A History of UNESCO

CW00330296

7

A History of UNESCO

Global Actions and Impacts

Edited by

Poul Duedahl
Professor, Aalborg University, Denmark

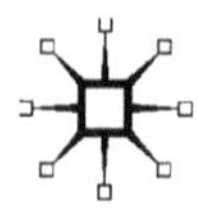

Selection, introduction and editorial content © Poul Duedahl 2016
Individual chapters © Respective authors 2016

Softcover reprint of the hardcover 1st edition 2016 978-1-137-58118-1

All rights reserved. No reproduction, copy or transmission of this publication may be made without written permission.

No portion of this publication may be reproduced, copied or transmitted save with written permission or in accordance with the provisions of the Copyright, Designs and Patents Act 1988, or under the terms of any licence permitting limited copying issued by the Copyright Licensing Agency, Saffron House, 6–10 Kirby Street, London EC1N 8TS.

Any person who does any unauthorized act in relation to this publication may be liable to criminal prosecution and civil claims for damages.

The authors have asserted their rights to be identified as the authors of this work in accordance with the Copyright, Designs and Patents Act 1988.

First published 2016 by
PALGRAVE MACMILLAN

Palgrave Macmillan in the UK is an imprint of Macmillan Publishers Limited, registered in England, company number 785998, of Houndmills, Basingstoke, Hampshire RG21 6XS.

Palgrave Macmillan in the US is a division of St Martin's Press LLC, 175 Fifth Avenue, New York, NY 10010.

Palgrave Macmillan is the global academic imprint of the above companies and has companies and representatives throughout the world.

Palgrave® and Macmillan® are registered trademarks in the United States, the United Kingdom, Europe and other countries.

ISBN 978-1-349-84528-6 ISBN 978-1-137-58120-4 (eBook)
DOI 10.1007/978-1-137-58120-4

This book is printed on paper suitable for recycling and made from fully managed and sustained forest sources. Logging, pulping and manufacturing processes are expected to conform to the environmental regulations of the country of origin.

A catalogue record for this book is available from the British Library.

A catalog record for this book is available from the Library of Congress.

The establishment of UNESCO was a direct response to the violent actions during World War II. Here is a street scene from Siegburg, Germany, in April 1945. A German woman runs through the streets with what belongings she is able to carry, as the American and German troops battle for control of the city. (Photographer: Troy A. Peters, US Army)

Contents

Figures and Tables

Figures

Tables

Contributors

Jens Boel is Chief Archivist at UNESCO, Coordinator of the UNESCO History Project and a member of the Global History of UNESCO Project funded by the Danish Council of Independent Research.

Samuel Boussion is Associate Professor of Education Science at the University of Saint-Denis-Vincennes, France, and a member of the Centre interdisciplinaire de recherche "culture, éducation, formation, travail" (CIRCEFT) laboratory.

Anabella Abarzúa Cutroni is a research member of the Research Program on Academic Dependency in Latin America (Programa de Investigaciones sobre Dependencia Académica en América Latina) at the National University of Cuyo, Mendoza, Argentina.

Josué Mikobi Dikay is Professor of History at the National Pedagogical University and at the Protestant University of Congo, Kinshasa, DRC.

Poul Duedahl is Professor of History at Aalborg University, Denmark, and Director of the Global History of UNESCO Project funded by the Danish Council of Independent Research.

Inés Dussel is a professor in the Department of Educational Investigations in the Center for Research and Advanced Studies at the National Polytechnic Institute, Mexico, and former Director of the Education Area at the Latin American School for the Social Sciences, Argentina.

Jaci Eisenberg is Assistant Editor of the Biographical Dictionary of Secretaries-General of International Organizations project. She holds a PhD in International Studies from the Graduate Institute, Geneva, Switzerland.

Mathias Gardet is Professor of Education Science at the University of Saint-Denis-Vincennes and a member of the CIRCEFT laboratory.

Aurélie Élisa Gfeller is currently a visiting research fellow in the Department of International History at the Graduate Institute in Geneva, Switzerland. She holds a PhD in History from Princeton University, USA.

Céline Giton is Manager of the Writers Retreat La Ferme des Lettres in Southern France. She holds a PhD from the Centre of History at the Paris Institute of Political Studies, France.

Miriam Intrator is Special Collections Librarian at Ohio University, USA. She holds a PhD in Modern European History from the Graduate Center, City University of New York, USA.

Edgardo C. Krebs is a research associate in the Department of Anthropology, National Museum of Natural History, Smithsonian Institution, Washington, DC, USA, and Adjunct Professor of Anthropology in the Department of Anthropology, University of Maryland at College Park, Maryland, USA.

Aigul Kulnazarova is Professor of International Relations and International Law at Tama University, Japan, and a member of the Global History of UNESCO Project funded by the Danish Council of Independent Research.

Celine Lai is Assistant Professor of Cultural Management at the Chinese University of Hong Kong, People's Republic of China.

Suzanne Langlois is Associate Professor of History at Glendon College, York University, Toronto, Canada.

Agnès Borde Meyer is a research fellow. She holds a PhD from Paris 1 Pantheon Sorbonne, France.

Thomas Nygren is a researcher and associate senior lecturer in the Department of Education at Uppsala University, Sweden. He is funded by the Knut and Alice Wallenberg Foundation.

Martine Ruchat is Professor of History of Education at the University of Geneva, Switzerland, and a member of the Laboratory of Social and Cultural History of Education.

Takashi Saikawa is a postdoctoral fellow of the Japan Society for the Promotion of Science, Tokyo, Japan.

Christian Ydesen is Associate Professor of History of Education at Aalborg University, Denmark, and a member of the Global History of UNESCO Project funded by the Danish Council of Independent Research.

Figure I.1 The Egyptian delegation outside UNESCO House in Avenue Kléber near the Arc de Triomphe in Paris, November 1946. (© UNESCO)

Introduction

Out of the House: On the Global History of UNESCO, 1945–2015

Poul Duedahl

In the era of globalization, there is a need for research which explains the cause and the importance of transnational phenomena that affect people's lives. International organizations are obvious objects of analysis in order to achieve a deeper understanding of some of the more prominent and organized transnational issues characterizing the 20th century because they are specific places – headquarters with offices, meeting rooms and conference facilities – where people meet beyond national borders and exchange knowledge.[1]

An organization that has attracted much attention in recent years is the United Nations Educational, Scientific and Cultural Organization (UNESCO), founded in November 1945. Its initial mission was to ensure peace and security by carrying out a considerable amount of mental engineering in the shadow of the aggression of World War II. As stated in the preamble to its constitution, "Since wars begin in the minds of men, it is in the minds of men that the defenses of peace must be constructed."[2]

Prior to the organization's anniversary in 2005, UNESCO was only subject to scattered attempts at writing its history. The publications were often sketchy reference works commissioned by the organization itself and written on the basis of published material produced by UNESCO, and not its unpublished, administrative documents.[3] In 2005 it then arranged the first of a series of conferences on its history and launched the UNESCO History Project, the objective of which was to encourage research on the history of the organization and use its holdings of archival material. The subsequent research has exposed UNESCO as an excellent prism reflecting ways of thinking that became popular on a global scale in the post-war period in the fields of science, education and culture. However, the research has still not convincingly revealed UNESCO's particular role in disseminating these thought patterns. In fact, one tendency has more or less dominated the research: that of making retrospective analyses focus on UNESCO's initiatives and their roots rather than their impact. This focus on "intellectual history" characterizes the larger overviews, the more detailed studies on the specific UNESCO departments' history as well as those of specific initiatives. The same strategy applies to the research on the headquarters' physical architecture and the employees' view on art; the organization's key concepts such as universalism, cultural relativism, multiculturalism, internationalization and cultural diversity, as they were formulated in-house; and the research on some of the organization's more influential personalities. In every case the center of attention is UNESCO House in Paris.[4]

Historians have in other words uncovered the roots of many of the soft power initiatives launched to construct sincere solidarity between people, but we still know very little about their subsequent impact. Consequently, UNESCO has come to appear as an organization reflecting contemporary intellectual trends rather than influencing them. It does, of course, make good sense that historians first try to understand the organization's initiatives before assessing their influence on the outside world, but after 70 years of history it also seems appropriate to take it a step further. The organization is, after all, much more than a physical building in Paris and a producer of piles of documents.[5]

The fact that the history of the reception of UNESCO's initiatives has not already been written has many reasons. It is not that it is considered irrelevant; quite the opposite. In fact, most historians will most likely acknowledge the importance of historical impact studies, even though they do not make them themselves, because of the fact that the organization's activities were launched in order to contribute to its overall mission of constructing peace in the minds of people outside UNESCO House. The main reason is that they are often relatively easy to begin with but become rather difficult to complete in a fully satisfactory way, which is again a matter of what kind of impact you are looking for.

Let me demonstrate both the easiness and the relatively difficult task of making historical impact analyses by highlighting two of UNESCO's most prominent initiatives in its history. The first came into being after the decision to build the Aswan High Dam in Egypt, which flooded the valley containing the Abu Simbel temples, a treasure of ancient Egyptian civilization. In 1959, after an appeal from the governments of Egypt and Sudan, UNESCO launched an international safeguarding campaign that mobilized international attention on a worldwide scale – from Jacqueline Kennedy to people in the poorest countries buying UNESCO stamps – to support archaeological research in the areas to be flooded and to remove the Abu Simbel and Philae temples and reassemble them elsewhere. The temples at their new locations are undeniably physical evidence of the potential impact of a UNESCO initiative. Also the impact of the subsequent world heritage declarations – the history of which is mentioned by Aurélie Elisa Gfeller and Jaci Eisenberg (Chapter 13) – is to a certain degree possible to show in physical ways, for both its positive, intended and its negative, unintended consequences.

It is much more difficult when we get even closer to what is at the heart of UNESCO's mission – namely, to change mindsets. One of UNESCO's early initiatives in that field was to launch a program that was supposed to combat racial inequality. Most scholars have so far primarily focused on the roots of the program, not least on how the four statements on race, made by experts within the field and issued by UNESCO from 1950 to 1967, came into being.[6]

The statements all stressed human equality. However, to eliminate thinking in terms of superior and inferior races among a large number of people takes more than issuing a statement; its existence has to be known of by people outside UNESCO House. The organization's Department of Social Science therefore sent the statements to a range of scientists, scientific journals, national commissions, newspapers, magazines and so on, and this created a lot of publicity. An inventory of the press clippings that UNESCO managed to collect in 1950 shows that it was mentioned in 133 news articles, 62 in-depth articles and leaders, and eight major news reports from all over the world. The text was also reproduced in full in three magazines, and it was estimated that there were an additional 50–75 articles that UNESCO staff had not tracked. In addition, there was some radio publicity and the distribution of thousands of copies of the statements. "Whenever it is, whatever form it takes, racism is an evil force, and to the extent that UNESCO can kill it by the truth, it will do good," *The New York Times* proclaimed.[7]

Also *The UNESCO Courier*, the organization's popular journal of the time, had an important role to play in the promotion of the race statements, as one can read in Edgardo C. Krebs' contribution (Chapter 1). Promoting such a viewpoint to a range of people and in huge numbers is in itself an impact, but a physical impact, and also an indication that there might have been a subsequent and even more far-reaching mental impact of the kind that UNESCO was put in the world to achieve. But it is not proof.

To reach an even wider audience and for a longer timespan, the Department of Mass Communications suggested that the organization should also engage a number of recognized researchers to write about different topics in relation to race, based on the viewpoints in the statements. UNESCO managed to launch three series of publications – *The Race Question and Modern Science, The Race Question and Modern Thought*, and *Race and Society* – each of them consisting of a number of small pamphlets in French and English. However, it soon turned out that the pamphlets had problems reaching the "man in the street" in most of the member states. This was first and foremost because they were written in languages that were foreign to people in many countries, but also, as a study from New York University showed, because they were too difficult to understand. The reader required at least a high-school degree to grasp the content. In addition, their layout was not very engaging.[8]

This fact indicates that the campaign was not as efficient as UNESCO had wished. In the long run, the publications were nevertheless able to infiltrate national education systems because they were written by recognized scientists, were discussed and used in leading scientific journals, and represented a steady bombardment of publications that at least physical anthropologists had to deal with. In the early 1950s they were among UNESCO's best-selling publications and represented a substantial proportion of all the new titles published in the USA in the field of anthropology, and by the late 1950s they had been translated into 13 languages and more than 300,000 copies had been printed.

That had consequences, of course, but not always the ones intended by UNESCO. In the USA, some of the publications were severely criticized, and in Los Angeles, all UNESCO publications were eventually banished from the public school system in 1953, which again led to apprehension among school administrators all over the USA concerning the use in public schools of any of UNESCO's publications, regardless of their content. South Africa even withdrew from UNESCO in 1956 as a consequence of the race pamphlets and what was felt to be the organization's interference in internal affairs.[9]

On the other hand, these incidents only added to the public's awareness of the issues, and in some cases we actually have proof that the publications not only reached the intended audience – people on the ground, or at least some of the leading figures – but also influenced them: "UNESCO came out with a study that said that blacks – at that time Negroes – were not inferior, and there was no fundamental genetic difference between blacks and whites. We were determined in our differences by social conditions," recalls, for example, the famous US Civil Rights activist Revd Jesse Jackson. He got to know the statements and pamphlets around 1960.

> We went around the South giving speeches, holding up the UNESCO study, saying that blacks were not inferior. A world body had studied and concluded that we were not inferior. It was a big deal. UNESCO, a world body – not some Southern segregated school, not some Southern governor, not even the President – UNESCO said we were not inferior.[10]

Also, as the cases on segregation reached the US Supreme Court, the outcome of UNESCO's race program would play a role. On several occasions during the 1950s and 1960s, experts affiliated with UNESCO and its race program were brought in as expert witnesses and the statements on race were highlighted as the newest available research, paving the way towards the eradication of state-approved segregation.[11] In 1967, for example, the US Supreme Court drew heavily upon them in its landmark decision to declare those laws unconstitutional that banned interracial marriages.[12]

There is therefore no doubt that the statements and the authority with which the experts spoke had a psychological impact, and that to some degree, due to the early date of the release of the statements, they paved the way for a new way of thinking. However, with regard to the exact extent and range of UNESCO's impact in relation to the race question, this remains unknown and is for future historians to figure out.

One could mention a lot of other activities outside UNESCO House of which it would be possible, whether easy or difficult, to study using historical impact analysis because, in fact, "direct action" in the shape of pilot projects, expert missions, experimental centers and regional offices – far away from the headquarters – was from the very beginning one of the organization's most prominent working methods, and UNESCO was – and remains – the only branch of the UN family with a network of national commissions.[13] Even though these commissions are often branches of the national educational or cultural ministries, and thus not part of the organization as such, their task is still to select and implement its policies in the world outside UNESCO House, in the member states, and to feed the headquarters with information about national viewpoints and local needs. These field operations and the collaboration with governmental and non-governmental organizations are very much a part of UNESCO's history, and their history deserves to be written down – not despite the fact, but rather because, they indicate that the impact of the very same initiative most likely varies from place to place.

The focus of this book is therefore on the routes rather than the roots of UNESCO's initiatives, and on the local interventions and their impacts rather than the global initiatives and the ideas behind them. It is important to stress that it is not a book about successes and failures. Impact is a neutral concept which can be observed, whereas the proclamation of successes and failures depend on the eyes of the beholder. The race statements mentioned above were, for instance, conceived as a success by politicians in Brazil, which saw their country as a racial melting-pot, whereas they were conceived as a gross failure by politicians in Apartheid South Africa. But it is hard to deny that the statements had an impact. Impact – the change attributed to UNESCO initiatives, whether good or bad, intended or unintended – is what this book is about. A number of scholars (of which I have already mentioned a couple) have for this purpose been invited to write to contribute to this first attempt at tracing the routes of various UNESCO initiatives from the center to the periphery – from the organization's headquarters in Paris to the member states – to assess its exact impact on mindsets in the wake of World War II.

To do this we have asked all of them to base their contribution to this volume upon one or more of the following research questions:

- How were ideas and initiatives transmitted in practice from UNESCO headquarters in Paris to the member states?
- How did the UNESCO-related national institutions work in practice, and what distribution channels did they have with regard to the different populations?

- Were UNESCO's initiatives implemented equally in all member states?
- What explains country-specific priorities?
- How were initiatives made acceptable to the different populations?
- What effect did the implementation of UNESCO's specific initiatives have on changing people's mindsets? And where this cannot be unambiguously determined, was knowledge transferred, domesticated and made available for the construction of "peace in the minds of men"?

Theoretical, conceptual and methodological framework

This book draws on globalization theories, which indicate that interventions of international organizations have in fact been shaping the lives of individuals everywhere in the world for at least half a century, and on impact assessment methodologies used by governments and international organizations to produce their own precise performance data.[14]

The focus of this volume on the routes rather than the roots of knowledge is on the very hypothesis which underlies these approaches – namely, the close relationship between ideas, initiatives, interventions and impacts, leaving it to the researcher to determine the exact results (impact) attained by an activity (intervention) designed to accomplish a valued goal or objective of a program (initiative) based on the reflections of its inventers (idea).

The above theories and concepts are very practical in the sense that they point at the kind of documents which could prove useful for historians wanting to study transnational interventions and local impacts, while the methods are designed to answer questions about whether a certain initiative actually made a difference – in this case according to the historical documents.[15]

Such a study requires the identification of a number of key factors, as detailed below.

Identification of global initiatives whose subsequent local impacts should be studied

In July 2013 the Danish Council for Independent Research allocated funds for a major research project on the global history of UNESCO, of which this book is part. In their unfolding of the theoretical and methodological framework of the project, Danish historians Christian Ydesen and Ivan Lind Christensen begin by asking two questions: "Impact of what?" and "Impact on what?"[16]

The answer to both questions, with regard to the research project as well as this volume, is that the contributors trace the impact of selected key UNESCO initiatives, and that, in consideration of UNESCO's constitution and overall aim, they want to trace the impact on the construction of peace in the minds of people after World War II.

In international organizations, initiatives are reflected in written decisions, formulated with the purpose of solving specific problems and promoting certain values, and, through a selection process among the more than 50 abstracts submitted for this book, written by researchers who have studied these documents, a pattern soon appeared. The pattern is a reflection of what initiatives historians of today focus on as central to UNESCO's work on mental engineering over the last 70 years. This volume will thus focus on the impact of these five types of initiative:

- Initiatives taken to disseminate knowledge via different media platforms from UNESCO headquarters to implement change in the surrounding world.
- Initiatives taken right after World War II as a first step towards the physical and mental reconstruction of a war-devastated world.
- Initiatives taken to help poor and newly independent countries during the period of decolonialism.
- Initiatives taken to break down hostile stereotypes and promote peace via local educational systems.
- Initiatives taken to implement a new vision of humanity as a unified entity by promoting the idea of a common heritage.

The first step for the contributors was to outline briefly – on the basis of the resolutions of UNESCO's General Conferences, which are accessible via the so-called UNESDOC online archive database – the organization's precise expectations regarding the outcome of the initiatives, which then formed the basis for tracing their subsequent routes and determining their possible local impact on people's ways of thinking and acting.[17]

With regard to what more precisely initiatives can impact on, historians can study the effects on two levels: institutional and beneficiary. The first study focuses on the effects on governmental and civil society institutions, private corporations, regional or local policies, strategic support to institutional actors and so on. The second focuses on the impact of interventions that directly or indirectly affect communities; in the case of international organizations in general, that could be the impact on communities of the local trade liberalization measures, medical treatments and microloan programs. This second level gets closest to what is the overall aim of this book, but the first level is undeniably also important because it is often the gateway into a community that makes it possible to have an impact on the beneficiary level.

Some of the authors in this volume focus on changes at the institutional level, such as Takashi Saikawa in his contribution (Chapter 5) on Japan's efforts as a defeated nation to re-enter the international community after World War II, but never without keeping an eye on how changes at the institutional level impacted the communities and led to the launch of several hundred UNESCO clubs, and how the widespread popular engagement in UNESCO affairs and values again impacted the institutional level and strengthened the Japanese officials' argument for membership of the organization. One could also highlight Miriam Intrator's contribution (Chapter 6) about UNESCO's policies and their local implementation, which is exemplary in its demonstration of the many different travel routes a UNESCO initiative can take from the headquarters to the national institutional level, in this case with the purpose of implementing new international standards for libraries around the world.

More contributors have focused on the beneficiary level, such as Agnès Borde Meyer in her contribution (Chapter 14) on archeology experts and their impact on museums, and on safeguarding heritage at specific locations in Iran and Afghanistan. However, most of the contributors at the same time have in mind that institutional changes often come before those on the ground – and that it is only due to governmental support that countries eventually open their borders for UNESCO experts to come to assist with the construction of schools, formulating curricula, producing textbooks and so on – and therefore need an explanation too. At the same time, it is the institutional level that has produced our primary sources – the documents. A sole exception is Josué Mikobi Dikay's contribution (Chapter 8) on the D.R. Congo (hereafter Congo), where there was hardly a local government. This paved the way for the direct involvement of the UN and UNESCO in Congolese affairs and made it possible to build up an entire educational system, with the consequence that local communities suddenly had schools where there had not been schools before, and that thousands of children were suddenly able to read and write where they had previously been doomed to illiteracy.

Counting how many books, films and radio broadcasts the organization produced throughout its existence, how many experts it sent abroad and how many schools it helped to create is indeed relevant because the numbers describe the scope and range of means and agents for change, and they represent physical changes and therefore a local impact in their own right. There is also no doubt that UNESCO's initiatives on fundamental education led to the physical construction of schools worldwide with the aim of eradicating analphabetism and empowering people in the local community. One can also assume that a relatively large number of the kids attending were actually able to read and write when leaving the building a few years later. However, the more ideologically profound UNESCO wanted the changes to be, the more difficult they are to document.

For example, it is possible to provide evidence that UNESCO's initiatives within the field of textbook revisions led to the production of several new books, as Inés Dussel and Christian Ydesen show in their case study of history textbook revisions in Mexico (Chapter 11), and we can even read and analyze their content. However, to assess to what degree UNESCO actually managed to eradicate nationalist prejudice is much more difficult to assess in such an accurate way.

At least we can estimate how many people were exposed to the books. That is more challenging in the case of UNESCO's Major Project on the Mutual Appreciation of Eastern and Western Cultural Values, which also had a physical outcome by leading to the translation, publication and transfer across continents of hundreds of classical texts. However, here it is close to impossible to estimate who and how many had access to them via libraries, and to know if they made a difference in people's perception of other cultures. UNESCO in fact made a few polls to document any changes, but the changes were not necessarily the outcome of UNESCO initiatives.[18]

The fact is that books can be remembered, misunderstood, ignored and forgotten, or used differently, and they have a variety of impacts depending on who the recipients are, what their social background is and where they live. One of the aims of this volume is therefore to try to address some of the more profound methodological challenges of conducting historical impact studies and to find a way to get as close to the mental impact as possible.

Identification of methods useful in historical impact studies

To conduct historical impact research is difficult. Not many interviews, questionnaires or written statements were made before, during and after the local implementation of UNESCO initiatives. At the same time there is no consensus on how best to conduct this kind of analysis methodologically.

However, many historical impact studies with an ideological dimension tend to use qualitative methodologies primarily. These include conceptual history and discourse analyses through which it should be possible to see, for example, whether the content of the concept of race changed in books, articles and magazines in the wake of the UNESCO statements on race of the 1950s and 1960s – that is, whether "race" changed from being perceived as a concept based on both physical and mental differences between large groups of people and which could be used to legally discriminate, to being conceived as only physical differences with limited possibilities for political abuse; and also whether there has been a shift in the use of concepts, such as a switch from the biologically rooted concept of "race" to the culturally rooted concept of "ethnicity", or whether the concept of "racism" had positive, neutral or negative connotations.[19]

However, analyses of discursive and conceptual change and continuity have, as demonstrated by Christensen and Ydesen, a problem, which is the question of causality: How can we be certain that the potential changes in discursive formations and conceptual architecture are in fact due to the impact of specific initiatives taken by international organizations? And, furthermore, does the international organization represent the starting-point in the construction of a discursive formation?"[20]

Sometimes it is relatively easy to demonstrate the direct link between UNESCO's media and their impact, and that is when UNESCO is mentioned by name. One could mention the testimony of the former South African president, Nelson Mandela, about the importance of the *Courier* to his worldview. As a prisoner on Robben Island, he had almost exclusive access to this particular magazine because the Afrikaans-speaking prison authorities for some reason considered it to be harmless reading material. It thus became a major information channel for him and the other prisoners to get to know what was going on in the world outside the prison and outside South Africa. Through the *Courier*, he later explained to UNESCO's director-general, Federico Mayor, they "learnt about so many subjects never before encountered, such as cultural diversity and mankind's common heritage, African history, education for development and so on. All these subjects did not exist in the apartheid lexicon, let alone in the solitary confines of Robben Island."[21]

However, Mandela's statement is a rare case. A few other good examples can be found in Jens Boel's contribution on fundamental education (Chapter 7), where he demonstrates how historians can sometimes be lucky to find letters from individuals writing to UNESCO to explain what the organization meant to them. For example, a letter from a woman from the Marbial Valley in Haiti who learned to read and write at the age of 42 thanks to a UNESCO pilot project on fundamental education. The presence of such material is the exception rather than the rule, but the less tangible nature of discursive and conceptual processes should not discourage us – just the opposite. It simply requires a large number of sources and from a longer timespan – before and after the initiative was launched.

A study of the impact of UNESCO's statements on race in Denmark shows that, with a background of conceptual history and by comparing the content of a variety of written sources, it is indeed possible to estimate their exact impact with regard to mentality change. In this case the analysis shows that UNESCO's anti-racist agenda did in fact make a difference, but also that it to a large degree impacted scientists more than ordinary people.[22]

Let me also mention an almost exemplary example in this book of how to undertake discourse analyses on UNESCO initiatives, and that is the contribution of Thomas Nygren (Chapter 10), in which he demonstrates how textbook changes were implemented in Sweden and actually made a difference in the long term according to a comparison of the topics taught by teachers and chosen by pupils as interesting to write about – most likely as a consequence of a change in mentality, and most likely as a consequence of UNESCO's efforts due to the similarities between the concepts and values promoted by UNESCO and the ones to be found in the Swedish curricula and the written outcome of the pupils' exams.

However, in order to get there we need to address the fact that observed local changes can be the outcome of competing initiatives. After all, international organizations consist of many different agents that are located in an international political context and draw on both ideological and institutional predecessors, and often almost similar initiatives exist out there so, as Christensen and Ydesen conclude, "the choice of the international organization as a starting point for the analysis is not the same as understanding the international organization as the starting point of the discourse".[23]

The question of "noise" – discourses that are present at the same time and similar to the ones promoted by the international organization – is in fact something that the impact evaluations of international organizations are often preoccupied with because they aim to demonstrate the "effectiveness" of a certain initiative, and are therefore inclined to eradicate the "pollution" of the multiple factors which can affect the observed changes.[24]

This volume does not necessarily have to show which effect was solely the result of a UNESCO intervention and which was only caused by external factors. Its approach is rather to investigate the interplay between various factors, and is an attempt to be open to both the effects intended by UNESCO and the possible unintended effects or local varieties – assuming that UNESCO initiatives were never conducted in complete isolation. However, to at least make sure that the observed changes were not achieved totally independently of the UNESCO initiative, we have chosen a research strategy where local activities are only seen as representing an impact if the local documents contain direct references to UNESCO. This might lead to a slightly conservative estimation of UNESCO's impact, but it will at least not overestimate it.

Identification of the time period on which the studies will focus

When evaluating their programs, international organizations tend to focus on the immediate rather than the long-term effects because the timespan of their programs is much shorter than the time significant effects would take to manifest, and because the member states want to be able to tell their populations what they get for their membership fee.[25]

However, if there is one thing we can learn from the contributions to his book then it is that there is a chronological delay between when the idea is fostered and the initiative is taken, and when the first institutional changes and physical outcomes of UNESCO's work can be seen, and till beneficiary changes and the subsequent mental impacts can be observed. That is often long after the experts have left the countries. In fact, most of the organization's initiatives would require an analysis with the perspective of one or two generations. For example, that would be necessary in order to study the effects of UNESCO initiatives aimed at breaking down stereotypes in early education, and the impact of changes in school curricula and textbooks. However, it differs from initiative to initiative. At the same time the contributions to this volume are made within the framework of the possible and reflect the competence, interest and timeline that made sense to the authors.

The chronological perspective therefore differs considerably, from Thomas Nygren's demonstration (Chapter 10) of how textbook changes were implemented in Sweden over a long timespan to Suzanne Langlois' and Takashi Saikawa's focus (chapters 3 and 5) on the immediate impacts.

The reader will also observe that there seems to be a preponderance of contributions from the first half of UNESCO's history. That inevitably gives the impression that the organization's earliest initiatives were the most important, which would not be a fair conclusion. Even though there were many urgent tasks for UNESCO of lasting importance in the wake of World War II and the decolonization, the weighting is also a consequence of the chronological delay from idea to impact, and the time it takes for an impact to manifest itself and become visible for the historian as something worth studying. Another reason of a more practical nature is that many administrative documents are not available for at least 30 years.

Identification of the places and spaces on which the studies will focus

Historical impact studies cannot be conducted from an armchair because impacts are local and must be studied locally. The local is the site of impact – a specific place, a geographical entity where changes can take place.

That has some implications, since UNESCO consists of 195 different member states and nine associate members. To write a truly global history of the organization, one would therefore either need a very large and expensive research group or to make randomized controlled trials, which can also be challenging because comparative studies that focus on national similarities and differences tend not only to assume but also to construct and reinforce boundaries that are crucial to the possibility of comparison. At the same time they tend to assume that initiatives came to them in the same way, often due to a local demand for them and not as supply-driven transfers of knowledge.[26] However, the interventions of international organizations are in fact often both demand- and supply-driven. UNESCO's race pamphlets were, for example, in great demand in Brazil because the country was often emphasized as one without racial tensions, and something to be proud of, whereas they were supply-driven in South Africa owing to the fact that the authorities there did not want them at all because they were seen as false and offensive, and as an interference in internal affairs.[27]

These differences in the reception of the same initiatives also force us to focus not only on "place" but also on "space", the international, regional and local contexts that cannot be reduced to geographic entities alone. The spaces consist of individuals, groups, networks, organizations and other connections with their own ideas and backgrounds, and living in an environment with specific national or religious narratives – a collective biography – that form the local meaning, which explains the different receptions. This has to be taken into consideration by researchers undertaking discourse analysis.[28]

Demand- and supply-driven moves across place and space are a focal point for historians who study transnational history or *histoire croisée*. They are preoccupied with the process of "transfer", which is the border-crossing movement via influential agents from, for example, the international organization to the local environment. These agents can be written statements and pamphlets, or visits by UNESCO experts, such as the many hundreds sent to South America as part of the UN's technical assistance program according to Anabella Abarzúa Cutroni's contribution (Chapter 9). The transfer can be close to direct, such as the adoption of a UNESCO convention into the national legislation, or the way described by Josué Mikobi Dikay (Chapter 8) in the case of the Congo, where UNESCO experts had almost a free rein to transfer and implement a Western-style educational system.

More often the initiative will first go through a "translation", which is a reinterpretation of the initiative to make it fit the local environment. That happens when the first people receive the initiative at the national ministries of education, culture and science, and again when it is transferred from there to the national commissions for UNESCO and from there to local communities. Translations happen both unconsciously (as an outcome of a local way to understand the initiative) and consciously (e.g. when an initiative encounters a different set of local political priorities and a translation serves a specific purpose of national interest). Textbook changes and education for mutual understanding were, for example, very hard to implement in Japan, where the country's relationship with South Korea and China, and disagreements about the correct interpretation of their relationship in the past, made politicians concerned and conscious about them, as Aigul Kulnazarova shows (Chapter 12).

That leads to the process of "transformation", where the initiative takes its final shape, as it is domesticated and entangled in the new, local setting: a shape which is often different from the intended version and sometimes turns into a completely unrecognizable and unintended form. In the aforementioned case of Japan, the initial skepticism has remained, and UNESCO's initiative regarding textbook changes has therefore been transformed into a version acceptable in a Japanese educational setting, and which does not mention sensitive historical events, or mention them in a certain way. Another case is demonstrated in Celine Lai's contribution about the impact of UNESCO's world heritage program in China (Chapter 15), where it has not only had an impact intended by UNESCO – namely, to safeguard sites and promote them as humanity's common heritage – but also had a range of unintended impacts, such as sites subsequently overrun by tourists or abused by local politicians for egoistic and nationalistic purposes (a pattern similar to that of many other member states).[29]

Finally, one should mention the "trading" of ideas, which indicates that there is also a movement of ideas and initiatives from the member states back to UNESCO, such as via national delegates at the general conferences, but this does not hold much interest for the contributors to this book, unless these agents bring other ideas and initiatives back with them, because the aim of this volume is to follow in the footsteps of the postman, fieldworker, mission expert and schoolteacher from UNESCO and out of the house.[30]

With these four concepts (transfer, translation, transformation and trading) in mind, a way to study the global history of UNESCO could be to select a number of case-study countries that would, for example, represent all continents – mainly large countries where an intervention would affect many people, and countries representing different historical, social, religious and ideological contexts that would most likely affect the national reception of UNESCO's initiatives. For example, in the overall research project, we have – for the same reason – chosen to make archival studies in ten countries: Ghana, South Africa, India, China, Japan, the Federal Republic of Germany, the USSR, the USA, Brazil and Indonesia – the latter, for instance, being the only South Asian representative, a former colony and newly independent country, a country with a huge problem and special needs regarding analphabetism, and at the same time the country in the world with the largest Muslim population.

Having chosen a number of case-study countries, there appear to be a range of more practical problems regarding writing global history: the fact that traveling is expensive and that member states have different traditions for preserving documents and providing access to them, making it difficult to ensure uniformity in the data-collection process. At the same time, historians conducting research on global, historical impacts will often also need assistance from people with local archive knowledge to retrieve the documents and to translate them. These practical problems are probably the main reasons why historians tend to give up long before they have even started. It is much easier to go to the UNESCO Archives in Paris, where they will get instant access to a lot of documents in only one place and with most of them in either English or French. However, there is then a danger that they will plunge right into the field of intellectual history and give up on the historical impact analyses.

None of the contributors to this book, unlike the American historian Matthew Connolly, have had the time, funding and opportunity to conduct research in 50 archives in seven different countries, in this case to tell the history of various international organizations' family-planning policies and their global impact history in the shape of sterilization camps in India and Chinese one-child policy. Instead they have done more like his colleague, Prof. Akira Iriye, by giving an estimation of the impact on the basis of the documents available to them.[31]

The contributions are, in other words, made within the framework of the possible. To address this we have chosen to at least make sure to include chapters that represent the different regions of the world, with the limitation that, for the reasons mentioned above, most of the contributions are confined to follow in the footsteps of an initiative and assess its impact in only one or two countries. We have as part of the selection process therefore chosen country case studies that can be seen as central and exemplary cases for a correct understanding of UNESCO's activities, such as the reintegration of Japan in the international community after World War II, the construction of an educational infrastructure in Congo in the wake of the decolonization process, and the importance of the world heritage concept today in the world's most populous country, China.

For guidance and inspiration for their own impact studies, the contributors have had several role models due to similar efforts made by scholars conducting research on other international organizations. One could mention the award-winning book by professor of American history Carol Anderson about how the UN went hand in hand with the US Civil Rights Movement and had an impact on the rights of the African-American population, and – from the other side of the globe – the work of Japanese historian Liang Pan on the UN's influence on Japanese foreign and security policy since World War II, as well as the uncovering by English historian Sunil Amrith of the World Health Organization's impact on disease control in India and Southeast Asia. They all take a close look at what happens when initiatives leave the headquarters of international organizations and are received, reshaped and executed in a national and regional setting. European integration researchers have in the same way focused on "Europeanization", or how European cooperation, first and foremost the European Union, has influenced and transformed national politics, administration, culture and society in the members states.[32]

This book is a contribution to this relatively new tradition and another step in a more profound attempt to move the research agenda further away from recognizing international organizations as purely political products and understanding them also as producers of local politics on a global scale.

Identification of documents in which interventions can be studied

As already mentioned, direct action was one of UNESCO's prominent working methods, often executed by regional or national UNESCO-related institutions which adopted responsibility for carrying the initiatives forward and through pilot projects, study grants, radio, film, books and other channels of knowledge transmission.

The documents produced by these intermediaries between UNESCO headquarters and the populations are important in order to clarify how the transmission of initiatives took place and how they turned into publications intended for specific national audiences. The documents, which play an important part in explaining the organization's success or failure nationally, are available at the UNESCO Archives in Paris. They consist of:

- country-specific UNESCO secretariat records, which include administrative files, project files and working files created by UNESCO in relation to each member state;
- topic-specific secretariat records, which include the same types of documents in relation to specific topics across national borders, such as national education for mutual understanding;
- records of national offices, which were official UNESCO agents and thus intermediaries with the task of transferring knowledge from the headquarters to the member states;
- archives of field offices, regional offices and temporary UNESCO expert missions which were other platforms for the delivery of UNESCO activities.

Identification of documents in which impacts can be studied

Comparing UNESCO initiatives with observed changes reflected in local documents with direct references to the organization provides us with the best picture we can achieve of UNESCO's impact. These local records consist of unpublished documents accessible at the various national archives and ministries of education, as well as published documents accessible via interlibrary loan:

- Annual reports of UNESCO's national commissions, which mention the work done within the member state and which reflect their priorities and not the organization's.
- Records of UNESCO's national commissions, which report in detail on the implementation of the nationally favored initiatives and contain assessments on their impact.
- Archives of non-governmental organizations (NGOs), which, according to the above documents, are identified as local collaborators.
- Local publications with references to UNESCO, which were active by placing national limits on people's understanding of the initiatives.

Overall content of this book

Given the size of UNESCO, and the number of projects it has initiated and taken part in throughout its existence, we have invited scholars to point at what they think has been central to the organization's work within the field of science, education and culture.

We have asked them to focus on how ideas and initiatives with which they are already familiar were transmitted in practice from organization's headquarters in Paris to member states, how the local UNESCO institutions and UNESCO-related national institutions worked in practice, and what distribution channels they had with regard to the populations. We have also asked them to assess – to the extent it is possible – the effect that the implementation of the organization's specific initiatives had on local people and on changing people's mindsets, or at least how knowledge was transferred, domesticated and made available for the construction of "peace in the minds of men". The answers might not be fully elaborated yet – it is, after all, the first step – but they give other scholars a hint of where to look for impacts and provide us all with a valuable contribution to a first estimation of what mattered, what worked and what made a difference.

Part I addresses the central distribution routes of knowledge or information channels out of UNESCO's headquarters in order to reach ordinary people, whether it was through the popular monthly, the *Courier*, as shown by Edgardo C. Krebs (Chapter 1), through the organization's many publications, primarily books and relatively easy-read pamphlets, as Céline Giton (Chapter 2) explains, or through new media, in this case via films, as described by Suzanne Langlois (Chapter 3).

Part II looks at the most urgent tasks preoccupying UNESCO in the very first months of its existence, such as the struggle to give orphans a home and hope, and to teach them to become good citizens, as shown by Samuel Boussion, Mathias Gardet and Martine Ruchat (Chapter 4). On the other hand, Takashi Saikawa (Chapter 5) tells the story of how people in the defeated and former fascist Japan worked hard to become part of the international community again. Finally, Miriam Intrator (Chapter 6) takes a look at how the organization helped to build up libraries in countries devastated by war and created common, international standards for libraries that would make it possible to get access to knowledge, also across borders, as a contribution to mutual international understanding.

Part III consists of case studies that exemplify central tasks for UNESCO in the following years within the field of technical assistance, particularly in the wake of the decolonization process, when the many new member states changed the organization's agenda. An important requirement for UNESCO to be able to spread its values was that people could read and write, and Jens Boel (Chapter 7) provides examples of its enormous work within the field of fundamental education (take also a look at figure I.2 and I.3). One of the former colonies with a profound need was Congo, and Josué Mikobi Dikay (Chapter 8) demonstrates how the organization came to play a vital role in building up everything from governmental administration to a functioning educational system from scratch as the former Belgian colonizer had barely left anything. Finally, Anabella Abarzúa Cutroni (Chapter 9) provides a statistical overview of the extent of UNESCO's fieldwork, in this case in the form of experts sent to South America.

In Part IV we get closer to what UNESCO, when all the basic educational facilities were present and working properly, saw as its main task – namely, to bring peace to people's minds, not least through schools, via new school curricula and through textbook changes. Here we get an almost comparative study across continents, and can see the similarities and differences, when Thomas Nygren (Chapter 10) looks at the implementation in Sweden, Inés Dussel and Christian Ydesen (Chapter 11) in Mexico and Aigul Kulnazarova (Chapter 12) in the newly defeated Japan.

In Part V we take a closer look at what became UNESCO's most famous projects ever. It shows how what began with an interest in writing mankind's common history and saving individual monuments gradually evolved into the creation of lists of the world's common heritage and efforts to protect it. Aurélie Elisa Gfeller and Jaci Eisenberg (Chapter 13) show how this work took place with the first really successful campaign in Egypt. Agnès Borde Meyer (Chapter 14) exemplifies how the organization protected world heritage in the early years in Iran and Afghanistan, while Celine Lai (Chapter 15) on the other hand explains how it looks today and how the competition to get on world heritage lists is sometimes so hard that it causes a number of unintended impacts, which threaten not only the world heritage sites themselves but also UNESCO's ideals.

This book is the first contribution by the Global History of UNESCO Project, a new major, international research project sponsored by the Danish Council for Independent Research. It does not reflect the final conclusions of the project but can rather be seen as an inspiration and a kind of state-of-the-art account of what we know about the impact of UNESCO in October 2015, when the papers for this publication were first presented at a conference in Paris on the occasion of the organization's 70th anniversary.

Acknowledgements

The Global History of UNESCO Project research team would like to thank everybody involved in the preparation of this book and the preceding conference: first and foremost the authors, the conference participants, the Danish Council for Independent Research and the Danish National Commission for UNESCO, and not least the commission's protector, Her Royal Highness Princess Marie of Denmark; its chair, Prof. Linda Nielsen; and its secretary-general, Jens Dalsgaard. We would also like to thank the center of attention itself, UNESCO. Even though the conference was arranged and the book has been written independently of the organization, the work has been followed with great interest and moral support from UNESCO, not least the director-general, Irina Bokova; the assistant secretary-general, Eric Falt; the senior program specialist, Marcello Scarone Azezi; and the chief archivist, Jens Boel. We add to our list the many people around the world who have helped the project team members as well as the authors to find documents and photos for the book, including Carole Darmouni, Fiona Ryan, Brenda Fong, Lois Siegel, Jane Gutteridge and Ralph Lucas. Finally, our gratitude goes to Palgrave Macmillan, Clare Mence, Lynda Cooper and Peter Cary for believing in the idea and for making it possible for us to give it a physical shape. May the book inspire researchers worldwide to undertake their own studies on the local, regional and global impact of international organizations.

Notes

1. Sunil Amrith and Glenda Sluga, "New Histories of the United Nations", *Journal of World History* 19:3 (2008): 251–274.
2. UNESCO, *Constitution of the United Nations Educational, Scientific, and Cultural Organization: Adopted in London on 16 November 1945* (London: Preparatory Commission of the UNESCO, 1945).
3. Denis Mylonas, *La genèse de l'UNESCO: La Conférence des Ministres Alliés de l'Éducation (1942–1945)* (Bruxelles: Émile Bruylant, 1976); UNESCO, *A Chronology of UNESCO, 1945–1987* (Paris: UNESCO, 1987); Michel Conil-Lacoste, *The Story of a Grand Design: UNESCO 1946–1993* (Paris: UNESCO, 1994); Fernando Valderrama, *A History of UNESCO* (Paris: UNESCO, 1995).
4. Roger-Pol Droit, *Humanity in the Making: Overview of the Intellectual History of UNESCO, 1945–2005* (Paris: UNESCO, 2005); Chloé Maurel, *Historire de l'UNESCO: Les trente premières années, 1945–1974* (Paris: L'Harmattan, 2010) and J.P. Singh, *United Nations Educational, Scientific, and Cultural Organization (UNESCO): Creating Norms for a Complex World* (London: Routledge, 2011), as well as *Sixty Years of Science at UNESCO 1945–2005*, ed. Patrick Petitjean et al. (Paris: UNESCO, 2006); Laura Elizabeth Wong, "Relocating East and West: UNESCO's Major Project on the Mutual Appreciation of Eastern and Western Cultural Values", *Journal of World History* 19:3 (2008): 349–374; Christopher E.M. Pearson, *Designing UNESCO: Art, Architecture and International Politics at Mid-Century* (London: Ashgate, 2010); Glenda Sluga, "UNESCO and the (One) World of Julian Huxley", *Journal of World History* 21:3 (2010): 393–418; R. Toye and J. Toye, "One World, Two Cultures?: Alfred Zimmern, Julian Huxley and the Ideological Origins of UNESCO", *History* 95:319 (2010): 308–331; Poul Duedahl, "Selling Mankind: UNESCO and the Invention of Global History, 1945–76", *Journal of World History* 22:1 (2011): 101–133; Teresa Tomas Rangil, *The Politics of Neutrality: UNESCO's Social Science Department, 1946–1956: CHOPE Working Paper No. 2011–08* (2011); Perrin Selcer, *Patterns of Science: Developing Knowledge for a World Community at UNESCO* (PhD dissertation, University of Pennsylvania, 2011), to mention a few.
5. On UNESCO as a norm-producing agency, see Michael Barnett and Martha Finnemore, *Rules for the World: International Organizations in Global Politics* (Ithaca: Cornell University Press, 2004).
6. See, for example, Claudio Pogliano, " 'Statements on Race' dell'UNESCO: Cronica di un Lungo Travaglio", *Nuncius. Annali di Storia della Scienza* 6 (2001): 347–441; Jean Gayon, "Do Biologists Need the Expression 'Human Races'? UNESCO 1950–51" in *Bioethical and Ethical Issues Surrounding the Trials and Code of Nuremberg: Nuremberg Revisited*, ed. J.J. Rozenberg (Lewiston: Edwin Mellen Press, 2003): 23–48; Jenny Reardon, *Race to the Finish* (Princeton: Princeton University Press, 2004); Perrin Selcer, "Beyond the Cephalic Index: Negotiating Politics to Produce UNESCO's Scientific Statements on Race", *Current Anthropology* 53 (2012): 173–184.
7. "The Myth of Race", *The New York Times,* 19 July 1950; "All Human Beings", *Time,* 31 July 1950; Letter from Douglas H. Schneider (Director, Department of Mass Communication, UNESCO) to Alva Myrdal (Director, Department of Social

Sciences, UNESCO), 16 October 1950; and letter from Douglas H. Schneider to Max McCullough (UNESCO Relations Staff, US Department of State), 4 January 1951, 323.12 A 102 (Statement on Race), UNESCO Archives, Paris.

8. Letter from Alfred Métraux (Division for the Study of Race Problems, UNESCO) to Harry L. Shapiro (Curator, American Museum of Natural History), 19 February 1952, 323.12 A 102 (Statement on Race), UNESCO Archives, Paris; Gerhardt Saenger, "The Effect of Intergroup Attitudes of the UNESCO Pamphlets on Race", *Social Problems* 3 (1955): 21–27; L.D. Reddick, "What Now Do We Learn of Race and Minority Peoples?" *The Journal of Negro Education* 34 (1965): 368; Maurice Freedman, "Some Recent Work on Race Relations: A Critique", *The British Journal of Sociology* 5 (1954): 342–354.
9. Michelle Brattain, "Race, Racism, and Antiracism: UNESCO and the Politics of Presenting Science to the Post-War Public", *American Historical Review* 112 (2007): 1386–1413; M. Daniel, "Race, Prejudice and UNESCO: The Liberal Discourse of Cyril Bibby and Michael Banton", *History of Education Researcher* 85:2 (2010): 74–84; Poul Duedahl, "From Racial Strangers to Ethnic Minorities: On the Socio-Political Impact of UNESCO, 1945–60" in *Current Issues in Sociology*, ed. Gregory A. Katsas (Athens: Athens Institute for Education and Research, 2012), 155–166; Anthony Q. Hazard Jr., *Postwar Anti-Racism: The United States, UNESCO, and "Race", 1945–1968* (New York: Palgrave Macmillan, 2012).
10. "Strategies and Tactics in the Struggle for Civil and Human Rights with Reverend Jesse Jackson" (18 November 2002): 2, John F. Kennedy Presidential Library & Museum, Boston, MA, USA.
11. Poul Duedahl, "UNESCO Man: Changing the Concept of Race, 1950–70" (paper presented at the American Anthropological Association: 107th Annual Meeting, San Francisco, CA, USA, 2008), http://vbn.aau.dk/da/persons/poul-duedahl(231811d3-5ba7-47f0-b879-e1b545132488)/publications.html (accessed 23 June 2015).
12. Loving v. Virginia: Brief of Amici Curiae: 388 US 1 (1967), US Supreme Court Records and Briefs, p. 28, Library of Congress, Washington DC, USA.
13. "List of UNESCO's Methods", UNESCO's Executive Board, 17 January 1950, 19 EX/5, UNESDOC.
14. Globalization theories: Akira Iriye, *Cultural Internationalism and World Order* (Baltimore: Johns Hopkins University Press, 1997); Akira Iriye, *Global Community: The Role of International Organizations in the Making of the Contemporary World* (Los Angeles: University of California Press, 2002), and *UN Ideas that Changed the World*, ed. Thomas G. Weiss et al. (Bloomington: Indiana University Press, 2009). Impact assessment methodologies: D. Hulme, "Impact Assessment Methodologies for Microfinance: Theory, Experience and Better Practice", *World Development*, 28:1 (2000): 79–98. B. Späth, *Current State of the Art in Impact Assessment* (Swiss Agency for Development and Cooperation, 2004); Franz Leeuw and Jos Vaessen, *Impact Evaluations and Development; NONIE Guidance on Impact Evaluation* (Washington, DC: NONIE – The Network of Networks on Impact Evaluation, 2009).
15. For example, in Thomas Nygren, *History in the Service of Mankind: International Guidelines and History Education in Upper Secondary Schools in Sweden*, 1927–2002 (PhD dissertation, Department of Historical, Philosophical and Religious Studies, Umeå University, Sweden, 2008); and in Poul Duedahl, "From Racial Strangers to Ethnic Minorities: On the Socio-Political Impact of UNESCO, 1945–60" in *Current*

Issues in Sociology, ed. Gregory A. Katsas (Athens: Athens Institute for Education and Research, 2012), 155–166.
16. Ivan Lind Christensen and Christian Ydesen, "Routes of Knowledge: Towards a Methodological Framework for Tracing the Historical Impact of International Organizations", *European Education*, 47:3 (2015): 274–288; Poul Duedahl: "Routes of Knowledge: The Global History of UNESCO, 1945–1975" (unpublished project description for the Danish Council for Independent Research, 2013).
17. All of the resolutions of UNESCO's various General Conferences can be retrieved via www.unesco.org/ulis (accessed 19 October 2015); Leeuw and Vaessen, *Impact Evaluations and Development*, 15.
18. Laura Elizabeth Wong, "Relocating East and West: UNESCO's Major Project on the Mutual Appreciation of Eastern and Western Cultural Values", *Journal of World History* 19:3 (2008): 349–374.
19. Poul Duedahl, "UNESCO Man".
20. Christensen and Ydesen, "Routes of Knowledge".
21. Annar Cassam (Director, UNESCO Special Program for South Africa 1993–1996): "Mandela: Reading the Courier on Robben Island", available at http://www.unesco.org/new/en/unesco-courier/single-view/news/nelson_mandela/#.VYv9Sk0cTcs (accessed 25 June 2015).
22. Poul Duedahl, "Fra overmenneske til UNESCO-menneske: En begrebshistorisk analyse af overgangen fra et biologisk til et kulturelt forankret menneskesyn i det 20. århundrede" ("From SUPERman to UNESCOman: A conceptual and historical analysis of the transition from a biological to a cultural view of man in the 20th Century", unpublished PhD dissertation, Aalborg University, Denmark, 2007).
23. Christensen and Ydesen, "Routes of Knowledge".
24. Leeuw and Vaessen, *Impact Evaluations and Development*, 21–34.
25. Leeuw and Vaessen, *Impact Evaluations and Development*, 15.
26. Roger Dale, "From Comparison to Translation: Extending the Research Imagination?" *Globalisation, Societies and Education* 4:2 (2006): 184; Martha Finnemore, "International Organizations as Teachers of Norms: The United Nations Educational, Scientific and Cultural Organization and Science Policy", *International Organization* 47:4 (1993): 565–597.
27. Poul Duedahl, "UNESCO Man".
28. B. Warf and S. Arias, eds., *The Spatial Turn: Interdisciplinary Perspectives* (New York: Routledge, 2009).
29. Robert Cowen, "Acting Comparatively upon the Educational World: Puzzles and Possibilities", *Oxford Review of Education* 32:5 (2006): 566.
30. Read more in Christensen and Ydesen, "Routes of Knowledge".
31. Matthew Connolly, *Fatal Misconception: The Struggle to Control World Population* (Cambridge: Harvard University Press, 2008); Akira Iriye, *Global Community: The Role of International Organizations in the Making of the Contemporary World* (Los Angeles: University of California Press, 2002).
32. Carol Anderson, *Eyes off the Prize: The United Nations and the African-American Struggle for Human Rights, 1944–1955* (London: Cambridge University Press, 2003); Liang Pan, *The United Nations in Japan's Foreign and Security Policymaking, 1945–1992* (Cambridge: Harvard University Press, 2005); Sunil Amrith, *Decolonizing International Health: India and Southeast Asia, 1930–1965* (New York: Palgrave Macmillan, 2006); Martin Conway and Kiran K. Patel, eds., *Europeanization in the Twentieth Century – Historical Approaches* (Basingstoke: Palgrave Macmillan, 2010).

Figure I.2 Young woman who participated in a UNESCO fundamental education class describes the books to village children in the Kumasi region, Ghana, 1959. (Photographer: Paul Almasy, © UNESCO)

Figure I.3 Fundamental education, Iraq, 1958 (Photographer: Reiman, © UNESCO).

Figure PI.1 Director-General Luther H. Evans makes a radio broadcast in October 1954 during which he presents a gift for the UN Korean Reconstruction Agency. The collected funds were used to establish a UNESCO children's ward at the Tongnae Rehabilitation Centre in Korea. (© United Nations)

Part I
Routes of Knowledge

The cornerstone in UNESCO's extensive attempt to construct peace after World War II has been its ability to communicate with the world outside of its headquarters.

The instruments for transferring and sharing knowledge have been many. The organization has issued binding declarations and normative statements. It has opened field offices and sent experts to its member states. It is the only UN agency with national commissions, schools and clubs in the member states as a way of linking directly with civil society. It has cooperated with numerous NGOs that have helped implement its programs. It has been preoccupied with specialized activities in the field of mass communication and it has worked on ensuring the free flow of information, including knowledge from the organization itself and to the surrounding world. Some of these activities have had a big impact, others hardly any, but all of them have been essential to UNESCO's communication strategy.

Right from the beginning, UNESCO has been particularly engaged in the use of new media and their ability to reach a wider public. That includes radio broadcasting, TV, film and other audiovisual media as means to spread knowledge – for example, as part of its fundamental education programs. It has published magazines of which the most widely distributed was *The UNESCO Courier*, which in the 1960s was published in 35 different languages and for which the print run was 1.5 million. Not least has the organization taken quite a considerable number of initiatives in which books were at the center. Not only have books constituted a natural extension of UNESCO's work on fundamental education but the prevalence and use of printed literature has over time made it the major tool used by the organization in its efforts to develop and influence cultural and educational awareness, and to create "peace in the minds" of men and women.

1
Popularizing Anthropology, Combating Racism: Alfred Métraux at *The UNESCO Courier*

Edgardo C. Krebs

I

On 18 May 1945 the Swiss-born ethnographer Alfred Métraux (1902–1964) wrote the following letter to his wife, Rhoda, from Tübingen, Germany:

> My darling, This afternoon I have been deeply shaken by the sight of a group of Jewish girls who were coming back from one of the death factories – Auschwitz. How to describe them? Imagine corpses who had emerged from the grave. There was around these ambulating skeletons something out of this world. A woman whom I thought to be about 50 turned out to be 23. As she collapsed and was obviously dying, she was taken away in a hurry. I talked with the others. No sooner one of them began to mention the horrors of the camp, the others started to cry and the girl became hysterical. They had their tagged numbers tattooed on their bodies. Darling, I have seen that – most of them had been branded like cattle on the throat or on their shoulders. They were taken to rest in a room with beds on which they threw themselves sobbing and laughing. The few things that they were able to tell (sur)pass the published reports. They were thrown to dogs, forced to witness the burning of other women ... They screamed when they mentioned what happened to the children. The whole incident was so awful that there was not a person present who did not have tears in his eyes. I had to leave because I was breaking down. Yesterday I saw men coming from the same place. Often they also cry when they start to tell about happenings there. The worst of all is that the Jews are still afraid and expect ill treatments or abuses. Alas, their fears are not entirely unfounded. Anti-semitism is strong among the people I have seen lately. When you see the logical result of anti-Semitism, you begin to sense it as a vicious and murderous attitude

which must be combated. Some of the injustices which I mentioned in my last letters are being corrected, but not all and I still witness a few incidents which revolt me. If things do not change quickly in Europe, more blood will flow very soon. I cannot stand the idea that all the suffering of these people has been in vain. Fascism is far from dead. It has poisoned the minds, even of those who come back from the German inferno.

The experiences of this trip have a deep effect on me. It will be difficult to return to library work and the S.A. Indians. I would like to help these friendless people who are coming out of 5 years of serfdom and humiliation to discover that nobody wants them and who find out that their saviors feel about them very much as the Germans did. I am beginning to think that Spaniards, Poles, and Italians are the martyrs of Europe, those to whom our sympathy should go. Then there is the German problem. Eastern Europe now appears somewhat like it was in the 13th century when the Mongols were loose or during the black plague. From Germany come millions of people with tales of horror, confusion, murder and looting … I am not gay, somewhat sad, but also immensely happy to be plunged in this drama and for once in my life I have an intimate contact with so many people within my own culture. I am amazed at the decency and simple greatness I found in peasants and workers. They are the only ones who are really human and sensible. The rest is often dried up or rotten. I wish I had an UNRRA (United Nations Relief and Rehabilitation Administration) job … Tell Margaret Mead about my experiences and ask her whether she could not recommend both of us for some relief work in central Europe. In the evenings, I walk in my medieval city and read Goethe. Your photograph is constantly under my eyes and I look at it often – I even talk to it.[1]

II

Together with an eclectic group of people, especially recruited for the task, Métraux was part of the Morale Division – a component of the US Strategic Bombing Survey (USSB). The USSB's overall mission was to estimate the damage done by Allied bombing, first in Germany and later in Japan. The categories and areas addressed by the survey were organized into 12 divisions. Most of them dealt with the material damage inflicted on industry, roads, railways and the general capacity of the Germans to keep their machinery of war in operation. The Morale Division was charged with estimating the human costs of strategic bombing or, in the official terminology, "The Effects of Strategic Bombing on German Morale". The recruits for this mission had to speak German, and to possess certain qualities as interviewers and writers. Fluent in German and a seasoned ethnographer, Métraux was perfectly suited to the job. In April 1945 the Morale Division set off for Europe, where it followed the advance of the Allied Forces in Germany, interviewing German civilians and soldiers.[2]

Some 13 years earlier, another war, fought very far from Europe, had also trapped Métraux as a witness. In 1932, Paraguay and Bolivia clashed over portions of the Chaco, backed by France and Germany, respectively.[3] Métraux was doing fieldwork at the time among Toba-Pilaga, Wichi and Chulupi Indians. His distress at what he observed – the plight of the Indians, who fell as collateral damage and were in a constant state of flight, crossing rivers and frontiers to avoid murderous raids by soldiers of both armies, and even bombing campaigns not indicted by any image as powerful as Picasso's *Guernica* – produced a crisis in Métraux. He disappeared for a couple of weeks while contemplating whether he should quit anthropology and become a missionary. The Anglican Mission in the Chaco was dealing with practical matters: fixing broken limbs, providing healthcare and providing a safe haven.[4] After those two weeks in the wilderness, Métraux decided he could never be a missionary. Instead he wrote a letter to General Juan B. Justo, then the president of Argentina, proposing the re-creation of an old colonial institution – that of Protector of Indians – and to place him in charge.[5] As the founding director of the Institute of Ethnology at the University of Tucuman, he had the standing and the contacts to propose such a thing. The letter, a real report on the conditions of the Indians in the Chaco, was never answered.

In 1934–1935, Métraux participated as an anthropologist on a French-Belgian research mission to Easter Island, charged with the purpose of studying its impressive statues and the small surviving population of descendants of the original inhabitants. The island was under Chilean sovereignty but managed by a British sheep-farming company. Métraux was dismayed by the poverty of the locals, and by the fact that they were confined to just one village, Hanga-roa. They could not access any other points of the fenced-off land, including the areas where the moai stood, facing the ocean or collapsed and partially buried on the ground. He was struck by the injustice, and very critical of the imposed political and economic system responsible for it. Free himself to roam the island, he managed to take his local assistants with him and photographed them posing at the top of the statues or standing in front of rocks inscribed with ancestral petroglyphs. Métraux's scientific mission was successful. He was able to clarify conclusively long-standing questions about the origin of the human presence on the island, and the construction and meaning of the moai. However, in the popular book he wrote in 1941 about the island, he also described the living conditions of the native inhabitants and stated that after the European definitive occupation in the second half of the 19th century, Easter Island "asked nothing more of us but the fulfillment of a simple human duty: that the persons and dignity of the descendants of the Polynesians who carved the great statues and engraved the tablets should be respected by their new masters".[6]

Even though Métraux's original expertise was in the history and ethnology of South American Indians, a short stay in West Africa in 1934 on the way to Easter Island had opened up for him what was to become his other scholarly passion: the study of African cultures in the New World.[7] This affinity found an outlet in the early 1940s when he visited Haiti, a country where he returned repeatedly and about which he produced what, for his colleagues, in particular the Haitians, still remains one of the best ethnographies of voodoo available.[8] The influential Melville Herskovits, a Franz Boas student who created the first Department of African Studies in the USA at North Western University in Chicago, promoted his younger colleague. He wanted Métraux to be the person hired by The Carnegie Corporation to do the research and write the report that would later be known as an *American Dilemma*.[9] Métraux was seriously considered, and it was one of the great disappointments of his life – one shared by Herskovits – that Gunnar Myrdal was chosen instead.

When, in 1947, Métraux joined the staff of the UN at its headquarters in Lake Success, New York, he had resolved a personal question which troubled him repeatedly while serving in Germany as a member of the Morale Division: "What will be my future? Again a scholar in a peaceful room or man of action in other fields. I am torn between the two desires, the eternal conflict of my life."[10]

III

When, in 1950, Métraux replaced the Brazilian scholar Arthur Ramos as Head of the Division for the Study of Race Problems in UNESCO's Department of Social Sciences, he brought to the task a wealth of relevant scholarly expertise and personal experience. The force behind the creation of this program was an imperative to address the ideology of racism. The extermination camps of the Nazis were the most barbaric evidence of the consequences of marrying racial prejudices with politics and biology. The UNESCO program, in particular its first statement on race of 1950, dismantled all pretenses that there was a biological basis for racism. Race and racism, it established, were cultural constructions.[11]

The program was a direct offshoot of the Universal Declaration of Human Rights. This landmark document was problematic from the very beginning. As a "declaration" it lacked the legal instruments to be enforceable. It was also, from a juridical perspective, an attempt at synthesizing beliefs of different cultures. Its main architect, the French jurist René Samuel Cassin, worked indefatigably behind the scenes, and in the spotlight, to will it into existence.[12] To provide a philosophical basis for this enterprise, in 1948, UNESCO organized an international symposium under the direction of the

French Thomist philosopher Jacques Maritain. He realized that "the problem of Human Rights involves the whole structure of moral and metaphysical (or anti-metaphysical) convictions held by each of us". Would a consensus be possible, he wondered, among men "who come from the four corners of the globe and who not only belong to different cultures and civilizations but are of antagonistic spiritual associations and schools of thought?" The solution for him was to agree that the declaration should be "given an approach pragmatic, rather than theoretical". On this grounding, a dialogue between cultures could be imagined and, most importantly, a new and deliberate category created upon which an internationally acknowledged taxonomy of rights could be refined.[13]

The urgency of some of the issues to be resolved, and the immediacy of the impact and trauma of the Holocaust, suggested more concrete ways of action to Raphael Lemkin, a Polish lawyer who, like Cassin, was of Jewish descent. Lemkin was more interested in producing an actionable legal document, a convention, which would bind the signatories to act, than in a declarative statement on the universality of human rights.

He argued that the international community should be able to intervene in the domestic affairs of a sovereign country if it was known to pursue policies of territorial expansion, persecution of civilians and ethnic cleansing. His *Axis Rule in Occupied Europe: Laws of Occupation – Analysis of Government – Proposals for Redress* from 1944, and in which the word "genocide" was used for the first time, is indirectly a sustained dismantling of the juridical theories of Carl Schmitt – sometimes referred to as "Hitler's Jurist" – which provided a strong executive branch with the right to interpret the needs of the country, unrestrained by its corpus of laws.

The Convention on the Prevention and Punishment of the Crime of Genocide, propelled by the relentless lobbying of Raphael Lemkin, was passed by the UN on 9 December 1948, beating by one day the adoption of the Universal Declaration of Human Rights, the more philosophical instrument imagined by Cassin. Knowledge of this backdrop of complicated and evolving history in the making, which still defines enduring aspects of international relations today, is necessary to properly understand the role played by anthropology and UNESCO immediately after the sanction of the declaration.

In redefining the concept of race, and putting racial prejudice in the spotlight, UNESCO was not only confronting squarely the political and social problem of apartheid; it was also inviting a disciplinary controversy between social-cultural anthropologists and natural scientists. The differences were not new. Social anthropology had always had a difficult time in museums of natural history. A member of the Smithsonian's Bureau of American Ethnology once remarked that his discipline was the "unwanted, foster child" of the institution.[14] Franz Boas, after working at the Field Museum of Natural History in Chicago, concluded that the taxonomic bias of biologists was not well suited to the development of anthropology. Later on, when he moved to the American Museum of Natural History in New York, he also had a position at Columbia University teaching anthropology. It was a joint appointment. It is well known that Boas's anti-racist thinking got him in trouble with the director of the museum, the paleontologist Henry Fairfield Osborn, and with the museum's board, where Madison Grant, the author of *The Passing of the Great Race*, a book admired by Hitler, was one of the more influential members. Osborn and Grant were close friends, and leaders of the eugenics movement in the USA. Similarly, the astronomer and physicist Samuel Pierpont Langley, who became the third secretary of the Smithsonian Institution in 1887, did not favor the kind of studies which the Bureau of American Ethnology was carrying out in the field, building an impressive ethnographic record of Indian cultures, and studying their languages, myths and rituals. The picture that emerged from these studies was complex, intellectually challenging and definitely not in line with an evolutionary conception of humanity.[15] Moreover, both Langley and his friend, Alexander Graham Bell, a Smithsonian regent, were proponents of eugenics – the Smithsonian hosted an international conference on eugenics as late as the 1920s. Langley tried to derail the Bureau of American Ethnology by bringing it under the financial control of his office. Foreseeing this eventuality and to guarantee its independence, the founding director of the bureau, Colonel John Wesley Powell, established it as an independent agency within the Smithsonian structure, appropriating and managing its own funds. When Powell died, Langley immediately undid that arrangement and effectively crippled the bureau.

One of the first organizations to take notice of the Universal Declaration of Human Rights – and the UNESCO statement on race – was the National Association for the Advancement of Colored People (NAACP), the "oldest, largest and most widely recognized grassroots-based civil rights organization" in the USA. In *Eyes off the Prize: The United Nations and the African American Struggle for Human Rights, 1944–1955*, the historian Carol Anderson chronicles the strategies pursued by the leadership of the NAACP to adopt the articles of the declaration as their own charter for action, and the systematic, and ultimately successful, opposition of the US administration to those designs which, had they been achieved, would have put the government on the spot, and under international pressure.[16]

In spite of the inherently subversive subject matter of social anthropology – taking other cultures seriously by trying to understand their belief and value systems can be a disquieting departure from the norm – its links with colonialism made it vulnerable to justified criticisms from intellectuals born in Africa and Asia. When the Uganda-born writer Okot p'Bitek, attending the Institute of Social Anthropology in Oxford in 1960 to read for a master of letters (M. Litt), went to the first lecture, the scholar giving it

> kept referring to Africans or nonwestern peoples as barbarians, savages, primitives, tribes, etc. I protested; but to no avail. All the professors and lecturers at the Institute and those who came from outside to read papers, spoke the same insulting language. In the Institute's library I detested to see such titles of books and articles in the learned journals as *Primitive Culture, Primitive Religion, The Savage Mind, Primitive Government, The Position of Women in Savage Societies, Institutions of Primitive Societies, Primitive Song, Sex and Repression in Savage Societies, Primitive Mentality* and so on.[17]

Although he completed the master's and had good things to say about Evans Pritchard and Godfrey Lienhardt, the life of Okot p'Bitek was inevitably marked by the task of translation – translating traditional African folktales and songs into English so that they could serve to define the identity of young people growing up in urban centers removed from their ancestral villages with its languages and myths. He also became a dancer and an actor, following a similar path to that of Wole Soyinka in Nigeria.

Claude Lévi-Strauss, a close friend and colleague of Alfred Métraux and one of the authors of the first UNESCO statement on race, was aware of these difficulties. In the lead article for the *Courier* in 1961 he wrote:

> for the so called primitive or archaic peoples do not simply vanish into a vacuum. They dissolve and are incorporated with greater or lesser speed into the civilization surrounding them. At the same time the latter acquires a universal character. Thus, far from diminishing in importance, primitive peoples concern us more each passing day ... The great civilization of the west ... is everywhere emerging as a hybrid.[18]

With the end of colonialism, a new situation emerged for anthropology. Lévi-Strauss referred to the "distrust" of the people who were formally the subjects of the anthropologist's fieldwork and added: "Might not anthropology find its place again if, in exchange for our continued freedom to investigate we invited African or Melanesian anthropologists to come and study us in the same way that up to now only we have studied them?"

As an anthropologist, Métraux was, in some ways, closer to Okot p'Bitek than to his friend Lévi-Strauss. A student of the great theorist Marcel Mauss, he rebelled against him and scampered to Gothenburg to finish writing his thesis with the Swedish ethnographer Erland Nordenskiöld. Nordenskiöld was a fieldworker, interested in the collection of material culture and myths, and a great advocate of establishing a partnership with the peoples he studied. The culmination of this practice occurred when he met Nele de Kantule, a Kuna leader, in Panama in 1929. Nele and Nordenskiöld considered each other as peers, and shared one common interest: the study of Kuna culture. For Nele it was very important to demonstrate, in his own terms, the value and complexity of Kuna cosmology and traditions. This for him had political value. It was as intelligent, intellectual people that the Kuna should be seen by the Panamanian state. He was successful in this, and part of the success was owed to his strategy of collaboration with anthropologists.[19] In 1931, Nele sent a young Kuna Indian, Ruben Perez Kantule, to Gothenburg to work with Nordenskiöld on classifying and interpreting the Kuna collection of material culture in the Ethnographic Museum, which the Swede directed. During that time he wrote a diary – still unpublished – which can rightly be described as a reverse ethnography, a study by a Kuna Indian of scientists at work in their museum-village.[20] Métraux met Perez Kantule and was aware of the possibilities of such collaborations, which he practiced himself in a fieldwork situation.

At UNESCO, and in Paris, Métraux was finally at the center of a very active laboratory of applied anthropology. During the 1950s he met several times per week with Lévi-Strauss, and assisted him in reinventing French anthropology. According to the late anthropologist Claude Tardits, Métraux was always willing and able to provide funding through UNESCO for a young colleague to do fieldwork in Africa or Asia. This happened very early on after he joined the organization. By the time Lévi-Strauss created the Laboratoire d'anthropologie sociale in 1958, the links between UNESCO and French researchers were well established.[21]

IV

It was a "collegiate adventure", as Peter Lengyel, a witness and participant of the early days of the UNESCO House at Avenue Kléber, described it. "Not yet heavily bureaucratized, small enough (about 700 members strong) for most people to know each other, it was a freemasonry so fascinated by its task and composition that it socialized intensively, both in and out of the office."[22] Nothing communicates better the dizzying, volcanic ambition that characterized the first decade of UNESCO's history than the journalistic record of its activities reflected on the pages of the *Courier*. The force behind this extraordinary publication was its founding editor, Sandy Koffler.

Regrettably, not much has been written about Koffler, a fate that inexplicably affects the *Courier* too. The best semblance of Koffler that I know of was penned by anthropologist Alan Campbell for a paper delivered at the Annual Meeting of the American Anthropological Association, which took place in San Francisco in 2008. Since it has not been published, I will quote from it here with the author's permission:

> Sandy Koffler was American. He had a Diploma in Ethnology and Anthropology from the École Libre des Hautes Études en Exil in New York, where he was taught by Claude Lévi-Strauss. He had graduated from New York's City University, and was at the Sorbonne, when his plans were cut short by the war.

His war experience was with the Psychological Warfare Branch in the American army.

He landed with the troops of Operation Torch when Vichy-French North Africa was invaded. There he started radio broadcasts in Rabat and Algiers, before becoming the "Voice of America" correspondent throughout the Italian campaign of 1944 and 1945.

Whenever a city was liberated, Koffler would set up a newspaper called the "Corriere" of wherever:

Corriere di Sicilia (in Catania),

Corriere di Roma,

Corriere del Piemonte (in Turin),

Corriere del Emilia (Bologna),

and *Corriere del Véneto* (in Venice).

So when he joined UNESCO after the war, he set up … the *Courier* (what else), with this astonishing creative drive he had.

Here's a passion for publishing, and publishing with a clear moral vision attached.

He said that the magazine was intended for an "enlightened" public, in particular teachers and students, and it did get its greatest readership through schools, colleges, and universities.

It set out its principal themes as:

- impact of science on human life,
- racial problems,
- art and culture,
- human rights,
- history and archeology,
- cultural differences and conflicts between peoples.[23]

Perhaps one of Koffler's most inventive editorial decisions, as it contributed greatly to the originality of the *Courier*, happened at the very onset: it was to deliver the news of what UNESCO was doing with a "straight face". In both the format of the presentation and the style of the writing, the *Courier* looked initially like a regular. newspaper . And this produced the effect, not devoid of mischief – deliberate or casual; I think it was deliberate – of placing news about culture and science, unapologetically, at the same level and rank as regular news. It assumed that there was an audience out there that would take the subject matter of the *Courier* seriously, and did not talk down to it.

If UNESCO's first years were characterized by the idealism and activism of a remarkable array of scientists, writers, artists and intellectuals from different countries and cultures, the *Courier* was its perfect mirror. This intense portrayal of humanism on the march was too good to be true, a document of an almost parallel world of bold thinking and generosity of spirit – the Republic of Letters reborn from the ashes of World War II, and with a megaphone.

Métraux was very close to Koffler and used the pages of the *Courier* to popularize anthropology. It is an extraordinary adventure to examine the collection of the magazine from the 1950s and early 1960s, when Koffler and Métraux collaborated.[24] The quality and variety of the articles and topics treated is nothing short of astonishing. In the first issue alone there were articles by Julian Huxley, Joseph Needham, the philosopher Jacques Maritain and the fiery John Grierson, who was advocating, as usual, for the value of documentary filmmaking. More remarkable still was that the flow of first-rate articles by prominent authors in many fields, as well as the announcement of projects in the arts and sciences continued unabated for years. In later issues, filmmaker Jean Rouch wrote on "the awakening of African cinema", Henry Cassirier on the effects TV has on reading, and Thurgood Marshall on "Brown vs. Board of Education". There was a long-lost letter from Gandhi to Tolstoi, articles by Miró and Calder, and the philosopher Bertrand Russell, several articles by Lévi-Strauss, and by historian Basil Davison criticizing the legacy of colonialism in Africa. There was a piece about a very young Felix Idubor, the Nigerian sculptor. There are others on "a woman's life in an African village" and on "why Japanese women won't marry farmers", with entire issues dedicated to the Universal Declaration of Human Rights, racism, the "American Negro", the treatment of foreigners, the rights and the protection of children, women's rights, "Orient-Occident: A study in Ignorance..." There is a remarkable issue entirely dedicated to "Race Relations in Brazil" with an editorial by Métraux and photographs by his close friend Pierre Verger. The only effort that compares in ambition, betting on great ideas and looking for talented people to provide them and carry them out, happened during the New Deal, when the administration of Franklin Delano Roosevelt employed writers, photographers, painters and filmmakers to document the lives of the poor and the down-and-out during the Great Depression.[25]

For the *Courier* to produce such a steady output of essays, illustrated by first-rate photographs, required an intense and equally sustained commitment, and certainly much imagination and talent. Popularizing science, particularly biology, became standardized by the turn of the 20th century. The storylines coming out of Darwin's theory of natural selection were easier to render and to elaborate upon.[26] Each generation since Darwin has had its very popular interpreter of the theory: T.H. Huxley; Julian Huxley, his grandson and UNESCO's first director-general; C.H. Waddington; Stephen Jay Gould; E.O. Wilson. Popularizing anthropology and definitions of culture is a much more complicated enterprise. The most successful practitioners – such as Sir James Frazer and Joseph Campbell – provided reductionist, teleological narratives that flatten ethnographic detail and hover well above the complexities of distinct cultures, difficult enough to understand and account for in their own right, as what the British anthropologist Rodney Needham has called "forms of life".[27]

Métraux contributed 23 articles to the *Courier*. He did not pull his punches, even though there were risks involved.[28] The cover of the August-September 1953 issue carried the picture of an African woman. She is looking at the camera with an expression of calm defiance. Next to the picture it read: "The Intellectual Fraud of Racial Doctrines". Métraux wrote the editorial with the title: "A Man with Racial Prejudices is as Pathetic as his Victim." It was a long piece that continued on the inside pages under the subtitle "Slavery Ended When Men Thought it Shameful: The Same Will Hold Good for Racism." Those were fighting words.

V

There are at least two very good reasons to wonder why the *Courier* has not yet been more assiduously mined by scholars interested in the history of UNESCO and the influence of its programs. As its "window on the world", the *Courier* was the publication of record for UNESCO and a privileged vantage point from where to look both outside and inside the organization. There is the question of the readership, conceived very ambitiously as international and multilingual. And there is also the question of what happened during waking hours, behind the scenes of the intense exercise of idealistic dreaming and projecting that UNESCO represented, when political pressures were felt, and clear differences of opinion and cultural perception brought down the spirited aspirations of universality that are UNESCO's *raison d'être*. On both questions the *Courier* archive provides an extremely rich quarry.

On the first point, Alan Campbell offers the following summary:

> The Courier first appeared in three different language editions: English, French, and Spanish. That was steadily increased over the years until by 1990 they had produced 35 different language editions, plus four different language editions in Braille.
>
> Some of these folded – the Pashto and Hausa editions were intermittent. In 1980 they did one in Serbo-Croat, but the Croatians weren't too happy, so they produced a parallel one in what they called Croato-Serbe.
>
> 1949 – 40,000 copies of the magazine were produced. At the beginning of the 1980s, 500,000. It was estimated through time that each issue would be read by more than four people, so you're getting a readership of some 2 million upwards.

> You could buy it in newsagents; you could subscribe; you could find it in libraries.
>
> The *Courier* was *never* well-known in the United Kingdom nor the United States. But if you ask people, more or less my age, brought up elsewhere, it's astonishing to find how many will say:
>
> "Oh yes, I remember the *Courier*. We used to get that".
>
> And even: "That's where I first learned about anthropology".
>
> In *my* experience I've heard that said by people from: Indonesia, India, Pakistan, Ghana, Brazil, Jamaica. If you start asking in your own circles, I'm sure you'll get similar responses.[29]

It would make a great and ambitious research project to trace the effect of the *Courier* among its different readerships – European, African, American, Latin American and Asian – interviewing people who subscribed to the magazine during the Cold War. Who read the *Courier* in the Soviet Union, for instance?[30] It would be an inquiry much like those favored by the Mass Observation project in the UK, but in reverse.[31] We know that the *Courier*'s issues were anxiously awaited by many, and in many countries – and I include myself on the list because my mother had subscribed me to it. I was totally fascinated by Aboriginal walkabouts, Japanese theater and African villages. Two anthropologists, both of them good friends of Métraux, and through him involved in UNESCO projects – Claude Tardits and Georges Condominas – told me that in the 1950s the *Courier* competed on Parisian newsstands with magazines such as *Time* and *Life*.

On the second point – the undercurrents of dissent and competition troubling the will for a harmonious, universalistic approach to culture that was put forward by UNESCO and the *Courier* – some aspects have been well researched. Teresa Tomas Rangil, for instance, has analyzed what she calls "the politics of neutrality" in UNESCO's Department of Social Sciences, from 1946 to 1956. She draws attention to the influence of the American sociologist Edward Shils in shaping the agenda of the department during its early years, and also the selection of personnel.[32] Shils was very much behind the so-called Tensions Project, aimed at explaining animosities between countries and cultures applying neo-Freudian models of psychological interpretation of what they called the "national character". These studies had no small ambition as they meant to be predictive of collective behaviors and thus useful in pursuing international peace.

Another proponent of this line of research was the head of the Tensions Project, the Canadian social psychologist Otto Klineberg. Métraux was very familiar with Klineberg and his research methods because, together with Herbert H. Hyman, Klineberg had been the principal architect of the questionnaires used by the USSB in Germany and Japan to determine the "morale" of their civilian populations during the war.[33] Métraux had to use those questionnaires in Germany during his service in the Morale Division. He did not like them one bit, and was not shy in expressing his frustration and criticizing them in the letters he sent to his wife, Rhoda. He naturally favored a more ethnographic approach to the gathering of social information. Therefore it must have been very grating for him that one of the offshoots of the Tensions Project further blurred, or crossed, the line with ethnography. The *Way of Life* book series, announced in the second issue of the *Courier*, was intended to produce volumes dedicated to explaining the cultures of countries. "A reading of one another's *Way of Life*", the article said, "should help to bring about a deeper understanding" between nations, and it went on to suggest that "when, for instance, a Brazilian member of the secretariat of one of the UN organizations or agencies is visiting Poland for the first time, the volume entitled *The Polish Way of Life* should be of value to him".[34]

The authors of the books were expected to follow the Tensions Project guidelines. This clumsy conception seemed to ignore the very existence of ethnography, or the value of literature for reaching levels of expression and interpretation only rarely attained by anthropologists and social scientists. It must have been hard for Métraux to sit through meetings discussing such initiatives, not just because of his personal experience with questionnaires prepared by social psychologists but also because of the link between neo-Freudian ideas and the "culture at a distance" and "national character" studies that Margaret Mead and Ruth Benedict – both of whom were friends of his – and Rhoda herself were strongly advocating. A few books in the series were published, and then the trail narrows.[35] A comparison with the New Deal's Federal Writers Project (FWP) and Farm Security Administration comes to mind again.[36] Even though nothing as memorable as Richard Wright's *12 Million Black Voices*, a study of black poverty in America, or Walker Evans' and James Agee's *Let Us Now Praise Famous Men* came out of the *Way of Life* series, the entire experiment is worth looking at for its colossal aims, questionable methodology and the "tensions" it created among UNESCO functionaries.

For Julian Huxley it must have meant something very distinct to become the first director-general of a UN organization that placed in the same sentence which defined its mandate the words "culture", "science" and "education". Different members of his family had been involved in decades-long arguments to define those terms and police the boundaries between them. His great-uncle, the poet Matthew Arnold, whose *Culture and Anarchy* from 1869 had launched a national discussion in England about the meaning of the term "culture" and the best ways to preserve it through teaching (he was also superintendent of schools), did not include science in his definition. He had a Ciceronian view of culture as the work of humanists and artists, passed on through great works of literature and the imagination. T.H.

Huxley – Julian's grandfather – disagreed. For him, science was a component of culture, and essential for developing a critical and enquiring mind. Years before joining UNESCO, Julian Huxley collaborated with H.G. Wells – a former student of his grandfather – in writing *The Science of Life*, which came out in three volumes from 1929 to 1930. They were books intended to be a grand interpretation and explanation of biology for the general public, and bridging the polemical gap that had separated Arnold from T.H. Huxley.

Not everybody was convinced. In his autobiography, Julian Huxley tells of a visit of writer D.H. Lawrence to the home where he and Wells were writing. Lawrence did not accept that life – the life of the mind – could be explained by biology, and he refused to look at the teeming world revealed by a microscope. Julian Huxley had strong views on this subject. He enlarged the meaning of evolution to encompass culture, proposing three stages in the process: inorganic, organic and psychosocial. He surely knew of Wells' idea, put forth in the book *A Modern Utopia* from 1905, that in the current psychosocial stage of evolution it was up to a class of ascetic and technocratic intellectuals, which he called the Samurai, to take the reins and rule the world, as some sort of post-modern philosopher-kings. It is a stretch to propose that Huxley understood UNESCO as the embodiment of the best hopes imagined by Wells, but not impertinent to consider the British polemical tradition he was a product of, and how it influenced his perception of UNESCO's role – precisely because UNESCO was not conceived as a provincial effort. It aspired to represent a heterogeneous grouping of civilizations. It must have been quite clear from the beginning, with historians of science such as Joseph Needham very much in the picture, that the task of promoting a dialogue between cultures and bringing about a world civilization was a formidable and daunting challenge.

Without leaving the confines of a single country, the UK, and while UNESCO was admirably attempting to define its universal mission and launch ambitious projects, the 19th-century debate about the meaning of the word "culture" took another, sometimes vicious, polemical turn. In 1956 the chemist and novelist C.P. Snow gave his famous Rede Lecture at Cambridge University on the "two cultures", describing and lamenting what he considered to be an antagonistic chasm separating the sciences from the humanities. Three years later the literary critic F.R. Leavis responded with his own public lecture at Cambridge, "Two Cultures?", accentuating the question mark and putting in doubt Snow's intelligence and his understanding of literature: "As a novelist (Snow) does not exist; he doesn't begin to exist. He cannot be said to know what a novel is. He is utterly without a glimmer of what creative literature is and why it matters."[37] The old British debate between Arnold and Huxley was back in force, against a context of decolonization and the widening of voices sitting at the table, pushing to be heard and have their own definitions of culture – and history – taken seriously into account.

Métraux was clearly sympathetic to Julian Huxley's notion of a "scientific humanism", and Huxley thought well enough of Métraux to define him as the embodiment of the "UNESCO man". Yet when it came to joining as an anthropologist the management of the Hylea Project in the Amazon – an ambitious conservation project very dear to Huxley – Métraux finally declined, realizing that political interests would stand in the way of any successful operation and, furthermore, that the concerns of biologists and anthropologists were certainly not identical, and another source of conflict.[38]

VI

Métraux deserves a good biography. His was an examined life, registered in clinically written diaries and letters that began in the late 1920s and ended with the Delphic note he scribbled at the Vallée de Chevreuse, a place that reminded him of the Amazon, his back reclined against a tree, looking out on the lake as an overdose of barbiturates slowly overcame him. The body was still in that contemplative position, the ethnographer's notebook resting on his lap, when it was found a few days later. The chronicler of Indians and of the African diaspora in the Americas passed away with a pen in his hand, witnessing his own death.

I do not know of any other anthropologist so relentlessly applied to the pursuit of the lives of others and the observation of his own. The private letters and diaries, as lucid and well written as the public texts of the scholar, are more than a self-portrait. They follow with intelligence the course of his discipline, anthropology, as it took shape and changed over more than three decades. Métraux's vantage point was privileged. He had been part of several intelligentsias, and always at the top – in Argentina, Chile, Brazil, the Caribbean, Mexico, Easter Island the USA and France.[39]

The cast of characters is heterogeneous and staggering. A quick sample could include Borges, Xul Solar, Breton, Drieu La Rochelle, Leger, Herskovits, Richard Wright, Bastide, Bataille, Pierre Verger, Jacques Roumain, Mme Rigaud, Margaret Mead, W.H. Auden and Lévi-Strauss.

He met Lévi-Strauss in Brazil in 1934. They walked up and down a beach in Rio, talking. "A taxi took us to a deserted beach, haunted for us by the ghosts of the Indians Jean de Lery and Hans van Staden had encountered, those Indians of whom Métraux was the undoubtable historian."[40]

Their friendship tightened during World War II, when Métraux worked at the Smithsonian as editor of the *Handbook of South American Indians*, and Lévi-Strauss took one of the last ships to leave Portugal and arrived in New York, fleeing from Nazi-occupied France. Métraux commissioned a couple of essays for the handbook from Lévi-Strauss. When he visited his friend in New York, they shared the bed in Lévi-Strauss' tiny apartment.[41]

After Métraux moved to Paris in the early 1950s to work at UNESCO, the lives of these two former students of Marcel Mauss came even closer, as professionally they began to diverge. Métraux had always found a totalizing theoretical approach to anthropology problematic, and certainly unsuited to his temperament. In the 1930s he had written somewhat fiery letters to his old *maître*, championing fieldwork over armchair speculations. The archive and history were important for Métraux. His type of anthropology owed more to Giambattista Vico than to *L'Année Sociologique*. On 9 April 1945 he wrote in his diary: "Abus de l'humanisme en France. Trop d'analise, trop de critique, trop de brillante, de menetration" (Abuse of humanism in France. Too much analysis, too much critique, too much brilliant interpretation).[42]

Throughout the 1950s, Métraux and Lévi-Strauss met frequently, every week, sometimes daily. UNESCO brought them together. Lévi-Strauss was building a formidable academic power base and sought to harness the torrent of projects that came out of UNESCO House, placing students and colleagues in them. Métraux was the man in the middle, always attempting to marry his demanding scholarship and penetrating ideas with practicalities that caused things to unnervingly disappear in a cloud of bureaucracy.[43] Around them, the intellectual stage provided by Paris was crisscrossed by historical fault lines and an abundance of journals vying to claim new horizons: Diop's *Présence Africaine*, Sartre's *Les temps modernes*, UNESCO's own *Diogene*, Mounier's *L'Esprit*, Lévi-Strauss' *L'Homme*. Métraux contributed to all of them, an "intersticial figure" in this varied archipelago.[44] The *Courier* was part of it, perhaps the most original part. It had invented a way to address an infinitely wider audience, with high standards, through articles written often by the same writers who published in the other, more restricted, journals.

Lévi-Strauss failed twice to be admitted to the College de France. That changed after the publication of *Tristes Tropiques* in 1955. In a letter to Rhoda on 13 November, Métraux wrote:

> Lévi-Strauss has also published a book, *Tristes Tropiques*, which may turn into a great literary event. The silent, the mysterious man has written a book which is a blunt confession and the revelation of an extraordinary sensitive and artistic personality. Everybody says that it is now my turn. I have signed the contract, but shall I dare to reveal myself to say openly what I have felt when I was in the field and what it meant to me to be an anthropologist?

In 2001 I interviewed Jean Malaurie, the arctic explorer and author who created the collection *Terre Humaine*, in which *Tristes Tropiques* appeared. He signed up Métraux to write his own personal book. The title was to be "La Terre Sans Mal", taken from one of the Tupi-Guarani myths that Métraux had studied. Malaurie recalled with regret that the project was sharply interrupted by Métraux's suicide. In his opinion it would have changed the course of French anthropology. At stake was the opposition between ethnography and ethnology, fieldwork and historical narratives versus theory and overarching interpretation. Malaurie approved of Lévi-Strauss' work, as it was original to him, but not of the work of his imitators. The "abus de l'humanisme" had gone too far. One afternoon, in Métraux's apartment, Malaurie was discussing with him "La Terre Sans Mal", going over the plan, chapter by chapter. At one point another guest, Georges Condominas, complained ruefully that he had wasted his life writing straight ethnographies that nobody seemed to value. Métraux stopped in mid-sentence, turned to Condominas and said: "You are very wrong. The hard-won facts that you gathered will still be useful when the theories now in vogue are long gone."[45]

The conversation between Métraux and Lévi-Strauss continued until the very end. In the suicide note recovered at the Vallée de Chevreuse there was a sentence dedicated to the old friend: "Lévi-Strauss, mon ami, je vous admire et vous aime, vous m'avez *inspiré*" (Lévi-Strauss, my friend, I admire you and I love you, you have *inspired* me).[46]

Notes

1. Rhoda Bubendey Métraux (1914–2003) was Alfred Métraux's second wife. She archived a series of letters sent to her by Alfred while he was serving in Germany with the Morale Division. The first one was written "at sea", on April 1945 en route to London and then Germany; the last one in London, on 15 August, when the mission was completed. Private collection.
2. Métraux also kept a diary during his mission. Both diary and letters, together with other documents, will be published as a book: Alfred Métraux, *The Morale Division: An Ethnography of the Misery of War*, ed. Edgardo C. Krebs (Oxford: Wiley-Blackwell, forthcoming).
3. For a book that reflects the impact of the Chaco War on the Indian populations spread over all sides of the frontiers between Bolivia, Argentina and Paraguay, see Nicolas Richard, *Mala Guerra: los indigenas en la Guerra del Chaco (1932–1935)* (Asuncion-Paris: Museo del Barro, 2008).
4. For a history of the establishment of Anglican missions in the Chaco, see Wilfred Barbrooke Grubb, *An Unknown People in an Unknown Land* (London: Seeley & Co., 1911).
5. Métraux wrote about the crisis and what followed in letters to Yvonne Odonne, and two in particular: the first one written in Mendoza, Argentina, on 27 April 1933 describes his fleeting attempt at quitting anthropology to start a Protestant mission on the River Pilcomayo to provide a refuge for the Indians; the second in Tucuman, Argentina, on 25 May 1933, outlines his plans for becoming inspector general de Indios in the Chaco. See Letters to Alfred Métraux and Rhoda Bubendey Métraux, 1932–1951, Yvonne Oddone's Papers, Gen. Mass. 350, Box 1, folder 1, Beinecke Library, Yale University.
6. Alfred Métraux, *Easter Island* (Geneva: Editions Ferni, 1957 [1941]), 75.
7. In a letter to Oddone, written in Doula, Cameroon, on 8 April 1934, Métraux described with exuberant enthusiasm his first encounter with the village life and landscapes of Africa, wondering what course his career would have taken had he decided to be an Africanist.
8. Alfred Métraux, *Le vaudou haïtien* (Paris: Gallimard, 1959).
9. Métraux and Herskovits had a mutually supportive professional and personal relationship, documented in a steady correspondence which is now archived at Northwestern University. The Carnegie Corporation's file on "Negro Study"

contains a note on "Personnel Suggestions through July 15, 1937", where it is stated that Herskovits put forward Métraux's name for the job.

10. Letter to Rhoda from Tübingen, 11 July 1945. Métraux was also considering the possibility of joining the faculty of Howard University in Washington DC, and of the New School in New York.
11. Poul Duedahl has thoroughly analyzed the struggle within UNESCO's race project between biologists and social anthropologists, each camp defending its own definitions of the concept. The victory, he concludes, went to the anthropologists, who were able to recast race in cultural, not biological terms – a perspective that had powerful real-world repercussions, and one which for the most part dominates our thinking about this issue today. See Poul Duedahl, "From Racial Strangers to Ethnic Minorities: On the Socio-Political Impact of UNESCO, 1945–64", in *Current Issues in Sociology: Work and Minorities,* ed. Gregory A. Katsas (Athens: Athens Institute for Education and Research, 2012), 155–166. See also Michelle Brattain, "Race, Racism and Antiracism: UNESCO and the Politics of Presenting Science to the Postwar Public", *American Historical Review* 112:5 (December 2007): 1386–1413, and Chloé Maurel, *"'La question des races': Le programme de l'UNESCO"*, *Gradhiva* 5 (2007): 114–131.
12. On René Samuel Cassin, see: Gérard Israel, *René Cassin,* 1887–1976: *La guerre hors la loi, avec de Gaulle, les droits de l'homme* (Paris: Desclée de Brouwer, 1990).
13. Jacques Maritain, *Human Rights: Comments and Interpretations* (London: Allan Wingate, 1950).
14. For a history of the Smithsonian's Bureau of American Ethnology, see Curtis Hinsley, *The Smithsonian and the American Indian: Making a Moral Anthropology in Victorian America* (Washington, DC: Smithsonian Press, 1994).
15. Hinsley, *The Smithsonian and the American Indian,* xiv. Jesse Green, ed., *Cushing at Zuni: The Correspondence and Journals of Frank Hamilton Cushing, 1879–1884* (Albuquerque: University of Arizona Press, 1990).
16. Carol Anderson, *Eyes off the Prize: The United Nations and the African American Struggle for Human Rights, 1944–1955* (Cambridge: Cambridge University Press, 2003).
17. Okot p'Bitek, *African Religions in European Scholarship* (New York: ECA Associates, 1990).
18. *UNESCO Courier* (November 1961): 12–17.
19. James Howe, *Chiefs, Scribes and Ethnographers: Kuna Culture from Inside and Out* (Texas: University of Texas Press, 2010).
20. I am currently translating Ruben Perez Kantule's diary.
21. Interviews with Claude Tardits and Isaac Chiva. In 2000–2002 I was able to conduct interviews in Paris with Tardits, Chiva and other colleagues and contemporaries of Alfred Métraux with a grant, HAP-22 (1999), from the Wenner-Gren Foundation for Anthropological Research. I should like to acknowledge here their generous support.
22. Peter Lengyel, *International Social Science: The UNESCO Experience* (New Brunswick-Oxford: Transaction Publishers, 1986),16.
23. Alan Campbell, "The 'UNESCO Courier', Race, and Anthropology" (unpublished manuscript).
24. In May of 2000 I interviewed Olga Redel in Paris – an editorial assistant at the *Courier* in the 1950s. She gave me a glowing recollection of the *Courier*'s staff meetings, which often took place in the cafeteria of UNESCO House, with Koffler and Métraux presiding. They were very impressive to Redel, who was very young

at the time, for the caliber of the guests that came to sit around the table, the topics discussed and the erudition of the participants. They were "real seminars", she said.

25. On this subject, see David A. Taylor, *Soul of a People: The WPA Writers Projects Uncovers Depression America* (New York: Wiley, 2009); Jarrold Hirsch, *Portrait of America: A Cultural History of the Writers Project* (North Carolina: The University of North Carolina Press, 2003).
26. For the relationship between scientific and literary narratives in the late 18th and early 19th centuries, see Richard Holmes, *The Age of Wonder: How the Romantic Generation Discovered the Beauty and Terror of Science* (New York: Pantheon, 2009). For the same topic in the Victorian Era, see Gillian Beer, *Darwin's Plots: Evolutionary Narrative in Darwin, George Eliot and Nineteenth Century Literature* (Cambridge: Cambridge University Press, 2009).
27. On Frazer and Campbell, see Marc Manganaro, *Myth, Rhetoric and the Voice of Authority: A Critique of Frazer, Eliot, Frye and Campbell* (New Haven: Yale University Press, 1992). On cultures as "forms of life", see Rodney Needham, *Exemplars* (Berkeley: University of California Press, 1984).
28. *Chloé* Maurel, *Histoire de l'UNESCO: Les trente premières années, 1945–1974* (Paris: L'Harmattan, 2010), xi. In a series of letters to Rhoda, Alfred Métraux described the problems besetting the writing and different publications of UNESCO's second statement on race. This is from one of the last letters on the subject, written in Paris on 22 April 1952: "It is possible and even likely that I shall have to go to New York very soon because of this fantastic affair of the pamphlet on race. It is a major crisis in UNESCO from which so far, I emerged safely, but a changed man. If my faith in UNESCO and in its heads is stronger than ever, I have lost my respect for many people and certainly this was also a crisis in my conscience ... Did you know that Luther Evans [UNESCO's director-general at the time] is a fine and courageous man – I would have never thought so. What a comfort to think that there are still some decent people in Washington ... To tell you the truth, I am glad to fight and I am glad that my chiefs in UNESCO gave cause to believe in their sincerity and courage." Private collection.
29. Campbell, *The "UNESCO Courier"*.
30. On the *Courier*, racism and the role of the Soviet Union in UNESCO, see Anthony Q. Hazard Jr., *Postwar Anti-Racism: The United States, UNESCO, and "Race", 1945–1968* (New York: Palgrave Macmillan, 2012).
31. On the Mass Observation project, see Tony Kushner, *We Europeans?: Mass Observation, "Race", and British Identity in the Twentieth Century* (London: Ashgate, 2004); Nick Hubble, *Mass Observation and Everyday Life: Culture, History, Theory* (London: Palgrave Macmillan, 2010); Judith Heimann, *The Most Offending Soul Alive* (Honolulu: University of Hawaii Press, 1999).
32. Teresa T. Rangil, *The Politics of Neutrality: UNESCO's Social Science Department, 1946–1956: Center for the History of Political Economy at Duke University Working Paper No. 2011–08* (April 2011), 15–21.
33. For Hyman's perspective on how the questionnaires were assembled and on the policy value of the study, see Herbert H. Hyman, *Taking Society's Measure: A Personal History of Survey Research* (New York: Rusell Sage Foundation, 1991).
34. Percival W. Martin, " 'Way of Life' Book Series Planned by UNESCO", *UNESCO Courier* 1:2 (March 1948): 6.
35. " 'The Way of Life' series gives an intensive analysis of the national character in various countries. This series includes 15 monographs, available in English

and French, on 15 countries. This set of monographs has been prepared according to a basic standard plan which permits each country to be studied as a whole" according to "Studies of Social Problems Likely to Create National and International Tensions" in *Report of the Director General on the Activities of the Organization From April 1951 to July 1952* (Paris: UNESCO, 1952), Chapter 7, 194. The countries represented in the series were Australia, Austria, Egypt, France, Greece, Italy, Lebanon, Mexico, New Zealand, Norway, Pakistan, Poland, Switzerland, the Union of South Africa and the UK.

36. The writers who participated in the FWP included, among others, Loren Eiseley, Saul Bellow, John Steinbeck, Ralph Ellison and Zora Neale Hurston – a very impressive list by any standard.
37. Leavis delivered the Richmond lecture, *Two Cultures? The Significance of C.P. Snow*, at Downing College, Cambridge, in 1962.
38. Harald Prins and Edgardo C. Krebs, "Vers un monde sans mal: Alfred Métraux, un anthropologue à l'UNESCO (1946–1962)", *60 ans d'histoire de l'UNESCO* (Paris: UNESCO, 2007), 115–125.
39. Aside from his published journals –Alfred Métraux, *Itineraires I (1935–1953): Carnets de notes et journals de voyage*. Compilation, Introduction et Notes par Andre Marcel D'Ans (Paris: Payot, 1978) – there are two main bodies of correspondence that cover the better part of three decades of Métraux's life. For the 1930s and part of the 1940s are the letters to Yvonne Oddone, which Michael Leiris thought wise to preserve far from Paris, at Yale University's Beinecke Library. For the late 1940s and throughout the 1950s, covering the UNESCO years and much else, are the letters Métraux sent to Rhoda, which are invaluable. D'Ans prepared the manuscript of *Itineraires II* but it was never published.
40. Métraux, *Itineraires*, 42.
41. Didier Eribon, *Conversations with Claude Lévi-Strauss* (Chicago: University of Chicago Press, 1991).
42. Manuscript of *Itineraires II*. A copy of the manuscript was given to me by Marcel D'Ans.
43. Letter from Alfred to Rhoda Métraux, 18 October 1952: "My work at UNESCO is fraught with so many frustrations and disappointments that I am wondering whether it is worthwhile giving to it so much of my time of my energy. We are consumed and devoured by the most insane bureaucratic machine."
44. Clifford Geertz came up with this definition of Métraux during the course of a telephone conversation we had a few months before his death.
45. Interviews with Jean Malaurie, Paris, May 2000.
46. Private collection. I would like to express my deepest gratitude to Daniel Métraux, who entrusted me with the classification of his father's papers, and to Rhoda and Vevette Métraux, both of whom I was able to interview before their passing. Guy Métraux shared his memories of his uncle. I am also pleased to acknowledge the support I received from The Wenner-Gren Foundation (HAP-22, 1999) to do archival research in Paris and interview Alfred Métraux's surviving friends and colleagues.

2
Weapons of Mass Distribution: UNESCO and the Impact of Books

Céline Giton

After World War II, for the first time in mankind's history, an intergovernmental organization had the ambition to launch a worldwide policy in the literary field. UNESCO's constitution says that the organization must assure "the conservation and protection of the world's inheritance of books", encourage "the international exchange of persons active in the fields of education, science and culture and the exchange of publications" and initiate "methods of international cooperation calculated to give the people of all countries access to the printed and published materials produced by any of them". UNESCO therefore put in place an ambitious book policy.

In order to evaluate correctly the impact of this policy, it is necessary to take into account at the same time its philosophy and goals, its actors and its concrete actions on the ground. Such an approach reveals, as we will see, a complex situation and an ambiguous policy, the effects of which on people's mindsets and peace are difficult to estimate. The growing number of member states over the years – and therefore the rising heterogeneity of international community – made it very difficult to collaborate in a spirit of peace and solidarity. Between the temptation of promoting a "universal culture" and the concrete diversity of cultures present in an intergovernmental organization, UNESCO worked in two directions: on the one hand, the research of minimum values in order to obtain a large international consensus; and on the other hand, the regionalization of projects and actions in order to gain in coherence and cultural homogeneity.[1]

UNESCO's book policy: A philosophy based on the sacralization of writing

Whereas many subjects have caused controversy since the creation of UNESCO, consensus was almost general concerning books and their benefits to human beings. Incarnating many hopes, the book was always presented as a technical and cultural support essential to development and mankind's happiness. However, this official position concealed quite different approaches for the founding members of the organization.

In a certain "French/Latin" position, the book was described as an emancipator, giving individual knowledge and stimulating greater reflexion. Thanks to these qualities, the book appeared to be an ideal support for dialogue and mutual understanding between peoples. Great books and writers played an important part in defining a common literary heritage, valorized through several programs: collection of representative works, commemoration of great writers, protection and valorization of old manuscripts and incunabula, creation and circulation of exhibitions, and setting of copyright to encourage literary creation and writers.

Meanwhile, from the Anglo-Saxon and especially US point of view, the book was above all a tool for educating people and encouraging economic development, leading to general wellbeing, and was also considered to be a useful support for communication. In this respect the main tasks were to collect and distribute books and periodicals to countries ravaged by World War II and to developing countries, to build public and school libraries, to favor the exchanges of scientific magazines, to publish and circulate textbooks and books for new literates, to encourage professional training and exchange scholarships, to organize training courses, and to use copyright to reward the distribution of books by publishers.

This double point of view reflected two visions: for France and its partners, the organization was the heir of the International Institute of Intellectual Cooperation (the League of Nations), whereas for the USA and Anglo-Saxon countries, UNESCO had to keep on with the educational activities undertaken since 1942 by the Education Conference of Allied Ministers, and especially by its commission for books and periodicals, created in January 1943 after a proposition by the British Council. However, UNESCO never formulated this ideological dichotomy underlying its speech about books. On the contrary, the organization always presented its activities as a multiform, but nevertheless coherent, book policy as a whole.[2]

What is sure is that the book, at UNESCO, was almost never questioned. Only its contents were discussed, never its qualities. For René Maheu, director-general from 1960 to 1974, the book was "the individual machine tool *par excellence*, the informant constantly available everywhere, the faithful companion of personal quest through the collective treasure of knowledge and wisdom passed on by past generations."[3] The act of reading was considered to be "a way of living, a way of asserting and developing one's personality" which was "part of the inner life, as well as the social life, of a human being".[4] And in his book *The Importance of Living*, the Chinese philosopher Lin Yutang, who was the first chief of the Division of Arts and Letters, spent two chapters talking about "the art of reading" and "the art of writing", and compared the act of reading to a journey in time and space, an excursion into a different world, a way of discussing with great people and escaping our narrow and routine lives.[5]

Some years later, in a fix film (photographs accompanied by texts) produced in 1962–1963, UNESCO affirmed again that "children need books to be happy" and that a book "is the best gift that can be put in the hands of a child".[6] Emile Delaveney, the head of the organization's Division of Publications, wrote that "the reader freely chooses his interlocutor, messenger of a thought, a wisdom, a sensibility which are part of common heritage of mankind. More and better than any other form of communication, the book is the great liberator of the human element in man."[7] Closely associated, in Western minds, to the notion of civilization, the book appears to be an instrument of freedom conveying thoughts, ideas, knowledge, symbols and dreams elaborated by other human beings.[8] Independent of passing time, it seems to be superior to other means of communication, printed or audiovisual; according to UNESCO, it creates bridges between human societies and cultures, and it can improve dialogue and mutual understanding.

This idea played an important role in the philosophy and goals of the organization: for Jaime Torres Bodet, director-general from 1948 to 1952, books constituted one of the major defenses of peace because of their enormous influence in creating an intellectual climate of friendship and mutual understanding.[9] "If UNESCO proposes to encourage the translation of the most important literary works in a great number of languages," he wrote, "it is that, because of its call to a sensitivity mixed with intelligence, because of the vibrant colors it gives to the feeling of human solidarity, literature is one of the most authentic factors of universal understanding."[10]

The book was considered as a factor of open-mindedness, emancipation and intercultural dialogue, and UNESCO wished to put forward the existence of a "common literary heritage", which should be used to bring cultures and people together. Already, Goethe had invented the concept of *Weltliteratur*, based on the existence of a common worldwide literature, but after World War II the concept was understood in a more universal way and claimed that worldwide literature was composed of all the texts produced by mankind. "There are countless forms of narrative in the world," wrote the French literary theorist Roland Barthes in 1966. "Narrative starts with the very history of mankind; there is not, there has never been anywhere, any people without narrative; all classes, all human groups, have their stories, and very often those stories are enjoyed by men of different and even opposite cultural backgrounds: narrative remains largely unconcerned with good or bad literature. Like life itself, it is there, international, transhistorical, transcultural."[11]

Since its inception, UNESCO had among its objectives the preservation, protection and distribution of a common literary heritage, postulating *de facto* both its existence and its legitimacy. This notion was regularly confirmed – for instance, by René Maheu when in 1959 he encouraged the International Federation of Translators to pursue its efforts which contributed to favor "a better distribution of common literary heritage."[12] However, this notion raised many questions and controversies because of the fact that the valorization of a common heritage implied looking back at the past and on countries with a prestigious literary heritage, and there was no way that the setting of qualitative selection could be absolutely unbiased.[13]

However, UNESCO proclaimed equality between literatures and its wish to get together the "great representative works produced by the genius of different peoples".[14] In fact, the organization did include indiscriminately, in this heritage, literature from each culture, each linguistic area and each civilization without hierarchy or preconception. However, in this attempt to define a common heritage and a universal pantheon of great writers, UNESCO embraced a memorial field largely occupied by many countries – mainly Western ones. The European member states, and especially the French, were almost obsessed with the fundamental importance of the past and wanted many works of literature on the list, as if they were nostalgic of a past which was irrevocably disappearing. The organization, which was still at that point dominated by Western member states, seemed in fact to be searching for these charismatic literary figures known to a range of people to personify its ideals and values, literary figures that could be used to affirm its international legitimacy and to create a feeling of shared heritage between peoples.[15]

In addition, the book at UNESCO was linked to the question of literacy as an important step in economic development. In the 1950s and 1960s, the UN as a whole rocked from functionalism to developmentalism, and UNESCO fully entered into the "era of development", especially through education. Literacy became an essential task of the UN and of several specialized agencies, among them UNESCO and the UN Children's Fund (UNICEF). In UNESCO's discourse, literacy, education, book and library went together, and book distribution and the creation of libraries were completed by an increase in the number of readers throughout the world. UNESCO insisted that access to books depended on social, economic and educational development, and most developing countries quickly considered literacy to be a priority.[16]

"Basic education", a concept launched by UNESCO in the early years, was closely connected to literacy campaigns. However, from the beginning the idea that teaching people to read and write can be useless also appeared, particularly in rural areas and countries where there is almost no texts in local language if new literates have nothing to read. That is what Edward J. Carter, head of the Library Division, pointed out in 1950 in a memorandum concerning UNESCO's fundamental education program.[17] He brought the subject up again two years later, saying that "in communities where there is no circulation of appropriate reading material and no stimulation to write, literacy in itself is not really significant. Experience shows that in such regions, new literates often fall again in illiteracy."[18]

In the 1960s, UNESCO made a strong link between the promotion of books and the social and economic question in developing countries.[19] This attitude of "book developmentalism" considered the book "not as an elite object of cultural exchange but as an agent of economic and political change in the underdeveloped world. This phase unfolded with the establishment of a new majority within UNESCO, the majority made up of the newly decolonized and the anti-colonial nations, many of which were part of the Non-Aligned Movement."[20]

In 1970, International Year of Education, UNESCO asked the former head of UNESCO's Applied Science Department, Herbert Moore Phillips, to write the booklet *Literacy and Development*, in which the access to books was presented as fundamental for development.[21] This theory, a fundamental part of the industrialization ideal advocated by the Western world, resulted quite naturally in valorizing non-fiction books: textbooks, scientific and technical books, and professional literature.[22] This is what UNESCO underlined in a report produced the same year for the UN's Economic and Social Council (ECOSOC).[23] With this vision concentrating on development, education took on growing importance in UNESCO's program, and its budget was always larger than those of the cultural and scientific sectors. For instance, two centers dedicated to the production of textbooks were created in Africa – at Accra in Ghana and Yaoundé in Cameroon – and through UNESCO, textbooks for African schools were printed abroad free of charge.[24]

Lastly, books were considered by UNESCO as tremendous communication tools. The US vision in particular considered books first and foremost as a medium for spreading knowledge. Whereas the US Information Agency coordinated US cultural propaganda abroad from 1953 onwards, the USA encouraged UNESCO to consider communication as a legitimate field of action. As early as 1945, Archibald MacLeish from the US delegation showed great interest in mass media and asked UNESCO to include them in its activities.[25] Indeed, with the acceleration of international exchanges and technical innovations in transport and communication, information became a crucial challenge of the modern world. At UNESCO, this interest was reflected in projects encouraging books and information distribution, data centralization and networking. The opening of an International Literary Exchanges Center, the collection of data concerning books, the harmonization of bibliographical statistics, the standardization of bibliographical standards and the large-scale adoption of Dewey classification played their part in this dual movement of gathering information through worldwide networks and redistributing it across the entire world. In this way the book was first of all considered a wonderful communication tool.[26] It was valued not for its esthetic, emotional and patrimonial dimension but essentially as a medium of information.

However, whereas the UN and UNESCO became day after day, membership after membership, the center of "worldwide public opinion", in the 1970s the latter denounced more and more firmly the rising of inequalities in the circulation of information, and asked for a New World Information and Communication Order – a sensitive subject which became one of the reasons explaining the USA's withdrawal from UNESCO.[27] This subject was all the more sensitive because there were also problems with censorship and state control of media in non-democratic states. Many of the programs supported by UNESCO were "designed to offset imbalances in the communication system" and were opposed to the Western doctrine of "free flow", masking in reality "a one way traffic between the dominating and the dominated".[28] This raging debate corresponded to a real debatable point regarding the place of information in the organization, given, for instance, that the Library Division, which was attached to the Cultural Sector and under French influence, and the Social Sciences Division, which was attached to the Scientific Sector and under US influence, quarreled for a long time about the responsibility for projects linked to documentary centers and information services.[29] Michael Keresztesi notes that the Division of Scientific and Technological Documentation and Information worked mainly to encourage a greater integration of the scientific community, mainly to the benefit of the advanced countries, and that "its mission-oriented and well-focused program sharply contrasted with the many-faceted and broad-gauged activities of the Department of Documentation, Libraries and Archives".[30]

Therefore the book raised expectations, stakes and sometimes excessive hopes, which UNESCO had to take into account. A mixture of these visions often appeared in speeches and publications, and they were used to explain and legitimize the choice and usefulness of the organization's projects.

On the ground

UNESCO's book policy took four main forms: normative action; preservation and valorization of worldwide literary heritage; encouragement of increased professionalism of people working in the book industry; and direct action to promote books and reading.

Of all the standard-setting instruments adopted between 1945 and 1975, five concerned the book: the Agreement on the Importation of Educational, Scientific and Cultural Materials, also called the Florence Agreement; the Universal Copyright Convention; the Convention Concerning the International Exchange of Publications; the Recommendation Concerning the International Standardization of Statistics Relating to Book Production and Periodicals; and the Recommendation Concerning the International Standardization of Library Statistics. Of these instruments, the most important were the Florence Agreement and the Universal Copyright Convention. Both of these resulted in confrontations between Western states as book producers and developing countries as book importers.

The Florence Agreement was adopted in 1950 and came into effect in 1952. After two meetings in Geneva, in 1967 and 1973, the General Conference of UNESCO revised it in November 1976 and adopted the Nairobi Protocol, which enlarged the scope of the agreement by extending its benefits to new technological supports. The protocol took into account some requests of developing countries – for instance, the fact that tax-free books for higher education institutions had to be explicitly adopted or recommended as textbooks by the institutions concerned in these countries. On the other hand, the protocol stated for the first time that customs exemption was extended to all books and was no longer limited to educational, scientific and cultural materials. In this way the Nairobi Protocol contributed to accentuating the imbalance in the flow of books between Western and non-Western countries.

Concerning the Universal Copyright Convention, it was adopted in 1952, came into effect in 1955 and was revised in 1971. It largely arose from diplomatic negotiations carried out in the interwar period in order to move closer the two already existing copyright systems in the USA and Europe. When it was adopted the convention widely reflected Western states' interests as the main book producers. Nevertheless, after a close collaboration with the Copyright Office of Washington in the 1950s to prepare the convention, in the 1960s the UNESCO secretariat became more critical and demanding, sometimes almost rebellious, in its relations with Western institutions, such as the Copyright Office, but also the United International Bureaux for the Protection of Intellectual Property in Berne.

In July 1961, for instance, the chief of the Copyright Division, Díaz Lewis, wrote out for the director-general a detailed description of the copyright situation in African countries after decolonization, in which he highlighted the fact that copyright laws were not adapted to African realities and that these countries needed more flexible and pragmatic legislation.[31] Indeed, African countries were mainly consumers of Western books, especially textbooks, and their intellectual production was very limited, so the existing legislation forced them to pay very high copyright fees to Western publishers and booksellers in order to have books for schools and pupils. Lewis put forward the importance of an international conference to be organized by UNESCO in Brazzaville in 1963 in order to advise African countries in copyright matters and to encourage legislation adapted to their situation. In letters and memoranda concerning the preparation of the working documents to be used during the Brazzaville Conference, several members of the secretariat also criticized US arguments used to impose copyright in developing countries, in particular the idea that the non-protected works were harmful to protected ones, that Western cultural industries needed copyright to survive, and that paying copyright fees was beneficial to developing countries' economies.[32]

The outcome of the Brazzaville Conference was a revision of the convention in order to take into account the needs of developing countries, and to prepare a draft copyright law adapted to African realities.[33] For this last project, UNESCO and the United International Bureaux for the Protection of Intellectual Property began to collaborate, but the Berne bureaux quickly proposed to exclude Ghana's representatives from the working group because that country had recently adopted a copyright legislation which was not favorable to Western interests. Refusing to give in to this kind of blackmail, the secretariat insisted on including Ghana's representatives, explaining that "the representation of an African country which has already adopted a legislation in this matter... seems very appropriate, because the lessons from the legislation enforcement can bring an effective aid to the committee's works".[34] Finally the meeting of African experts had a considerable influence on the Berne convention's revision in 1967 and the revision of the copyright convention in 1971.[35]

However, the change to the convention in favor of developing countries, especially with the adoption in 1971 of an appendix declaration that recognized "the temporary need of some States to adjust their level of copyright protection in accordance with their stage of cultural, social and economic development", caused some disappointment in Western countries. These countries then decided, with the diplomatic Conference of Stockholm in 1967, to begin a huge structural and administrative reform of the Berne bureaux in order to transform them into a World Intellectual Property Organization.[36] Created in 1967, this came into force in 1970 and became an organization within the UN in 1974, largely stripping the Universal Copyright Convention of its importance in real life.[37]

In the area of literary heritage, UNESCO also launched several programs with a global impact, even though they were not all equally successful.

One outcome was the inventory, safeguard and valorization of archives and old Arab, Coptic and Burmese manuscripts. For instance, between 1956 and 1961, UNESCO set up a microfilm unit which worked on a million pages of archives in eight Latin American countries. In the field of old manuscripts, the key project of UNESCO took place between 1960 and 1984 in Egypt, and consisted of inventorying, spelling out, microfilming and publishing a facsimile edition of the 13 Coptic manuscripts of the Nag Hammadi Library. UNESCO also focused on the collection and preservation of oral heritage, mainly in Africa after decolonization, and contributed to the written transcription of several African languages between 1963 and 1968. Thanks to Amadou Hampâté Bâ, the organization launched a huge collection program of African oral traditions between 1964 and 1974, within the framework of the General History of Africa project, and in 1968 participated in the creation of a Regional Centre for Research and Documentation on Oral Traditions in Niamey (Niger). Over the years, this center collected thousands of oral traditions and stories before their disappearance from popular culture due to urbanization, and economic and social changes on the African continent. It offered a unique resource center regarding African culture and traditions, open to all researchers and potential publishers, and then participated in the preservation and promotion of African culture, both in Africa and throughout the world.

The translation and circulation of works considered as "worldwide classics" with the Collection of Representative Works was another outcome and represented the key project of the Arts and Letters Division. Between 1948 and 1994 it published 866 books from all over the world, written in 91 different languages – in Bengali, Korean, Greek, Hungarian, Italian, Pali, Romanian, Turkish and so on. The aim was "to encourage the translation, publication and distribution in the major languages – English, French, Spanish and Arabic – of works of literary and cultural importance that are nevertheless not well known outside their original national boundaries or linguistic communities".[38] Taking the example of the Japanese author Yasunari Kawabata, Jens Boel remarks that "in many cases, the translation of outstanding literary works from 'small' languages has helped to achieve both international recognition for the author and wide distribution of his or her works".[39] However, owing to the lack of a substantial budget, UNESCO resorted to external editors, which limited considerably the symbolic and concrete impact of the project. Even if the collection "was influenced by the idea that distinct locations could learn a lot about each other – and hence build the framework for lasting world peace – if they read each others' representative literature in translation", the project concerned in reality first and foremost a selected Western or Westernized elite and encountered many difficulties in its efforts to reach the general public, especially because most people on earth couldn't read the languages used for publication.[40]

In its early stages, UNESCO also launched a program called Commemoration of Great Men in order to put forward artists, writers, intellectuals, educators and scientists from all over the world on the occasion of their "birthday". Between 1946 and 1965, 12 writers were celebrated: Johann Wolfgang von Goethe, Alexander Pushkin and Edgar Allan Poe in 1949, Honoré de Balzac and Confucius in 1950, Adam Mickiewicz in 1955, Sholem Aleichem in 1959, Anton Chekhov in 1960 and Rabindranath Tagore in 1961. The absence of writers from Latin America, Africa and the Arab world was due to complex reasons, not only historical ones, such as the late development of written fiction in these regions, or political ones, such as the cultural imperialism of the great powers, but also ideological and cultural. For example, through this program, many countries chose to celebrate famous educators, philosophers, scientists and revolutionaries more than writers. Moreover, the lack of budget and the strong opposition of some states, especially the USA, prevented this program from giving birth to a worldwide literary Pantheon. This failure resulted from a deep ideological disagreement between states concerning UNESCO's role, the existence of a "common" literary heritage and the place given to writers in society. For two centuries, collective memory phenomenon had been monopolized by nation states or even smaller particular groups, so it was impossible for UNESCO to create at once a sense of "world citizenchip" through culture and literature.

A third area of action consisted for UNESCO in facilitating and supporting the professionalization of the book world by creating or subsidizing schools and training centers, and giving scholarships and organizing events such as seminars and training courses for librarians, writers, translators, booksellers and editors/publishers. With its studies, research projects and numerous publications, UNESCO also played the role of a resources center.[41] The organization "produced and sponsored more coverage of the situation of book production and consumption across the globe than it did at any other point, in the process forming the first archive of postcolonial book history".[42] The financial help given to professional networks, the creation of pilot libraries, the promotion of modern Western methods and standards, the production of materials, books, periodicals and films for book workers, contributed not only to professionalizing people but also to bringing into the foreground a true *esprit de corps*. With the creation of international consultative committees, UNESCO also wanted to bring together the different groups working in the field of books and the cultural NGOs – for example, it managed to move librarians and documentalists closer, and the organization of International Book Year in 1972 made contacts and collaborations possible between NGOs dedicated to book promotion.

Finally, UNESCO directly encouraged books and reading through three types of activity. First, it favored a rebalancing of book circulation around the world by encouraging literary exchanges and by launching a book donation program called the Book Coupon Scheme. In 1954 some 33 countries participated in the project. UNESCO was then in contact with about 3,500 libraries in 78 countries, and it published a booklet listing all the libraries' projects and needs for potential donors. However, the scheme achieved mixed success and only collected a few donations, mainly from the USA, France, Japan and the UK. Its impact was quite limited because of the attitude of the USA, which saw it as a competing project for its own fundraising campaigns linked to development aid and reconstruction, thus giving many more books through its own programs than through the Book Coupon Scheme, which planned for books to be selected by receivers, not by donors.[43] During its first years, UNESCO gave modest aid to different countries – for instance USD620 to Bayreuth Library in 1951, USD600 to Rheydt Public Library, USD200 to the library of the Munich Quaker Student Centre, USD350 to the Budapest National Library, USD120 to the library of Iwate in Morioka in Japan, USD350 to Osaka University Library, USD1,000 to Miyazaki University Library in Japan, USD620 to the Doshisha University Library in Kyoto, USD550 to the Philippine University Library in 1952, USD550 to the library of the Prague National Museum, USD500 to the Belgrade National

Library and USD400 to the Patras Common Reading Room in Greece in 1953.[44]

In 1948 the US Government invited UNESCO to become a partner in the program called Cooperative for American Remittances to Europe (CARE), when it was extended to include book donations to libraries. Launched in 1949, it included several US institutions that wanted to encourage a better understanding of the USA around the world and targeted schools, universities, and technical and medical societies in Europe and Asia.[45] The books given were mostly American and all were written in English. They were chosen by a commission directed by the Librarian of Congress, Luther H. Evans (before he became director-general of UNESCO), with the help of the American Library Association and some important libraries. The aim was to give preference to recent technical and professional books and manuals.[46] Like other book donation programs, the CARE program was designed largely to distribute US and UK books abroad, and most research suggests that it "actually stifled local production rather than encouraging it – by, for instance, offering titles at highly subsidized prices and thus driving more expensive local books out of the market".[47] UNESCO quickly felt uncomfortable with this project and asked the Americans to improve communication, to include non-US books and to let the librarians choose the books they wanted to receive.[48] Some members of the secretariat also feared that the Americans took advantage of the name and image of UNESCO for their national propaganda and considered books first and foremost as "invaluable tools for indoctrination and training of the vast indigenous personnel required for overseas operations".[49] Disapproving of the methods and goals of the CARE program, UNESCO finally broke the partnership in December 1952 and after that concentrated its efforts on its own Book Coupon Scheme. In 1956 the total budget of the project, since the beginning, was USD9 million, a quite insufficient sum compared with the huge needs of the program.[50]

UNESCO also encouraged reading with two specific programs. One concerned libraries, which were a major area of activity during the first years of the organization, reflecting the importance of libraries in Anglo-Saxon societies. UNESCO insisted in particular on the role of the public library, "a product of modern democracy and a practical demonstration of democracy's faith in universal education as a life-long process... a living force for popular education and for the growth of international understanding, and thereby for the promotion of peace".[51] Influenced by the activities of the British Council and the Carnegie Foundation during the interwar period, UNESCO gave grants, scholarships and books to many libraries throughout the world and created several important pilot public libraries in Delhi, Bogota, Enugu and Dakar. However, an important administrative reorganization in 1967 and the organization of a series of regional conferences on library and documentation planning, such as in Quito in 1966, Colombo in 1967, Kampala in 1970 and Cairo in 1974, were a turning point in this field, with the cultural aspects of libraries included henceforth from that time on in a larger reflexion concerning technical and scientific documentation.

UNESCO also showed great interest in the challenge of giving new literates around the world enough books to read, in particular by including libraries in literacy projects.[52] The organization worked out tools and material, distributed two booklets to encourage literacy and tried to promote innovative methods.[53] From 1955, UNESCO launched a huge program entitled Reading Materials for South-East Asia. The idea was to produce and distribute texts in local languages for new literates in this region, where English and French were not spoken by the populations. This project took off with the creation in August 1958 of a regional center in Karachi, and between 1958 and 1967 this center helped in the publication of more than 500 books in 22 languages – such as Urdu, Hindi, Burmese, Bengali, Kannada, Kashmiri and Pendjabi – concerning concrete subjects such as aviation, radio, TV, the history of food and the peaceful use of atomic energy. Because of their scientific and practical content, these publications clearly had an educational aim and never proposed fictional narratives. However, even if "UNESCO multiplies missions and centers, supported by extensive campaigns," claims Philippe Moreau Defarges, "results are disappointing (notably in Africa); literacy cannot be imposed and, to progress, needs very precise conditions".[54] Despite extended efforts, in 1966, René Maheu denounced the existence of 700 million adults and more than 100 million young people who were illiterate in the world.[55]

Finally, UNESCO encouraged reflexion on the book and its promotion, by organizing conferences, symposiums and different kinds of event with writers and book professionals, by helping its member states to launch national book policies, by creating regional centers for book promotion, and by coordinating the symbolic International Book Year. The organization thus tried to supervise its member states, especially the developing countries, giving them advice to implement book policies, taking into account the different sectors of the book industry. Nevertheless, despite UNESCO's efforts, only 20 or so countries had a national body in charge of the book question in 1972. Concerning the book regional centers created by UNESCO in Karachi and Bogota at the beginning of the 1970s, they progressively took over the organization's book policy in Asia and Latin America.

The many actors

UNESCO's book policy was not exactly the same as a "normal" national public policy, and one of its most challenging aspects was the multiplicity of actors involved. Internally that would be the member states, the General Conference, the executive council, the director-general and the secretariat, and outside the book professionals and organizations, US foundations, other UN organizations and international organizations, NGOs and so on. Generally speaking, UNESCO's book policy depended a lot on personal commitments, and the role of civil servants was of crucial importance to encourage, promote and carry out most of the projects.

From the executive council, some representatives were particularly important for the book policy. Between 1945 and 1974 some 52 out of 166 representatives at UNESCO had what can be called a "literary background": 37 of them were writers, poets and intellectuals and thus in a majority, compared with other professions – mainly directors of archives/libraries, journalists, teachers, senior officials in charge of book questions and publishers. All in all, 24 countries appointed, at one time or another, a "literary" figure to represent them at UNESCO, and six of them almost constantly chose literary profiles: Brazil, France, Mexico, India, Turkey and the USA. Some members of the executive council were particularly active in the book policy, such as Amadou Hampâté Bâ, who was directing the collection of oral traditions in Africa; Julien Cain, who was involved in the development of libraries, book exhibitions and International Book Year; Luther H. Evans, who participated in the preparation of the Universal Copyright Convention before he became director-general; Josef Grohman, involved in the development of libraries and International Book Year; and Ventura Garcia Calderón, involved in the Collection of Representative Works. The most prestigious of these representatives was probably the famous Chilean poet Pablo Neruda, holder of the Nobel Prize for Literature, who participated in the executive council in 1972–1973, before the assassination of Salvador Allende in September 1973.

In addition to the executive council the director-general doubtless had a certain influence on UNESCO's book policy, but one which is difficult to quantify. Between 1946 and 1974, four directors-general had close links with books: Julian Huxley, Jaime Torres Bodet, Luther Evans and René Maheu.[56]

In fact the secretariat as a whole played a fundamental role in UNESCO's book policy concerning operational decisions and the concrete impact of the policy. These operational decisions consisted of financial gifts, such as donations of books, mobile libraries, equipment and microfilm material, and the setting-up and implementation of technical assistance projects, such as expertise and advice missions, and the establishment of libraries. They also involved the launching and management of collective projects, such as help for the production of reading material in Asia and the creation of regional centers for book promotion, plus diplomatic interventions, such as financial negotiations with governments for the projects, and presentation and mediation on copyright questions. The secretariat played a crucial role in these choices, thus exercising considerable influence on the concrete actions on the ground.

Globally, the book policy was implemented by three departments and five divisions at UNESCO's headquarters. In each of these administrative divisions, some civil servants were very influential, such as Jean Thomas,

Roger Caillois and François Hepp (French members), Edward Carter (British member) and Julian Behrstock and Everett Petersen (American members). In contrast, UNESCO had many difficulties in attracting the support and friendship of prestigious figures from the intellectual and artistic world outside its headquarters. In fact, diplomats, administrators and managers quickly replaced intellectuals and scientists, who were relegated into the background within the International Council for Philosophy and Human Sciences, and many intellectuals, such as Dubuffet, Ionesco and Benedetto Croce, thus showed skepticism about cultural interventionism and considered UNESCO to be an "erroneous venture" full of contradictions between ethical objectives and political considerations.[57] Whereas Torres Bodet regretted the virtual absence of African, Asian and South-American intellectuals at UNESCO, the French historian Chloé Belloc argues that communism also took many European intellectuals away from the organization, because they considered it to be a representative of US economic liberalism.[58]

However, UNESCO managed to bring together a small circle of book professionals and intellectuals who carried out expertise and consultation missions for the organization and influenced its book policy. They used their intellectual resources and networks to serve this policy, and the variety of their profiles contributed to maintaining a certain geographical and thematic balance. These personalities were not civil servants, so they had considerable room for maneuver compared with the secretariat and could be considered as "unofficial ambassadors" of UNESCO's book policy in the field.

The various programs on books also needed an active collaboration with a lot of other partners, quite different from each other, and this included hundreds of temporary experts and consultants, national delegations and commissions, other UN organizations and the UN Development Programme (UNDP), NGOs in the book field, such as the International Pen Club, private actors, such as translators, publishers, editors and printers, and some *ad hoc* consultative committees created by UNESCO. The organization also had to take into account the work accomplished in book development by other actors, notably national structures such as the Washington Copyright Office, the British Council and the French National Library, US foundations, such as the Ford, Carnegie and Rockefeller foundations, and the Franklin Book Program. Among the players were also other international organizations, such as the Arab League Organization, the European Council, the European Economic Community, the Organization of African Unity and the Organization of American States. One of the great ambitions of UNESCO was to channel, concentrate and centralize the disparate efforts made by all these actors in order to set up a real "worldwide book policy", but this task was quite difficult because of historical, economic, political, cultural and ideological reasons.

Did UNESCO's book policy change people's mindsets and encourage peace?

Evaluating the effect of UNESCO's book policy in changing people's mindsets appears to be a difficult exercise. Indeed, if this policy had a positive impact by encouraging education and reading, its effects on the "peace" aspect is more complicated to evaluate.

From their creation, the UN and UNESCO saw their origin in the humanist philosophy of the Enlightenment and the world peace concept emerging at that time, which imagined "a sort of permanent congress of monarchs, which should have a right of arbitration".[59] The creation of the UN corresponded to a wish to unify mankind in its symbols and ambitions, with the idea of imposing the main features of the Western political and social system at an international level.[60] To many people, war appeared to be an absurd and outmoded phenomenon. The UN system originated in part from the functionalist theory explained in 1943 by David Mitrany in his book *A Working Peace System*. This theory implied the possibility of creating a universal identity by constructing an "international civil society" and insisted on economic wellbeing, technical cooperation, and the fundamental role of technicians and specialists faced with political powers.[61]

The UN Charter referred to cultural cooperation as a factor of mutual understanding, and UNESCO became responsible for this task. Its constitution says that peace must "be founded, if it is not to fail, upon the intellectual and moral solidarity of mankind". The underlying idea is that economic and political exchanges between peoples need to take into account cultural, social and human aspects.[62] UNESCO subordinated its educational and cultural goals to political goals such as safeguarding peace and international understanding, moral goals such as the protection of human and children's rights, of the fundamental liberties and the defense of justice and law, and economic and social goals such as the contribution to development permitted by education, culture and science for the material and moral expansion of mankind; "cultural cooperation therefore appears, more than a goal in itself, as a mean to achieve higher objectives".[63]

However, the idea of promoting peace, security and international understanding only with education, culture and science is a very complex one, and UNESCO quickly became the most politicized of the UN specialized agencies. Year after year the organization was more and more heterogeneous from a political, ideological and cultural point of view.[64] Whereas Western countries gave their preference to intellectual activities and the circulation of knowledge, most of the developing countries asked UNESCO to take on urgent problems such as poverty, illness and illiteracy, through education and technical assistance. This indirect way of contributing to peace finally became the major part of UNESCO's program, and the evolution of its budget until 1974 shows a focus on the field of education, which seems to be a deliberate choice due to the rising importance of "technical assistance" in the UN system.[65] In 1949, Jaime Torres Bodet estimated that "if UNESCO wants to serve the cause of peace, it must concentrate on the concrete needs of mankind. It must not be an academy preaching virtues of theoretical pacifism without thinking of the means to achieve peace, nor an institution which, in the name of the primacy of intellectual life, considers culture as an end in itself, keeping it artificially away from social and economic factors concerning its development."[66]

Attaching great importance to the role of individuals for the establishment of peace in the world, learning from the mistakes of its elitist predecessor, the International Institute for Intellectual Cooperation, UNESCO claimed: "culture is not the monopoly of an intellectual minority, but progress in culture, education and science must benefit all mankind".[67] What really matters to achieve peace is solidarity between peoples and not between governments. During this period the ideal of mutual understanding between ordinary people replaced the aristocratic ideal of intellectual exchanges between the elites, "and this, all the more because culture is not any more the privilege of a few advantaged minorities but becomes accessible to everybody".[68] UNESCO worked therefore to encourage "mutual understanding" but could never prove scientifically by what process a better objective and intellectual understanding of others would naturally lead to collaboration and peaceful relations with them.[69] This limit, more generally, is part of the difficulty in linking, by a relation of cause and effect, education and culture on the one hand and peace on the other.

For some research workers, the official goals of UNESCO – peace, security and human rights – are not the real objectives of the organization. For them, UNESCO was set up first of all to favor the implementation of the necessary infrastructure for the development of a capitalist economy in developing countries.[70] This objective is not denied by UNESCO. On the contrary, economic development, through technical assistance, was always presented as a way to improve the wellbeing of humanity.

In any case, the conceptual presupposition that favoring education and culture should favor peace evolved over the years and was progressively challenged at UNESCO in the 1970s. Many personalities, such as Sulwyn Lewis, Paul Lengrand and Jacques Havet, expressed their doubts about the efficiency of international cultural exchanges in promoting understanding and peace.[71] In 1975, Jacques Rigaud said it was vain to hope that culture could reconcile people whereas "it is the expression of contradictory aspirations and contains the seeds of everything that can divide us . . . Culture, that should unite and fulfill us, divide and ruffle us."[72]

The idea of using literature to promote peace, which was at the heart of UNESCO's book policy, was therefore increasingly denounced as utopian. The word "book" indicates a medium, not its contents, and many intellectuals consider that books don't automatically encourage peace. UNESCO itself was conscious of, and never really solved, this question. Gaston Bouthoul explains, for instance, that "even popular poetry glorifies military exploits and heros that accomplish them, much more than peace... Sometimes regretting peace is even expressed in glorious stories."[73] As the French writer Pascal Bruckner claims, "There is never a direct link between a work of art and life. With a book, I can forget my prejudices, commune with the universe of a Chinese or South-American writer, fully enter another age, others customs, but it doesn't change anything in my open-mindedness when I leave the literary field. Sceptical, ironical when reading, temporarily liberated from thousands of links attaching me to my community, I become again sectarian, partial, quick-tempered as soon as I return to my century facing my fellow men. Alone, a work of art cannot eradicate the barbarian background of mankind."[74]

Jean-Baptiste Duroselle adds that "peace is a blessing in itself, an essential blessing. But it is not an absolute value. Human dignity, freedom, are higher values. One can conceive, at the ultimate level, violent rebellion against oppression."[75] Nevertheless, with its particularities and its importance for Western culture and development, the book has always been one of the pillars of UNESCO's actions. Its specificity was closely linked to its ambiguous position after the Renaissance: from that time, the book became both a potential tool for education and emancipation of the masses, and an instrument used by politics and diplomacy. As Canadian literature professor Sarah Brouillette remarks, "promoting the book wasn't about promoting it in any form and by any means – it was about unearthing the total interdependence of economic and intellectual systems, and recognizing that the book had become a specific kind of tool: a tool controlled by a small part of the world's population, but needed for participation in a global conversation about what kind of global order would unfold in the wake of colonialism".[76] The book quickly became a major tool for cultural foreign policies of states, which wanted "above all to modify, for the better, their image abroad and therefore change others' vision of them, transform their imagination".[77] The book seemed less frightening than a true weapon, less aggressive than a military expedition, but was nevertheless a very useful diplomatic tool.[78]

In the first decades, the great Western powers managed to use UNESCO's book policy not only to legitimize a Western vision of the book in society but also within the context of their national cultural and linguistic policies. UNESCO served in particular as a go-between for US conceptions concerning copyright, the promotion of the public library as a place for cultural activities, the demonstration effect of "pilot projects", the promotion of library planning conception, the discrediting of the traditional "erudite librarian model", library architecture, the donations of books written in English throughout the world, and the promotion of the Dewey system of classification.

More generally, UNESCO's book policy met difficulties in avoiding the pitfall of Western centrism. Many examples illustrated this problem, from the fact of imposing Western ideas in developing countries, such as the primacy of literacy, copyright or the public library concept, to the focusing on Western writers in the program of commemoration of great men and the use of the three main languages (English, French and Spanish) for practical and financial reasons for almost all UNESCO events. UNESCO then certainly contributed to the standardization and Westernization of practices and mentalities by advocating for children the reading of adapted books which were often translations of Western books, by separating clearly first-degree courses from in-service training, by compartmentalizing books and culture in closed places such as schools and libraries, by depreciating oral traditions with regard to writing, by seeing the only "true" culture in the Western means of communication such as books, but also radio and TV, and by expressly reducing the book to its educational aspects. In these conditions, UNESCO's book policy has probably encouraged a feeling of frustration and aggression in developing countries, which is quite the opposite of encouraging peace.

However, this policy was, of course, an interactive phenomenon which provoked reactions from the different actors concerned. The book experienced a real revolution in the 20th century, and its form, content and use changed. On a global scale, the modern printed book, conveying ideas, culture and imagination, has been appropriated, distorted and used by many non-Western peoples. By an acculturation and reappropriation process, a growing number of countries and new literate populations entered the worldwide literary scene, which is well illustrated by the evolution of the Nobel Prize for Literature from the 1960s. Thanks to the newly decolonized and anti-colonial nations forming a majority in the General Conference of UNESCO, which denounced imperialist attitudes and projects, these changes were accepted by the organization and contributed toward progressively modifying its book policy between the 1950s and 1970s.

On the other hand, both Western and non-Western countries – or, more precisely, their economic and political elites – were reluctant to promote all the books for everybody. Non-democratic states especially have been fearful of their potential power of subversion. The choice of UNESCO's book policy to favor educational books, such as specialized and technical books, instead of more "literary" books, such as essays, fiction and philosophical books, then seems to correspond to general educational conceptions and fears of political elites faced with the subversive and emancipatory potential of the book. For the American philosopher Martha Nussbaum, "educators worried about economic growth are not content with ignoring the arts. They are fearful of them. A growing sympathy is a particularly formidable enemy of closed-mindedness, whereas a morally obtuse mind is necessary to apply economic development programs ignoring inequality."[79] For the elites, ordinary people must be advised in their readings by schools and libraries:

> Almost all the mass literacy campaigns conducted during 20th century at national or worldwide level – by UNESCO for instance – in developed countries or in ex-colonies, concerned above all the development of reading and not writing. Obviously, this choice is the conscious result of the educational vocation of institutions which, everywhere, elaborated learning ideologies and methodologies... And there is something more at the root of this universal choice, shared by all the authorities and powers: the idea that reading was, before the age of television, the best vehicle for spreading values and ideologies, and therefore the easiest to regulate, once the process of production, distribution and preservation of texts can be under control.[80]

From this point of view, the wish to democratize the book and to encourage the production of books in local languages seems an ambiguous choice for an intergovernmental organization such as UNESCO. Developing these aspects could indeed lead to challenging the economic model of industrialization, the political model of a centralized nation state and the cultural model of constructing a nation around a unique language to the detriment of regional and local senses of identity. This complex environment explains the relative "failure" of humanist and universalist projects and the "success" of technical and legal projects, such as standardization, professional training and material equipment. However, the global impact of UNESCO's book policy in people's mindsets remains difficult to assess. By concentrating its efforts on diversified operational tasks, UNESCO gave legitimacy to its book policy and solved its fundamental contradiction: the wish to maintain peace – that is say, the *status quo* resulting from World War II – in a world experiencing a complete upheaval at all levels, not least decolonization and globalization.[81]

UNESCO's book policy had positive effects in many respects: libraries were created everywhere, thousands of people around the world learned to read and write, book distribution improved, many book professional were trained, precious archives and manuscripts were microfilmed and safeguarded and so on. For Michael Kereztesi, UNESCO seminars, meetings and conferences were meticulously prepared and provided valuable educational experiences to hundreds of participating librarians from all over the world, and "through its regulatory function, UNESCO has achieved standardization in many professional areas, as in statistics, bibliographical description, terminology and others".[82] Of course, the book policy suffered from the now well-known general dysfunctional aspects of UNESCO, mainly the administrative burden, red tape and bureaucratic inefficiency, the low budget, the wasting of money and communication difficulties. Its policy also encountered difficulties in reaching the general public, specific minorities and even some member states among the poorest and the less strategic from a Western point of view. Yet Richard Hoggart, who was assistant director-general from 1971 to 1975, wrote in his book *The Uses of Literacy* about the indisputable interest of many of UNESCO's programs, such as literacy campaigns, and concluded that "in spite of incredible, baroque and disconcerting failures, UNESCO remains one of the most promising institutions created in this ambiguous century".[83]

By spreading the book in the world, UNESCO contributed to transforming it into a tool of emancipation, support for the denunciation of Western ethnocentrism, a way to share imagination and to perpetuate oral literature. Its book policy therefore achieved results quite different from the official goals concerning peace and stability. In fact, it encouraged democracy and individual emancipation through books that "question and shake preconceived ideas, challenge our certainties, and show... grey area, dark side of our societies, sometimes better than experts".[84] In 1964, Brazilian biochemist Prof. Paulo E. de Berredo Carneiro observed that world peace seemed "distant and vanishing". Eight years later, the Indian writer Prem Kirpal described a similar situation, regretting that UNESCO had not managed to ensure peace and cooperation in the world. UNESCO's book policy was a good illustration of the difficulties and ambiguities of the organization as a whole. It showed that even when its action was based on noble and humanistic ideals, the multiplicity of actors, cultures and challenges, reflecting mankind's diversity, made it difficult to conduct a worldwide policy on any subject. The book policy could not have dramatic results. Nevertheless, it allowed many countries to express a symbolic opposition to the cultural and linguistic imperialisms of the great powers and American foundations, and acted as an intermediary to proclaim and defend the cultural, literary and linguistic diversity of mankind.

Notes

1. S. Nihal Singh, *The Rise and Fall of UNESCO* (New Delhi: Allied Publishers Pvt. Ltd., 1988), 46.
2. Emile Delavenay, *For Books* (Paris: UNESCO, 1974).
3. Lecture given by René Maheu on 5 June 1964, reproduced in René Maheu, *La civilisation de l'universel* (Paris: Laffon-Gonthier, 1966), 211.
4. Lecture given by René Maheu on 14 February 1964, reproduced in *La civilisation de l'universel*,103.
5. Lin Yutang, *The Importance of Living* (Paris: Editions Philippe Picquier, 2007 [1937]), 378–395.
6. "School Libraries (texts for a fix film)", undated, file 02:31 A 323, UNESCO Archives.
7. Delavenay, *For Books*, 9.
8. Norbert Elias, *La dynamique de l'Occident* (Paris: Calmann-Lévy, 1991), 247–248 and André Chamson, "Langage et images" in *La culture est-elle en danger? Débat sur ses moyens de diffusion* (Geneva: La Baconnière, 1955), 83.
9. Ronald Barker and Robert Escarpit, *The Book Hunger* (Paris: UNESCO, 1973), 149.

10. Opening speech of Torres Bodet for a meeting of the International Committee of Experts for translation of classics (21–25 November 1949), 21 November 1949, file 803 A064 -56, UNESCO Archives.
11. Roland Barthes, "An Introduction to the Structural Analysis of Narrative" in *Communications* (1966): 1.
12. Letter from René Maheu to Bothien, 12 June 1959, file 4 A 337/01 IFT -66, UNESCO Archives.
13. Pascale Casanova, *The World Republic of Letters* (Cambridge: Harvard University Press, 2007).
14. Expression used by the director-general of UNESCO in his official report concerning 1951–1952, p. 185.
15. Tzvetan Todorov, "La vocation de la mémoire" in *Cahiers français* 303 (2001): 3 and Jacques Juliard, "Que sont les grands hommes devenus?", *Politique internationale* 82 (1998–1999): 21–22.
16. Lecture given by René Maheu, 26 August 1964, reproduced in Maheu, *La civilisation de l'universel*, 97.
17. Memorandum from Edward Carter to Lloyd Hughes, 14 September 1950, file 375:02, UNESCO Archives.
18. *Report on Social Situation in the world, 18 December 1951*, quoted by Edward Carter in a memorandum to Elwin, 6 February 1952, file 375:02, UNESCO Archives.
19. Letter from Petersen to Carnovsky, 9.11.1961, file 02 (6) A074 (669) "62" TA, UNESCO Archives.
20. Sarah Brouillette, "UNESCO and Book Development", MLA, "Print Culture and Global Development" panel, 29 December 2009, p. 1.
21. Herbert Moore Phillips, *Literacy and Development* (Paris: UNESCO, 1970).
22. "Books without Chains", *UNESCO Courier* 6:6 (June 1953): 18.
23. "Development of Information Media: Book Development in the Service of Education: Report by the UNESCO Secretariat", 1970, p. 11 (United Nations document E/4958), file 04 A 066 72 AIL, UNESCO Archives.
24. Julian Behrstock, "UNESCO and the World of Books", *UNESCO Courier* (September 1965): 21.
25. Carl Doka, *Les relations culturelles sur le plan international* (Geneva: La Baconnière, 1959), 120–126.
26. Robert Escarpit, quoted in Marie-France Blanquet. "Robert Escarpit", April 2008, pp. 1–2, accessed 24 March 2015, website http://www.cndp.fr.
27. On "worldwide public opinion", see Jean-Baptiste Duroselle, *Histoire diplomatique de 1919 à nos jours* (Paris: Dalloz, 1993), 939.
28. Brouillette, "UNESCO and Book Development", 4–5.
29. Stephen Parker, *UNESCO and Library Development Planning* (London: Library Association, 1985), 259.
30. Michael Keresztesi, *The Contribution of UNESCO to Library Education and Training: The First 25 Years (1946–1971)* (Michigan: University of Michigan, 1977), 261.
31. Memorandum from Díaz Lewis to DG, 5 July 1961, file 347.78 A 06 (672.4) 63, UNESCO Archives.
32. Letter from Finkelstein to Bogsch, 26 April 1962. file 347.78 A 06 (672.4) 63, UNESCO Archives.
33. Letter from Bodenhausen to DG, 16 September 1963. UNESCO Archives, file 347.78 A 06 (672.4) 63.
34. Letter from Bodenhausen to Gomes Machado, 25 February 1964, file 347.78 A 06 (672.4) 63, UNESCO Archives.

35. Daniel Gervais, *La notion d'œuvre dans la Convention de Berne et en droit comparé* (Droz, 1998), 133 and Circular letter send to Member States on 30 December 1966, DG/6/126/397, UNESCO Archives.
36. Arpad Bogsch, *The First Twenty-Five Years of the World Intellectual Property Organization from 1967 to 1992* (Geneva: International Bureau of Intellectual Property, 1992), 10 and Isabella Löhr, "La Société des Nations et la mondialisation du droit d'auteur entre les deux guerres" in *La diplomatie par le livre: Réseaux et circulation internationale de l'imprimé de 1880 à nos jours*, ed. Claude Hauser, Jean-Yves Mollier and François Valloton (Nouveau Monde éditions 2011), 195.
37. Tabrizi Salah, *Institutions internationales* (Paris: Editions Dalloz, 2005), 254.
38. *UNESCO Collection of Representative Works* (Paris: UNESCO, 1994), 9.
39. Jens Boel, "Historical Perspectives on UNESCO Books", International colloquium on the History of the Book, Sydney, 12 July 2005.
40. Sarah Brouillette, "Literary Diplomacy: UNESCO's Collection of Representative Works," *Modernist Studies Association*, November 2007, 4.
41. For instance *Books for All: A Study of the International Book Trade* (1956), *The Revolution in Books* (1965), *Books for Developing Countries: Asia, Africa* (1965), *The Book Hunger* (1973), *Books in the Service of Peace, Humanism and Progress* (1974), *UNESCO-Sponsored Programmes and Publications* (1974), *For Books* (1974) and *Promoting the Reading Habit* (1975).
42. Brouillette, "UNESCO and Book Development", 7.
43. "*Memorandum Summarising Current UNESCO/CARE Relations*" (1950), 1–2, file 332.55:02, UNESCO/CARE, UNESCO Archives.
44. Correspondence files 36 A 653, UNESCO Archives.
45. US Office of Education, Education and Orientation Section of the Military Government, Library of Congress, American Library Association, US Department of State, National Education Association, National Council of Voluntary Agencies; CARE member agencies and US Book Exchange.
46. "Operations of the UNESCO-CARE Book program", undated, file 332.55:02 UNESCO/CARE, UNESCO Archives.
47. Brouillette, "UNESCO and Book Development", 3–4.
48. Letter from Carter to Stanley, 3 January 1950, and Letter from Carter to Stanforth, 4 September 1950, file 332.55:02 UNESCO/CARE, UNESCO Archives.
49. Brouillette, "UNESCO and Book Development", 4. See also Memorandum from Barger to Carter, 17 April 1950, file 332.55:02 UNESCO/CARE, UNESCO Archives.
50. George N. Schuster, *UNESCO: Assessment and Promise, Council on Foreign Relations* (New York: Harper and Row, 1963), 37.
51. "Books without Chains", *UNESCO Courier* 6:6 (June 1953): 2.
52. Memorandum from Carter to Elwin, 6 February 1952, file 375:02, UNESCO Archives.
53. *Learn and Live: A Way out of Ignorance for 1,200,000,000 People* (Paris: UNESCO, 1951) and *Men against Ignorance* (Paris: UNESCO, 1953).
54. Phillippe Moreau Defarges, *Les organisations internationales contemporaines* (Paris: Seuil, 1996), 28.
55. Maheu, *La civilisation de l'universel*, 36.
56. F.C. Cowell, "Planning the Organization of UNESCO, 1942–1946. A Personal Record", *Cahiers d'histoire mondiale* 10:1 (1966): 224.
57. Jean Dubuffet, *Asphyxiante Culture* (Paris: Editions de Minuit, 1968), 15–16.
58. Chloé Belloc, *Le CIPSH (1947–1955), Idéal et réalité d'un engagement scientifique et intellectuel* (Paris: Université Panthéon- Sorbonne-Paris I, 2005), 38.

59. Maurice Vaïse, *La paix au XXe siècle* (Paris: Editions Belin, 2004), 10.
60. Karl W. Deutsch, *The Analysis of International Relations* (Englewood Cliffs: Prentice-Hall, 1968), 12 and Georges Abi-Saab, "La notion d'organisation internationale: essai de synthèse" in *Le concept d'organisation internationale*, ed. Georges Abi-Saab (Paris: UNESCO, 1980), 21.
61. Maurice Bertrand, *L'ONU* (La Découverte, 2009), 18 and Charles Pentland, "Functionalism and Theories of International Political Integration" in *Functionalism. Theory and Practice in International Relations*, ed. Arthur Groom and Paul Taylor (London: University of London Press, 1975), 15.
62. T.V. Sathyamurthy, *The Politics of International Cooperation, Contrasting Conceptions of UNESCO* (Genève: Librairie Droz, 1964), 24.
63. Louis Dollot, *Les relations culturelles internationales* (Paris: Presses universitaires de France, 1964), 104.
64. Charles S. Ascher, *Program-Making in UNESCO, 1946–51* (Chicago: Public Administration Service, 1951), 1 and Sagarika Dutt, *The Politicization of the United Nations Specialized Agencies: A Case Study of UNESCO* (New York: Mellen University Press, 1995), 12, 22, 78–83, 109.
65. Robert W. Cox and Harold K. Jacobson, *The Anatomy of Influence: Decision-Making in International Organization* (New Haven: Yale University Press, 1973), 386.
66. Quotation in S. Singh, *The Rise and Fall of UNESCO*, 35.
67. Masanobu Konishi, *Les rapports de l'UNESCO avec l'ONU et les autres institutions spécialisées* (Paris: Université de Paris, Faculté de droit et des sciences économiques, 1970), 30. See also Walter H.C. Laves and Charles A. Thompson, *UNESCO, Purpose, Progress, Prospects* (Bloomington: Indiana University Press, 1957), 350 and Jean-Jacques Renoliet, *L'Institut international de coopération intellectuelle (1919–1940)* (Paris: Université de Paris I-UFR d'histoire, 1995), 1075.
68. Belloc, *Le CIPSH (1947–1955)*, 36.
69. Laves and Thompson, *UNESCO, Purpose, Progress, Prospects*, 222–223.
70. Daniel A. Holly, *L'UNESCO, le Tiers monde et l'économie mondiale* (Montréal: Presses de l'Université de Montréal, 1981), 11.
71. Chloé Maurel, *L'UNESCO de 1945 à 1974* (Paris: Ecole doctorale d'histoire de Paris I, 2005), 160.
72. Jacques Rigaud, *La culture pour vivre* (Paris: Gallimard, 1975), 12–13.
73. Gaston Bouthoul, *La paix* (Paris: PUF, 1974), 21–22.
74. Pascal Bruckner, "Faut-il être cosmopolite?", *Esprit* 12 (December 1992).
75. Jean-Baptiste Duroselle, *Itinéraires. Idées, hommes et nations d'Occident (XIXe-XXe siècles)* (Paris: Publications de la Sorbonne, 1991), 56.
76. Brouillette, "UNESCO and Book Development", 7.
77. Conclusion de Robert Frank, in *Les relations culturelles internationales au XXe siècle. De la diplomatie culturelle à l'acculturation*, ed. Anne Dulphy, Robert Frank, Marie-Anne Matard-Bonucci and Pascal Ory (Bruxelles: P.I.E. Peter Lang, 2010), 674.
78. Conclusion de Jean-Yves Mollier in *La diplomatie par le livre,* ed. Hauser, 456–457.
79. Martha Nussbaum, *Les émotions démocratiques. Comment former le citoyen du XXIe siècle?* (Paris: Climats/Flammarion, 2011), 35.
80. Armando Petrucci, "Lire pour lire. Un avenir pour la lecture", *Histoire de la lecture dans le monde occidental*, ed.Guglielmo Cavallo and Roger Chartier (Paris: Seuil, 1997), 405.
81. Pierre de Senarclens, *La crise des Nations Unies* (Paris: Presses Universitaires de France, 1988), 173.

82. Michael Keresztesi, *The Contribution of UNESCO to Library Education and Training*, 254.
83. Richard Hoggart, quoted in Nicholas Sims, "Servants of an Idea: Hoggart's UNESCO and the Problem of International Loyalty", *Millennium: Journal of International Studies* 11 (1982): 67.
84. Interview of Alexandre Prstojevic by Gisèle Sapiro, 7 June 2011, concerning his book *La Responsabilité de l'écrivain. Littérature, droit et morale en France (XIX*[e] *– XXI*[e] *siècle)* (Paris: Seuil, 2011), http://www.vox-poetica.org (accessed 24 March 2015).

3
And Action! UN and UNESCO Coordinating Information Films, 1945–1951

Suzanne Langlois

The history of the relationship between UNESCO and the UN coordinating body for film encapsulates the challenges and tensions of the founding phase of the UN and its specialized agencies.[1] The choice of information films as a lens through which to view the early history of UNESCO is based on the fact that the organization was founded at the very time when cinema-going was at its peak in the 20th century, just prior to audience fragmentation in the 1950s. Making your mark in 1945 meant appearing on the world's screens. The stakes were huge. Well aware that half of the world's population were illiterate,[2] UNESCO was particularly attracted to this medium as it would complement its worldwide delivery of information to all, literate as well as illiterate audiences, and work as a powerful tool for modernization. This commitment also followed the sustained interest in the uses of film for educational purposes since the days of the League of Nations.[3] In 1945, UNESCO was planning a film and radio information service while, at the same time, in New York, the UN was setting up its Department of Public Information, including a division for visual information, in the form of both still and moving images. Surveys and memoranda from 1945 and 1946 stressed the need to avoid duplication, but the central concern was to have a strong visual presence. Who would see to it? How, and with what resources?

Envisaged as early as 1946, the UN Film Board (UNFB) was established in New York in January 1947. Its mandate was to encourage and coordinate the production and distribution of visual information by the UN and its specialized agencies. Lines of action were mapped out. They included the planning of a large-scale information campaign to publicize UN institutions, promoting problem-solving through international cooperation, and fostering the free flow of educational and information films. The film services recorded meetings and conferences, produced descriptive visual documents and – a much lesser-known activity – commissioned documentary films. Here my focus will be on the founding years when UNESCO defined its specificity within the UN system and fought hard to stand on its own feet. However, the organization's declared aims with regard to its freedom of action concerning film for public information and mass communications were hampered by internal and external difficulties. In addition, some geographical areas remained closed to UNESCO, and it was not entirely clear how the ever-increasing ideological polarization during the first years of the Cold War would be detrimental to its action. Only in 1951 did UNESCO finally succeed in creating its own film service. Such a gap from 1945 to 1951, precisely at a time when urgent action to occupy world screens was needed, calls for an explanation. This chapter seeks to provide answers by highlighting some of UNESCO's early ambitions concerning film activities: it examines the objectives pursued, the lines of approach followed within the UN system, and the obstacles faced in the media reach of a large international organization engaged both in immediate action and in reflection on education, reconstruction and the post-war world during the brief and perilous transitional phase from war to peace.

The understanding of the role of information in the overall mission of the United Nations Organization (UNO) has attracted some attention but so far, and contrary to printed material, this series of films and the network of professionals involved have not received much scholarly attention from historians despite the fact that this medium played an important role in the uses of mass media at the service of UN/UNESCO missions.[4] Scholars in communications studies have been more active.[5] Others studying the educational cinema movement or the visual strategies used by the UN are focusing on either the interwar years or the current period.[6] There is continuity in the importance given to the role of film for international organizations, beginning with the League of Nations, to the UN – both as the war alliance and the later peacetime organization – and to UNESCO. However, I will limit my discussion to the years covered by this chapter, 1945–1951. The *United Nations Bulletin* and *The UNESCO Courier* regularly discussed film production and distribution, subjects echoed in various general and specialized publications, especially in Europe, the Americas and the British Empire,[7] but the early films were not accessible to research. It is only recently that technology and cultural interests have combined to bring some of them back to light.

Although historical scholarship is still limited, it is encouraging that historic UN films are now beginning to reach broader audiences as a selection were presented at the UNESCO Institute for Lifelong Learning in Hamburg in August 2004. The program included a short film produced to fight illiteracy. It had been specifically commissioned for UNESCO by the UNFB in 1948–1949.[8] Interest is therefore growing. A film event was organized at UNESCO's headquarters in Paris in 2007,[9] while recent work on digital technology calls for the preservation of this unique cinematographic heritage.[10] In collaboration with the Institut national de l'audiovisuel (INA), in Paris, UNESCO has selected 70 hours from its archival film collections for digitization and access by the general public using the INA website. Also, the Center for the Moving Image, founded in Edinburgh in 2010, is planning a project on the history of the Edinburgh International Film Festival (EIFF). Established in 1947, the EIFF was dedicated to international documentary films during the early years of its existence and most UN films have been shown there.[11] In 2014, the UN Audiovisual Library in New York produced a short video of UN material from the past 70 years "to raise awareness of the urgent need to digitize and preserve the United Nations' audiovisual heritage for future generations".[12] Lacking detailed inventories of their collections, UN agencies, in New York and Geneva, left many historic films unused. The short video is evidence that the situation is changing, and the preservation of the early films will lead to more studies on this material. Important work has also been done in the UNESCO Archives in Paris, along with the INA development project, a fundraising campaign is under way for safeguarding and promoting its documentary heritage, but some undated titles remain in its listings. It is clear that work on these collections presents the dual objective of research and preservation – something like a "search and rescue" mission.

Why film?

Film was one of the pre-eminent mass media from the years before World War II until the beginning of the 1950s. During the war it was used intensively by all of the warring nations. At the same time, the conflict considerably increased the popularity of films among ever-growing audiences eager for news and entertainment. It was obvious that motion pictures would continue to be a powerful means of communication in the post-war years. There was also a clear understanding of their social function, with mass education more than ever on the agenda. The cinematographer, wrote film-maker Marcel L'Herbier in 1946, was the "anchorman of humanity".[13] Film then proved as necessary for peace and the construction of the post-war world as it had for the war effort.

By 1945, however, the word "propaganda" had become problematic. Its modern conceptualization during World War I and the Russian Revolution, its practice during the interwar period, especially by dictatorial regimes – from the Left as well as the Right – and its purpose at the service of all belligerents during World War II were all factors that encouraged the successor organization to the League of Nations not to use this term and to prefer the expression "public information". At the time of the League of Nations and the International Institute for Intellectual Co-operation (the interwar organization that was the ancestor of UNESCO), the word "propaganda" was commonly used, including for the activities of the League of Nations. The worlds of publicity and propaganda were then closely connected.[14] In 1946 and 1947, in the first organization charts explaining the structure of UNESCO, one tends to find the expressions "media and mass communications" and "public information" as separate section titles. The word "propaganda" is still suitable to refer to what these international organizations intended: to inform and educate certainly, but also to convince and change attitudes, and to achieve their goals.

In 1946, at a meeting in New York of the Consultative Committee on Public Information, the representative of UNESCO, Lloyd Free, indicated that his agency had great ambitions and intended to organize its mass communications – including motion pictures – using two complementary perspectives: public information or public relations, and a substantive promotion of the objectives of UNESCO for education, science and culture.[15] In 1950, discussing the great permeation and penetration power of the motion picture, "which speaks so powerfully to the mind and heart", the *Courier* presented cinema as an instrument of persuasion.[16] A year later the director-general of UNESCO, Jaime Torres Bodet, spoke of the need to mobilize world opinion in favor of the Universal Declaration of Human Rights and the fight against prejudice by incessant campaigns using all means at its disposal – books, press, radio, cinema, exhibitions and education.[17] To promote? To persuade? To mobilize? Some 70 years after the founding of the UN and UNESCO, one could probably use the term "propaganda for peace, education and international cooperation" without too much apprehension. The two complementary missions stressed by Free in 1946 clearly set the agenda for the uses of both information and propaganda.

Objectives and first steps

The specific role of UNESCO within the UN system was not immediately clear to everyone nor recognized, and there were many areas of overlap. In 1946 a draft subsidiary agreement linked the two institutions, but their links had to be organized. UNESCO needed quickly to strengthen its specificity and to affirm its primary competence in matters of concern to the UN in education, science and culture.[18] However, international organizations are not fully formed from the date of their foundation, even as they are the end results of long processes in conceptualizing a different world order and cooperation. One needs to be careful when projecting clear distinctions that did not yet exist in the 1940s. Three years after UNESCO had been founded, the chief of the Non-Governmental Organizations Section asked the Films and Visual Information Division of the UN to produce a filmstrip about UNESCO, the specialized agency in which NGOs had expressed the most interest, although it was the least known. He added that many well-informed people thought it was a department of the UN, or vice versa.[19] Such confusion did not prevent UNESCO from being called on in the very first weeks of its existence, in November and December 1945, to join the UN Relief and Rehabilitation Administration (UNRRA) in providing aid to areas devastated by the war.[20] At the time, however, UNESCO was unable to pinpoint the exact nature and scope of its responsibilities in that area. Nor could the organization, itself in a preparatory phase and without an established program, commit any resources.

During the interwar years, the International Educational Cinematographic Institute had already established that film should be a tool to build bridges among peoples, helping them to know and understand one another better and thus becoming an instrument of peace.[21] UNESCO had similar ambitions, but it was not until 1947 that a film program was drawn up to be handed over to production companies in various countries, in the hope that the films would be made in 1948. The plan called for 48 films to be made as part of the regular output of 19 member states.[22] This ambitious goal ran into several obstacles, including the internal organization of the UN's information film service, which partly determined the media reach that UNESCO could hope soon to achieve through film.

Lines of approach and the dialogue between Paris and New York

The first line of approach, decided on in 1944 for the propaganda films of the United Nations civilian mission, was to commission films from recognized production companies. Films supporting UNRRA emergency relief missions in the liberated zones of Europe and Asia were produced in this manner, which allowed for some control over content without incurring any direct production costs. The principle of international cooperation could thus be applied, subject to an excellent network of contacts in production and distribution agencies.[23] The pressing need at the time was to be visible on the world's screens. The situation continued unchanged for a few years owing to a shortage of technical staff, limited financial resources and a concern for efficiency. This *modus operandi* also suited the UN Films and Visual Information Division, which sought to avoid any dispersal of resources.

On 25 June 1946, Jean Benoit-Lévy, director of the Films and Visual Information Division of the UN, was ready to begin discussions about the establishment of a coordinating body for film and visual information media. He met Valère Darchambeau, the UNESCO representative to the UN in New York, who, however, objected both to the goals and to the operation of the UNFB and strongly protested against such political powers being vested in the UN Public Information Department. UNESCO was free, he said, and did not need to take instructions from anyone.[24] Benoit-Lévy then turned to a new negotiating partner for UNESCO, William Farr, whom he met twice at the end of July. In early August 1946, the plan to establish the UNFB advanced. In New York it was discussed by Benoit-Lévy, Farr and Florence Reynolds, director of information films at the Food and Agriculture Organization (FAO).[25] The subsidiary agreement between the UN and UNESCO in December 1946 clearly stated that competition and duplication were undesirable and that it was primordial to establish and to use common technical services, and an article in the agreement again recognized the special functions of UNESCO.[26] The agreement was not finalized until 1947, after UNESCO had reported the difficulties it was experiencing. "Another field where cooperation between UNESCO and the United Nations has proved somewhat difficult is that of information," a confidential memorandum admits. In the negotiations of this agreement, "UNESCO has found it difficult to secure recognition of its specific function, in view of the too pre-eminent role that the United Nations would like to assign to its own Department of Public Information. UNESCO feels that particularly on the question of mass media of information it should be accorded a role of central importance."[27]

UNESCO wished to keep the initiative with regard to film production, either encouraging production or, if the occasion arose, producing films itself, which was exactly what the UNFB wanted to avoid.[28] Benoit-Lévy was convinced that UNESCO did not possess the necessary resources or contacts to produce films properly.[29] UNESCO nevertheless reiterated its position during the meeting of the Consultative Committee on Public Information held in October 1946 at Lake Success but did not want to see such clashes degenerate into "bureaucratic battles of jurisdiction".[30]

On 24 January 1947, at the UN's headquarters at Lake Success, the first meeting was held of the UNFB, the new body whose mission was to coordinate joint services for all visual information from the UN and its specialized agencies. The meeting brought together Gerald Carnes and Lloyd Free of UNESCO, Tor Gjesdal and Jean Benoit-Lévy of the UN and Florence Reynolds of the FAO. Representatives of the International Labour Organization (ILO), UNRRA, the International Monetary Fund and the Interim Commission for the World Health Organization (WHO) also attended as observers. The executive office was headed by Benoit-Lévy, and it was decided that his assistant director for the European branch of the UNFB would be Farr from UNESCO's Department of Mass Communications.[31]

Having returned to Europe, Farr was not in New York for the inaugural meeting.[32] Free explained his absence by the fact that UNESCO's position was not clear at the time because of drastic budget cuts, but he believed that UNESCO would be able to fulfill its obligation to provide personnel for the executive office in New York and Paris. This situation did not prevent Free from proposing projects involving UNESCO. At that time there was only one film planned by the UK Government, in addition to the UNESCO conference in Paris, which had been filmed. Free then enumerated no fewer than seven projects proposed by the Department of Mass Communications, already approved by UNESCO, which could all, in his opinion, be carried out by the UNFB: an international ideas bureau, a project office for all media, a survey of technical needs, the establishment of national bureaus, cooperation with international film associations, the use of visual tools to overcome the language problem and the use of German-language educational films.

In response, Benoit-Lévy said that such plans were premature since they had to be approved by the administrations of the UN and UNESCO. Time and solid planning were required before film production could commence. He took the opportunity to outline a sustainable production policy favored by the UNFB, whereby there would be few films on the workings of the organization in order to spare the yawns of viewers. Instead, the emphasis would be on broader issues, such as food, human rights, education and science, thus creating a link between individuals and the UN.[33] This focus can be interpreted as positioning the UNFB in the transition from previous UN film propaganda for immediate and urgent help – which was done by UNRRA from 1944 to 1947 – to trying to address deeper social and economic issues, and reaching peoples without appealing to nationalist sentiments.

In the meantime, in the USA, there was growing impatience to see Farr. A month later, Benoit-Lévy, rather annoyed, prodded UNESCO on the subject, since the specialized agencies were already being asked to submit concrete plans for film production by the following meeting, scheduled for 18 March 1947.[34] The Department of Public Information had secured a budget of USD450,000 for film production. One aspect of the problem seemed to be the financial arrangement, an obstacle that appeared to have been overcome by the middle of 1947, but in March 1948 the UNESCO representative had still not made an appearance.[35] The financial issue also plagued the UN in New York where Benoit-Lévy, only weeks after the founding of the UNFB, had to battle to save his projects when faced with budget cuts decided by the General Assembly.[36]

Institutions and offices are one thing, but let us also look at the people involved. Benoit-Lévy, executive director of the UNFB, and his family found refuge in the USA in 1941.[37] Highly regarded in France during the interwar period, he directed and produced educational films. He already had acquaintances in New York when he arrived, since he had been there in October 1935 and maintained ongoing contact with people at the Rockefeller Foundation, at the Film Library of the Museum of Modern Art in New York, and with some US universities interested in modernizing teachers' training. He knew John Grierson, who entered UNESCO in 1947 – the two had met a few times in New York and in Ottawa during the war. They would meet around the same table at the UNFB during the 14 months that Grierson spent as the head of UNESCO's Department of Mass Communications. Benoit-Lévy had also been involved in the International Institute of Intellectual Co-operation and the French Committee of the International Educational Cinematographic Institute at the time of the League of Nations. Questions about the use of film for peace had already been raised at the League of Nations, but production had not been on the cards; the league was fairly active in international discussions exploring, understanding and organizing access to educational material using the powers of the new mass media, film and radio. Politically, Benoit-Lévy was a liberal, at the center-left of French politics. While holding traditional views on some social institutions, such as the family, he was very modern in his understanding of the impact of audiovisual material for educational purposes. When Benoit-Lévy took up his post at the UN in 1946, he could draw on solid experience, a good international professional network and a vision of what needed to be done, and he was more than eager to at last see an international organization fully use all the resources of film in the pursuit of its mission. He remained in office until July 1949.

UNESCO's attempt to strike a major blow in the spring of 1947 can be understood with reference to the difficulties it had been experiencing regarding film and other means of mass communication. In January, Julian Huxley wrote to John Grierson, the renowned Scottish film producer, filmmaker and theoretician of the British documentary movement, a fervent internationalist and the first commissioner of the National Film Board of Canada, to offer him the post of director of public information and mass communications with a salary of USD12,000 for a one-year contract.[38] The men knew each other well from the interwar period. Grierson arrived with a great reputation as a documentary-maker and propagandist but, branded a leftist, he had experienced political difficulties in both Canada and the USA since the end of the war. He remained at the head of the Section of Mass Communications of UNESCO from February 1947 to April 1948. Already in September 1947, however, in a long personal letter to Julian Huxley, Grierson stated his intention to leave UNESCO at the end of the year; indeed, he had accepted the post only because the offer had come from Huxley.[39] In 1946, Grierson had feared the worst for UNESCO and for Huxley's real chances of finding the appropriate administrative structure to make such an organization function properly. A year later he considered that the process was well under way despite a glaring lack of resources and that, all in all, the bulk of his work was over, insofar as a program had been established, a team formed and the Commission on Technical Needs had been launched.

In October 1947, Grierson put together some thoughts about UNESCO's Bureau of Public Information and its program for 1948: there was a need to be realistic about what a team of 20 people at most, including stenographers and clerks, could accomplish with a budget of about USD250,000. Even more, he added, those involved had to know where they were going and how they were going to get there because public information in a big international organization such as UNESCO could not mean advertising, lobbying or misinformation – UNESCO had to be shown to be an organization involved in efforts towards concrete goals.[40] The idea of targeting particular interest groups for specific projects and following up on concrete results over the long term had been at the center of Grierson's thinking about public information for years. That had been the spirit in which he had worked at the National Film Board of Canada. He had already suggested a similar program for the ILO as far back as 1938.[41] In his observations of

October 1947, he remarked that UNESCO did not have to be on everyone's lips; rather, it should, without any illusions, do no more than publicize a realistic program of activities. "I assert this in spite of the occasional disappointment expressed that the public does not think we are as big as we think ourselves," he wrote. "We did not know we were even anything till three or even two months ago when the whole work began to come into concrete focus. We had, till then, to demonstrate that we were in fact an important instrument in public affairs; and outside skepticism was both natural and proper."

As for those around him, those to whom he would entrust the department after his departure, he wrote to Huxley that Philippe Desjardins, William Farr and René Maheu, who were respectively in charge of radio, film and the press, made a good team and worked well together in keeping with the objectives to be reached. The outstanding results of the Commission on Technical Needs were a testimony to their abilities.[42] Let us take a closer look at this commission, whose existence grew out of the need first to collect vital information and then to cope with the reality of political differences in the context of the Cold War.

Media reach in the context of the Cold War

The extensive survey of technical needs was a considerable undertaking and one that was extremely useful for the creation, rebuilding and modernization of information and educational resources. The project was launched at the first session of the General Conference of UNESCO in November 1946, and Grierson had made it a priority upon arriving at the organization.[43] However, as a pragmatic person, he did not believe in the concept of the free flow of information, to which, although an objective of UNESCO, he was not committed, at least with regard to the elimination of censorship. In a letter from December 1947, he explained that the principle of the free flow of information, which came from the USA, was bound to arouse fierce opposition elsewhere in the world, not only from the Left and those accused of being "totalitarian", but also from all those who, like himself, believed in the European conception of the state, those who saw the need to adjust information and education to political and economic necessities, and to provide a framework of responsibility as the foundation of freedom.[44]

The concept of the "free flow of information" encompassed the principle of freedom of information without censorship. Freedom of information was not achievable because governments would always have the right to control what was, or was not, permitted.[45] Furthermore, the concept made no sense in the case of people who had no say and, in addition, did not possess either the infrastructure or the technical means to communicate among themselves, let alone to make themselves heard outside their country. This situation had first to be remedied on the ground, hence the urgent need to find out as much as possible about the resources available to the press, radio and film in the countries devastated by the war, and then elsewhere. For Grierson, in 1947 and 1948, the issue was not yet the ideological polarization at the heart of the Cold War. To his mind the main problems remained the absence of the USSR and the Soviet-bloc countries, with the exception of Czechoslovakia and Poland, and the largely inadequate budgets of UNESCO. He even wondered if the UN was not in a better position than UNESCO to establish an information service, since the USSR remained out of UNESCO's reach.[46] Although the generous funding for progressive documentary film-making before and during the war had dried up soon after, Grierson still believed that a spirit of reform would survive that terrible experience and spread throughout the world. At the beginning of 1946 he was convinced that there was an enormous market for peace films.[47] He would later go back on this optimistic view of the situation "let me admit that for a short period I missed the point," he wrote, "[...] the war was on before the peace was started".[48]

Concretely, the survey of technical needs had the advantage of avoiding quarrels about content between Cold War opponents by focusing instead on technology and equipment. Recent studies on the ideological origins of the UN and of UNESCO agree in seeing in the nomination of the British biologist Julian Huxley as head of UNESCO, instead of the classicist Alfred Zimmern, a new focus on science, which could cross the ideological boundaries of the Cold War.[49] This extensive study by UNESCO in 1947 would facilitate access to places that were politically out of reach. It "would be difficult to make UNESCO known unless it was possible to spread knowledge of its aims behind the iron curtain".[50] In the end, among the Soviet-bloc countries, only Bulgaria, Czechoslovakia, Hungary and Poland appeared on the list drawn up in 1949.[51] Yet there were many other frustrations and restrictions, UNESCO having identified "three groups of critical relationships" – namely, "East and West, Eastern Europe and the Western World, and the occupied zones and the rest of the world", which were detrimental to its role to promote peace through international understanding.[52] All attempts to meet the military governor of the Soviet Zone of Occupation of Germany failed, and UNESCO was left to make arrangements with just the commanders of the Western occupied zones of Germany, a situation harshly criticized by the Polish representative to UNESCO in 1948.[53] The USSR did not hold a seat at UNESCO, and neither did the Soviet republics of Ukraine and Byelorussia, despite having received UNRRA aid until the end of emergency relief operations in the winter of 1947. Incidentally, it had been necessary to wait until late 1946 for UNRRA to secure access for two teams of cameramen into Ukraine, Byelorussia and Soviet-controlled territories in the East.[54]

At its meeting in December 1947 at Lake Success, the UNFB took stock of what it had achieved in its first year of existence. In the review, Grierson, who at the time still held his post at UNESCO while also heading the UNFB, underlined the main problems.[55] Despite the considerable progress achieved, the UNFB believed that film and other visual media were not being effectively used to their full potential in pursuit of its members' goals. Obstacles identified included a financial shortfall and problems stemming from the constitutional position of the UNFB's executive office. More specifically, the fact that the UNFB was responsible for all commissioned films meant, in Grierson's view, that it was paralyzed when national initiatives were suggested.

In reality, two fundamental problems divided New York and Paris. First, the UNFB wished to maintain complete authority in all matters regarding the work of various agencies in the production, promotion and distribution of films and other visual information, whereas UNESCO, for its part, wanted to exercise its prerogative in its areas of competence. The first problem certainly had an effect on the second, which involved the promotion of films according to "international" or "national" criteria. It was UNESCO's idea that each nation would make known to others its accomplishments in culture, science and education, which therefore meant making films of a more national nature.[56] The UNFB, on the other hand, was working to coordinate the production of films on situations for which the most effective solutions could serve as examples elsewhere, or clearly depended on international cooperation. In other words, there was an attempt to problematize universalism and internationalism. Different conceptions of the use of film and of the subject matter suited to that purpose thus emerged.

In June 1948, Tor Gjesdal asked UNESCO a series of questions about exactly why it was being led to sidestep the procedures laid down for joint information services for members of the UNFB. He was concerned about the consequences of UNESCO's initiatives, which, in seeking to increase film production, would do so by flying in the face of UN principles.[57] Since UNESCO was not in a position to monitor projects submitted by private production companies, there was a risk that a film might receive approval from UNESCO and make unrestricted use of the name of the UN. The clashes between UNESCO and the UNFB were high on the agenda then when Benoit-Lévy made a five-month tour of Europe during the summer and autumn of 1948.

On 25 October Benoit-Lévy found himself in the company of Tor Gjesdal and William Farr from UNESCO's Department of Mass Communications, in the office of Walter H.C. Laves, deputy director-general of UNESCO. The partners repeated that there was no disagreement between UNESCO and the UN Department of Public Information regarding the role of the UNFB. Various points were made that confirmed the authority of the UNFB and clarified differences between "national" films and "international" films, which were moved outside UNESCO's orbit. In the end, international films would have priority when it was not possible to promote both types of film in a particular country or for a particular producer.[58] Laves completely agreed with this view.[59] In part, it can be deduced that alternatives were not available, for want of resources.

The first phase of production and distribution coordinated by the UNFB came to a close in the winter of 1949. Some 17 short documentary films, considered as prototypes by the UN, were produced in 11 countries and disseminated in eight languages. While all the films had an international content, they were marked by the "national character" of their directors, something Benoit-Lévy accepted since individual style, rhythm and ways of introducing issues varied.[60] He had to reckon, however, with increasing difficulties. The situation would have to change if there was to be an increase in the size of their audiences in the world and if those audiences were to obtain the information they required. It was thus realized that there was probably no subject on which a film or film magazine could be produced that would satisfy all the peoples of the UN.[61]

The adoption of a second line of approach coincided with the launching of a new long-term program in April 1949. The new areas to be given priority were Latin America, the Far East and Eastern Europe, especially as part of the reconstruction effort. The scripts would be written at the UN's headquarters, with writers working closely with specialists in the field concerned and under the supervision of the Films and Visual Information Division of the UN. Filmed material would be compiled by cameramen working from a specific script and guidelines, and taking orders directly from the division. This method of direct production would also usher in a new development, in the form of a newsreel magazine, the first two editions of which came out in March 1950. This radical change was due mainly to the lack of financial and organizational resources, but it also opened the door to nationally framed films, something UNESCO had wanted from the beginning.

Benoit-Lévy, who retired from the UN in the summer of 1949, was not in agreement with such changes. In February 1949 he had clearly expressed his concern that the original scheme for the UNFB had become very shaky. The budget, reduced to USD250,000, was inadequate, contacts with foreign filmmakers were insufficient, and distribution, especially in the USA, had become uncertain.[62] The first line of approach had certainly been more open to the diversity of cultures, film traditions and political points of view than the second, which was put into place during the winter of 1949. Benoit-Lévy was also worried about energies being spread too thinly if each agency started producing its own films. Funds had been running dangerously low since early 1948 in both New York and Paris.

From 1949 onwards, UNESCO concentrated its efforts on the dissemination of films, including science films that called for more extensive international distribution. The organization also continued to gather information about technical needs and worked to facilitate the exchange of audiovisual material around the world, organize specialized seminars on the training of educators and disseminate select films in order to raise funds for fellowships, while carefully skirting major ideological debate.

UNESCO had been featured in independently produced films, such as *Children of the Ruins*, produced in 1948 by the Crown Film Unit for the Central Office of Information of the UK.[63] This was where Grierson worked after leaving UNESCO. The aims and purposes of the organization were also promoted by films produced with its support, such as *This is their story* from 1949, which examined the work of the World Student Service Fund in the countries ravaged by World War II.[64] The film *The Re-education of Children in Countries Directly Affected by the War* looked at UNESCO's reconstruction efforts. The organization's areas of competence were also supported internally despite a system that required projects to go through the UNFB as the coordinating body. UNESCO was interested in three films commissioned by the UNFB in Belgium, Czechoslovakia and France, but the only one to be commissioned by the UNFB at the request of UNESCO during the first phase of production was the short film *That All May Learn* of 1949.[65]

The script had been submitted to the UNFB in 1947 since it had been planned that literacy would be one of the topics covered by the international program of film production.[66] The vigorous literacy campaign undertaken by Mexico in previous years, including for indigenous languages, was a legacy of the Mexican Revolution; it had convinced the UNFB to have the film made in that country. *That All May Learn* is a hybrid documentary film demonstrating the power of literacy in fighting against injustice and spoliation. It combines the story of a poor peasant, exploited and dispossessed of his land because he cannot read, with archival footage of China, India and the USSR, showing the large-scale initiatives for accessible mass education that had been implemented there, and citing favorably the historical precedents of the Russian and Chinese revolutions. This relative indifference to difficult international relations at a time when the world was becoming increasingly polarized by the Cold War reflects the gap between the commissioning and making of films, the rapid deterioration of international relations, but also a will to open the film experiment up to differing options of internationalism. The UNFB maintained its choice. This suggests that the situation was not as rigid as one is led to believe when studying the post-war years and attests to a resolve to bring about fundamental reforms leading to major social changes. It may also qualify the common view of the UN as the univocal mouthpiece of Western imperialism.

Through the activities of the UNFB, UNESCO had been involved in information films and filmstrips on the UN and its specialized agencies, as well as in narrative films on subjects closely related to its mission. Beginning in 1948, however, UNESCO and the UNFB suffered sorely from a shortage of the resources needed to meet their objectives by truly developing their media potential worldwide. In 1949 and 1950, UNESCO was not well served by the new approach, which sought to meet the challenge of coordinated production. In 1950, UNESCO itself made a series of filmstrips on human rights, intended to contribute to the teaching of the fundamental principles set forth in the Universal Declaration of Human Rights adopted in 1948.[67] The UNFB, in its original form, was dissolved in February 1950. UNESCO had a critical role to play in this change and it created its own film service in 1951. Yet UNESCO's film capacity remained limited despite its assertion that it would from then on stand on its own feet where film was concerned. The organization would later develop a series of films, but, during the early years, its film services mostly focused on the film recording of conferences, speeches, ceremonies and a news magazine. The list of films presented at UNESCO's headquarters in 2007 confirms that there was no direct production from 1949 to 1951, the rare films being by external producers.[68]

The Korean War, which began in the summer of 1950, was the one major historical event that dealt a blow to the rise of international organizations dedicated to peace. This tense situation partly accounted for UNESCO's decision to turn away from the making of films centered on mutual understanding and to concentrate mainly on cultural achievements, literacy and scientific information. "To begin with, documentary-makers backed UNESCO heavily," the documentary filmmaker Paul Rotha noted critically five years after the founding of the UN and UNESCO. "Alas, beyond conducting surveys (valuable certainly) and holding interminable committees, where has UNESCO gone in the documentary purpose? With a miserable budget, the United Nations Film Board under Jean Benoit-Lévy made a bright show with a series of internationally commissioned films, but where is it now?"[69]

Rotha, along with Basil Wright, had been deeply committed to producing film material for UNESCO. His disappointment was real. In Benoit-Lévy's opinion, member states had not grasped the importance of film production by the UN.[70] This problem resulted from the general policy described earlier – that is, the emphasis placed on creating links between individuals and the UN. The Cold War exacerbated the situation.

Conclusion

The difficult conceptualization of a new world order in the context of the Cold War eventually caused problems for film professionals involved in international organizations and production companies that supported the ideals of such organizations. The ideological conflict was detrimental to the initial ambitions of both UNESCO and the UN by polarizing and poisoning relations among member states, by blocking access to UNESCO propaganda in some geographical areas, and by extinguishing the hope of international cooperation born out of the war, particularly for film projects, at a time when pre-war relationships still meant something and the different film agency personnel knew one another; and by discouraging the specialized personnel brought together to found a new, post-war world.

While it is true that the whole production sector suffered cruelly from insufficient resources and that the lack of money constantly hindered concrete achievements and crushed some projects, it is also true that the fact that member states rechanneled resources was partly due to the Cold War, which had both a direct and a delayed impact on the subject under consideration here. Concerned to see a major shift in the alliances shaped by the war and ideological polarization on an international scale, UNESCO nevertheless remained proactive, as did the UN, hoping, if not to escape, at least to circumvent the bipolarization of the world and the fraught relations between opposing camps. But those hopes dwindled over time. From the end of the 1940s onwards, the general public heard mostly about the US Marshall Plan films, which, more marked by the Cold War, focused on the colossal reconstruction efforts, the growing influence of an explicitly modern cultural model and an array of political weapons to fight the expansion of communism. The UN and UNESCO had a specific multilateral dimension. Democratic ambitions and regulated capitalism were present in their worldview, but alarmist anti-communist fear did not permeate the work of the UNFB. Even though the multilateral cooperative approach did not prevail during the years of transition from war to peace, the output of the UNFB and UNESCO demonstrated that there were other ways of understanding the world than through the lens of the Cold War, which continued to impinge on all post-war political history.

Seen from Paris, the UN suffered from the huge disadvantage of having its headquarters in the USA, a fact that was not unrelated to UNESCO's low opinion of the superpower's influence on culture. During the preparations for the Committee of Experts that would meet in Paris in the spring of 1951, UNESCO had strong words for Hollywood: "If the meeting is called by UN-UNESCO, it should not be called in Hollywood. In film industry politics – which are bitter – Hollywood is regarded as the capital of U.S. imperialism, and in cultural circles as the capital of mass communications vulgarity."[71]

A final remark needs to be made about the transitional period from war to peace. World War II had a limited timespan, it was finite, whereas peace opened up a very different time dimension, both individually and collectively. Peace propaganda through film was a new and mighty challenge. If peace is built through the peaceful resolution of tensions and conflicts, then it is essential to address the causes of wars and living conditions, not only during the precarious post-war years but also over the long term. The magnitude of the crisis in the 1930s, the outburst of violence triggered by World War II and the urgent need to act had all given an exceptional boost to documentary film-making during the 1930s and the war years. The transition from war to peace was particularly difficult for progressive documentary films. Before they had achieved the purpose envisaged for such films by international organizations, a new war engulfed the world. Propaganda for peace and a better world suffered from the backlash of the Cold War at the very moment when the topics for internationalist film production planned by the UNFB were largely selected from social and economic concerns, and pointed to the evidence that inequality in all its manifestations was the root cause of wars. These issues were finally being put on the agenda for world distribution and viewing, but the momentum was halted before progress, necessarily a long-term phenomenon, could be set in motion and documented on a regular basis. The window that had been opened between 1945 and 1951 was slammed shut.

My aim at the beginning of this chapter was to shed some light on UNESCO's ambitions, its use of film and the restrictions between 1945 and 1951. In addition to limited budgets and resources, different views about the production system and the perspective of visual propaganda for world audiences, as well as the increasing instability of international politics, the post-war years were a time of great hope but also of great uncertainty. The issue was how to do propaganda while distancing it from what Nazi Germany had done? The discussion within the UN system about national and international films was informed by this immediate historical and violent nationalistic experiment. How should the UN and UNESCO be positioned? How should internationalism, peace and education be problematized? How should it transition from short-term urgencies to long-term objectives? How should it gain access to and disseminate visual material for information and propaganda in non-member countries? Answering such questions took time. Some clues to understand the gap mentioned at the outset were provided by John Grierson's private remarks to Julian Huxley. First he feared the worst for UNESCO, unsure it could find an adequate administrative structure. There followed his guarded confidence that a program had been established and a process was under way. He admitted, however, not to have immediately understood the extreme polarization of Cold War politics which were also in a flux. Jean Benoit-Lévy did not either,[72] but he was convinced that he had created the tool (the UNFB) that allowed film to play its international role. He was proud of having commissioned what he called "prototype" films,[73] many of them inspired by variants of social liberalism. In this conclusion I should like to stress these dimensions of time and uncertainty, in the present and the future, in terms of organization and capacity to succeed in this crucial transition from wartime to peacetime. The international organizations examined here owed a lot to their pre-war predecessors. They were nevertheless trying to invent themselves for a new era.

Notes

1. This research was made possible by a grant from the Social Sciences and Humanities Research Council of Canada. I thank Jens Boel and Mahmoud

Ghander of the UNESCO Archives in Paris, and Angela Schiwy (former chief archivist at UNO, New York), Lily Chau and Miguel Gonzalez (UNO, New York, multimedia resources unit) for their generous assistance.

2. *World Illiteracy at Mid-Century: A Statistical Study* (Geneva: UNESCO, 1957, Monographs in Fundamental Education XI), 10, 13; the world illiteracy rate was estimated at 62 percent in the 1920s and at 44 percent in 1950, the latter percentage based on available census data since 1945.
3. See the enlightening discussion by Zoë Druick, " 'Reaching the Multimillions': Liberal Internationalism and the Establishment of Documentary Film" in *Inventing Film Studies*, ed. Lee Grieveson and Haidee Wasson (Durham: Duke University Press, 2008), 66–92.
4. Mark D. Alleyne, *Global Lies? Propaganda, the UN and World Order* (Basingstoke: Palgrave Macmillan, 2003).
5. For excellent works on the historical development and issues concerning documentary film, see the important contributions of Bill Nichols, and of Jack C. Ellis and Betsy A. McLane. Through her case study (*Museum Movies: The Museum of Modern Art and the Birth of Art Cinema*, Berkeley: University of California Press, 2005), Haidee Wasson has examined the change of paradigms which brought popular movies into the sphere of high culture. Closer to our preoccupation with the role of film and communication discourses and practices in the UN system, Zoë Druick has made essential contributions in addition to her article, "Reaching the Multimillions", mentioned above; see, in particular, her articles "Visualising the World. The British Documentary at UNESCO" in *The Projection of Britain: A History of the GPO Film Unit*, ed. Scott Anthony and James G. Mansell (Basingstoke: Palgrave Macmillan, 2011), Chapter 23; and "UNESCO, Film and Education. Mediating Postwar Paradigms in Communication" in *Useful Cinema*, ed. Charles R. Acland and Haidee Wasson (Durham: Duke University Press, 2011), 81–102. Also, since 1998, Jasmina Bojic, film critic and lecturer at Stanford University, has organized one of the oldest international documentary film festivals in the USA (the United Nations Association Film Festival) at Stanford University.
6. Kenneth H. Garner, *Seeing Is Knowing: The Educational Cinema Movement in France, 1910–1945* (PhD dissertation, University of Michigan, 2012). In the UNESDOC database of research about UNESCO, Garner's work is the only one dedicated to film culture. See http://www.unesco.org/archives/new2010/en/research_on_unesco.html (accessed 25 March 2015). Another angle of approach comes from the work of François Debrix, "Deploying Vision, Simulating Action: The United Nations and its Visualization Strategies in a New World Order", *Alternatives. Social Transformation and Humane Governance* 21:1 (January–March 1996): 67–92.
7. For example, the enthusiastic Basil Wright who wrote about how the British could contribute to the UNESCO film programme: "Films and Unesco" in *Informational Film Year Book 1947* (Edinburgh: Albyn Press, 1947), 38–41.
8. *That All May Learn* (Carlos Jiménez, Mexico, 1949, produced by E.M.A., sound, black-and-white, 17 minutes).
9. Treasures from the UNESCO Film Archives Festival, 20 October 2007–3 November 2007.
10. Flora Losch, *Les Archives audiovisuelles de l'UNESCO, un patrimoine en danger d'oubli. Étude pour la définition d'un plan de sauvegarde et de numérisation appliqué aux collections audiovisuelles du Bureau de l'Information du Public*, Bry-sur-Marne, INA-Sup, Enseignement supérieur 2e cycle, Spécialité Gestion de patrimoines audiovisuels et numériques, 2010.

11. The Center for the Moving Image (Edinburgh): http://www.edfilmfest.org.uk (accessed 22 June 2015).
12. *70 Years of Archives Treasures* (1 January 2014; 00:06:13): http://www.unmultimedia.org/avlibrary (accessed 22 June 2015).
13. Marcel L'Herbier, *Intelligence du cinématographe* (Paris: Corrêa, 1946), 31.
14. See the issue of "Dossier: Publicité et propagande", *Vingtième siècle. Revue d'histoire* 1:101 (2009).
15. "Consultative Committee on Public Information of UN and Specialized Agencies", restricted, 17 October 1946, 3, intervention by Mr Free, DPI Records/S-0540-0056, UN Film Board, UN Archives.
16. "Films – A Universal Medium to Propagate a Universal Declaration", *The UNESCO Courier* 3:11 (December 1950): 8. See also the slightly different wording in the French version, *Courrier de l'UNESCO*, of December 1950, 8: "Considéré au début du siècle comme un simple 'jouet scientifique', le cinéma est devenu un instrument de persuasion – disons: 'd'éducation' – d'une singulière importance."
17. *United Nations Bulletin* 9:11 (December 1951): 450.
18. "Draft Subsidiary Agreement between the United Nations and the United Nations Educational, Scientific and Cultural Organization", Paris, 7.12.1946, 2 and Executive Board 1946–1947, vol. I, 1st session, 26 November 1946–10 December 1946, UNESCO Archives.
19. Memo from J.B. Orrick to Maurice Liu (Films and Visual Information Division), 22 September 1948, copy of Jerome Oberwager, filmstrip editor, Films and Visual Information Division, Office of Director/S-0540-0055, UN Archives. A filmstrip is a short strip of photographs or other images on a roll of 35 mm film. Inexpensive and adaptable, filmstrips were widely used in educational settings and in associations.
20. See documents from January to June 1946, AG3 Preparatory Commission, London-Paris 1945–1946, vol. VI, Sub-committee on the needs of devastated areas, UNESCO Archives. For recent research on the contribution of UNESCO to post-war reconstruction, refer to Chloé Maurel, "L'action de l'UNESCO dans le domaine de la reconstruction", *Histoire@Politique* 1:19 (2013): 160–175.
21. Christel Taillibert, *L'Institut international du cinématographe éducatif. Regards sur le rôle du cinéma éducatif dans la politique internationale du fascisme italien* (Paris: L'Harmattan, 1999), 273. See also Jean-Jacques Renoliet, *L'UNESCO oubliée. La Société des Nations et la coopération intellectuelle (1919–1946)* (Paris: Publications de la Sorbonne, 1999), 306; and Zoë Druick, "Reaching the Multimillions".
22. "Overseas News", *Documentary News Letter* 7 (February 1948): 16–17, cited in Jack C. Ellis, *John Grierson: Life, Contributions, Influence* (Carbondale: Southern Illinois University Press, 2000), 234.
23. See *The Union List of U.N.R.R.A. Film. A Guide to Motion Picture Records Produced by Agencies throughout the World on the Activities of the United Nations Relief and Rehabilitation Administration 1943–1947* (Lake Success, New York: Archives Section, Communications and Records Division, United Nations, March 1949). In addition to showing film titles and subjects, this guide listed the 27 North-American and European production companies that hold UNRRA-related film footage.
24. Memorandum "First Discussion of draft proposal with Darchambeau 25 June 1946", DPI Records/S-0540-0056, UN Film Board, UN Archives.
25. "Notes on discussion with Mr Farr of UNESCO, Monday August 5th, 1946", DPI Records/S-0540-0056, UN Film Board, UN Archives.

26. "Draft Subsidiary Agreement between the United Nations and the United Nations Educational, Scientific and Cultural Organization", Paris, 7 December 1946, 2, UNESCO Archives.
27. "Relations between UNESCO and the United Nations", confidential memorandum, Paris, 31 March 1947, 2, Cons. Exec./9/1947, UNESCO Archives.
28. "Revision of UNESCO/Prep.Comm/51 III E, Projects in media of mass communication", 8, article C.2, Production of Films, n.d., probably August 1946, DPI Records/S-0540-0056, UN Film Board, UN Archives.
29. Memorandum "First Discussion of draft proposal with Darchambeau 25 June 1946", DPI Records/S-0540-0056, UN Film Board, UN Archives.
30. "Relations between UNESCO and the United Nations", confidential memorandum, Paris, 31 March 1947, 3, Cons. Exec./9/1947, UNESCO Archives. See also "Consultative Committee on Public Information of U.N. and Specialized Agencies", restricted, 17 October 1946, intervention of Lloyd Free, UNESCO representative, DPI Records/S-0540-0056, UN Film Board, UN Archives.
31. "United Nations Film Board. New Coordinating Body Provides Joint Services", *United Nations Weekly Bulletin*, 4 February 1947, 98.
32. Telegram from William Farr (UNESCO, Paris) to Jean Benoit-Lévy (UN, New York), 22 January 1947, informing him that he cannot attend the meeting, Central Registry (RAG-1) S-0472-096-(1002-8-2), UN Archives.
33. "Summary Record of First meeting of the United Nations Film Board", 24 January 1947, 1–3, S-0540-0073, UN Archives.
34. Telegram from Jean Benoit-Lévy to William Farr, 28 February 1947, RAG-1/S-0472-096-(1002-8-2), UN Archives.
35. "Draft summary record of the fifth quarterly session of the United Nations Film Board", Washington D.C., 8 March 1948, 3, S-0540-0073, UN Archives.
36. Private letter from Jean Benoit-Lévy to Jean Renoir, 23 March 1947. UCLA, Arts Special Collections, Coll. 105 Jean Renoir Papers, Box 7, fldr 6, Correspondence March 1947.
37. See my entry on Jean Benoit-Lévy in *Encyclopedia of Documentary Film* 1, ed. Ian Aitken (New York and London: Routledge, 2006), 108–109.
38. Telegram from Julian Huxley to John Grierson, 15 January 1947, G5:11:32, John Grierson Papers, University of Stirling (Scotland). In the UNESCO hierarchy, only Huxley had a higher salary. Grierson's salary was the same as that of the deputy director-general of UNESCO, and much higher than that of all the other section directors.
39. Letter from John Grierson to Julian Huxley, 14 September 1947, incomplete document, G5:11:37, John Grierson Papers, University of Stirling.
40. "Public Information UNESCO. General Observations", 9 October 1947, G5:4:5, John Grierson Papers, University of Stirling.
41. Ellis, *John Grierson*, 114.
42. Letter from John Grierson to Julian Huxley, 14 September 1947, G5:11:37, John Grierson Papers, University of Stirling.
43. Transcription of an interview conducted by James Beveridge with William Farr (UNESCO) about Grierson's work, GA:10:23, John Grierson Papers, University of Stirling.
44. Letter from John Grierson to Zechariah Chafee (Law School of Harvard University), 30 December 1947, G5:11:38, John Grierson Papers, University of Stirling.
45. Ellis, *John Grierson*, 234.

46. Ellis, *John Grierson*, 238.
47. "Information policy – London", points raised by J. Grierson, 4 January 1946, 2, DAG-12/1.0 box 2, Office of Public Information, Sub-Committee of Technical Advisory Committee on Information, UN Archives.
48. Preface by John Grierson for the third edition of Paul Rotha, Sinclair Road and Richard Griffith, *Documentary Film: The Use of the Film Medium to Interpret Creatively and in Social Terms the Life of the People as It Exists in Reality* (London: Faber and Faber Ltd., 1951 (1st edn 1935)), 21.
49. Mark Mazower, *No Enchanted Palace. The End of Empire and the Ideological Origins of the United Nations* (Princeton: Princeton University Press, 2009), 23. See also John Toye and Richard Toye, "One World, Two Cultures? Alfred Zimmern, Julian Huxley and the Ideological Origins of UNESCO", *History* 95:319 (July 2010): 310; the authors see in this episode a struggle between literary intellectuals and scientists for cultural leadership.
50. Summary report of the 7th meeting, 13 April 1947, dated 7 May 1947, 2–3, Cons. Exec./2e Sess./S.R.7/1947 (Rev.), UNESCO Archives.
51. Temporary International Council for Educational Reconstruction, General Conference, Technical needs, 20 February 1949, TICER/Conf.2/4, UNESCO Archives.
52. 7 EX/SR.2 (rev.), 30 June 1948, 3, UNESCO Archives.
53. 6 EX/19, 13 February 1948, 1 and 7 EX/5, 23 March 1948, 2, followed by a supplement on 31 March 1948 and cable from Jan Opocensky (Poland) about UNESCO's program in Germany, in 7 EX/12, 1 April 1948, UNESCO Archives.
54. Visual Media Branch Report of activities for November–December 1946, 1–2, PAG-4/1.5.0. UNRRA-OPI, UN Archives.
55. Letter from John Grierson, President of the United Nations Film Board for the year under way, to the Secretary-General of the United Nations (Trygve Lie), 2 February 1948, RAG-1/S-0472-096-(1002-8-2), UN Archives.
56. "Recommendations of Sub-Committee 4 Films and Visual Information, Restricted", 1 June 1948, DPI Records/S-0540-0056, UN Film Board Sixth Session, Advisory Committee of information experts, UN Archives.
57. This was the essence of the plan proposed by Basil Wright, who had been asked to submit suggestions to UNESCO on how to increase the production of films concerning its areas of action. See "Mr. Gjesdal's Questions addressed to UNESCO at UNFB Meeting on 5 June 1948", DPI Records/S-0540-0056, UN Film Board Sixth Session, UN Archives.
58. Letter from Tor Gjesdal to H.W. Laves, 4 November1948, DPI Records/S-0540-0057, UNFB Seventh Session, UN Archives.
59. Letter from Tor Gjesdal to H.W. Laves, 5 November 1948, DPI Records/S-0540-0057, UNFB Seventh Session, UN Archives.
60. Statement by Jean Benoit-Lévy at a press conference, 24 March1948, following the Motion Picture Academy Award for the United Nations documentary *First Steps*. Academy of Motion Picture Arts and Sciences Library (Beverly Hills, Calif.), Special Collections, Margaret Herrick Library, fldr "Jean Benoit-Lévy Biographical clipping file".
61. *United Nations Bulletin*, 15 April 1949: 405.
62. Notes by John Marshall (Rockefeller Foundation European officer) from interviews with Frances Paine, 14 February 1949, and Jean Benoit-Lévy, 24 February 1949, file 2946, box 437, series 100 UN Film Program, RG2-1949, Rockefeller Foundation Archives.

63. *Children of the Ruins* (Jill Craigie, Great Britain, 1948, produced by the Crown Film Unit for the Central Office of Information and the Foreign Office, 35 mm, black-and-white, 11 minutes).
64. In the documents of the UNFB the film's title is *For Their Service*. See "Report on Films and Filmstrips", 3, DPI Records, UNFB, 7th Session, December 1948, Annex 4, UN Archives.
65. *That All May Learn* (Carlos Jiménez, Mexico, 1949, produced by E.M.A., sound, black-and-white, 17 minutes). The first, provisional title was *The Fight Against Illiteracy*. The Spanish title mentioned in the *United Nations Bulletin* 15 December 1948 is *Mentes Sedientas*, but this is the translation of the title of another short film, *Hungry Minds* (National Film Board of Canada, 1948), about another important issue for UNESCO, the essential task of rebuilding educational resources in war-torn areas. The Spanish translation of the title of Jiménez's film is in fact *Saber es poder* (knowledge is power), which better fits the spirit of the film. Incidentally, this Spanish title was used for the presentation of historic UN films at the UNESCO Institute for Lifelong Learning in Hamburg in August 2004. *That All May Learn* was released in 35 mm and 16 mm formats, with commentary in English, French and Spanish. UNESCO did not claim direct responsibility for the film although it had been specifically commissioned for the organization. The English version is available on the website of UNESCO, thanks to the commendable decision to make this historic film accessible to all. See http://www.unesco.org/archives/multimedia/?s=films_details&pg=33&id=16#.VLGkpnvo7d8 (accessed 25 March 2015).
66. "Summary Records", UNFB meeting 8 March 1947, 1, S-0540-0073, UN Archives, and "United Nations Film News", *United Nations Bulletin*, 15 October 1948: 840. Production was planned to be completed by the end of October 1948.
67. The first filmstrip, *Milestones*, examined the historical development of the concept of human rights. Each of the other five – *Abolition of Slavery, Emancipation of Women, Freedom of Thought, The Right to Education* and *Arts and Life* – dealt with a single article in the Universal Declaration of Human Rights. See *The UNESCO Courier* 3:11 (December 1950): 9.
68. Treasures from the UNESCO Film Archives Festival, 20 October 2007–3 November 2007. The database on UNESCO's film collection indicates the same type of material, with a few exceptions.
69. Paul Rotha, "Foreword to Third Edition" in Paul Rotha, Sinclair Road and Richard Griffith, *Documentary Film. The Use of the Film Medium to Interpret Creatively and in Social Terms the Life of the People as It Exists in Reality* (London: Faber and Faber Ltd., 1951), 31.
70. John Marshall's notes of an interview with Jean Benoit-Lévy, 24 February 1949, file 2946, box 437, series 100 UN Film Program, RG2-1949, Rockefeller Foundation Archives.
71. Memorandum from Douglas H. Schneider (Mass Communications) to the Director-General (Jaime Torres Bodet), 16 August 1950, "Expert meeting on Production & Exhibition of Films Serving Purposes of the Organization – UNESCO House April 1951", AG 8/2.2 REG, Box 134, 307: 778.5 A 064(44) "51" Part I up to 28 February 1951, UNESCO Archives. Torres Bodet refrained from using such terms when he wrote to Benjamin Cohen, who was in charge of the Department of Public Information, on 28 August 1950 to inform him of his wish to hold the meeting in Paris at UNESCO's headquarters.

72. Letter from Jean Benoit-Lévy to Jean Renoir, 24 March 1947. UCLA, Arts Special Collections, Coll. 105 Jean Renoir Papers, Box 7, fldr 6, Correspondence March 1947.
73. Jean Benoit-Lévy mentioned this at his press conference of 24 March 1948, following the Academy Award for the UN documentary *First Steps*. The "prototype" idea is also stressed in a letter to Jean Renoir, 17 April 1949. UCLA, Arts Special Collections, Coll. 105 Jean Renoir Papers, Box 9, fldr 7, Correspondence April 1949.

Figure PII.1 Terezska, a girl at a special school for war-handicapped children in Warsaw, Poland, drawing a picture of "home" in 1948, not knowing what a home is after a childhood spent in a concentration camp. The photo was first published in *Children of Europe*, a UNESCO publication chronicling the situation of children in five countries devastated by World War II. (Photographer: David Seymour, © Magnum Photos)

Part II

Rebuilding a World Devastated by War

The establishment of UNESCO was a direct response to the violence during World War II. Emergency action was therefore among its first and most urgent tasks.

The organization promptly engaged in a range of post-war reconstruction activities in war-devastated countries in Europe and Asia, and on both sides of the former enemy lines. It coordinated the activities of voluntary work camps and supported the creation of well-functioning communities for orphans to provide them with not only the most basic physical needs but also fundamental education and a democratic mindset. At the same time it contacted the Allied authorities in order to promote UNESCO's work in Japan and Germany, due to their role as defeated aggressors. Here it paid special attention to changing the attitudes and general conceptions of the youth by promoting democracy, human rights and international understanding, and paved the way for the countries' re-entrance to the international scene.

UNESCO co-founded the Temporary International Council for Educational Reconstruction in 1947, which consisted of 31 international and more than 700 national organizations willing to help. It also launched its own worldwide campaign to gather funds, and *The Book of Needs* of 1948, a catalogue of needs country by country, was distributed to potential donors. Many countries took part in the activities, sent textbooks and offered fellowships for the training of teachers and experts. UNESCO itself donated science teaching materials and laboratory equipment, and provided more than a hundred fellowships from its own resources, while several hundred were given by its member states. It also took on the task of reconstructing the many war-devastated libraries.

The reconstruction program after World War II was the most extensive, but it was not the last emergency action UNESCO engaged in. Later, similar campaigns were launched for the reconstruction of Korea and Palestine, among others.

4

Bringing Everyone to Trogen: UNESCO and the Promotion of an International Model of Children's Communities after World War II

Samuel Boussion, Mathias Gardet and Martine Ruchat

Introduction

The end of World War II gave rise to a number of communities of child war victims in France, Greece, Italy and Switzerland. The communities were more than a material link – they were an idea and a spirit that became popular in the post-war era. They belonged to an educational tradition in which the child was the center of an education that needed to be adapted to their needs – in other words, the spirit of new education. They also belonged to what was sometimes called the "spirit of Geneva": liberalism, pacifism and internationalism around the League of Nations.[1]

The children in the communities were primarily "dispersed" and "displaced" refugees, often orphans, half-orphans or "homeless" children, whose parents had been killed or deported. Their cause soon mobilized a large number of people, knowledge networks, official bodies and NGOs, all facing the challenges of reconstructing post-war Europe. For UNESCO, the children's communities were both establishments and representations of "international understanding", promoted by the organization to ensure lasting peace. In fact, it was an education in international civics and life in the global community.[2]

UNESCO's involvement began in July 1948, when the organization took the initiative to arrange and hold a conference of directors of children's villages in the Pestalozzi Children's Village of Trogen, Switzerland, which gave rise to an International Federation of Children's Communities. Its sponsors were Bernard Drzewieski from Poland, the head of UNESCO's section for reconstruction, and Thérèse Brosse from France, in charge of the organization's war-handicapped children program. The unusual backgrounds of these two figureheads and prime movers of UNESCO's involvement – the first a primary schoolteacher and educationalist, the second a doctor and spiritualist, both members of the resistance and senior civil servants in their respective governments, before they were recruited by UNESCO in 1946–1947 – give some idea of the development of the international organization's fascination with what were then called "children's villages". Where these two senior officials came together was from then on the meeting-point of two foundations of UN policy in the immediate post-war period: reconstruction, which for UNESCO meant educational reconstruction, and help for war victims, and in this case for children.

The communities, with the experimental approach to children's self-government, lasted barely ten years, from 1945 to 1955, in which UNESCO first and foremost played the role of international coordinator, and promoted facts and awareness about the communities, but the afterlife – the impact of the experiments and practical cases on education – was of lasting importance.

Building and extending a culture of international understanding

The choice of the children's community as an "international" model was an opportune one because it was connected with a tradition of caring for child war victims given the "urgent needs" of "millions of homeless children". The term "community" could be used for a home in a settlement or village, a ward, a garden city, a villa, a military camp or a center. The "home" was an attractive model and a symbol of security. However, the common point was to be found more in the spirit of the place than its buildings – namely, education in groups (not too large family-type groups with coeducation of all ages) and an environment where a culture of international understanding could be built, based on a "new teaching" for child war victims.

In practice, the community often took the form of a family that gave meaning to children's lives and was the only condition for the healthy development of their "egos". The stated objectives were freedom, initiative and autonomy, sometimes called "active methods", such as in the orphans' home at the Cité Joyeuse in Belgium. For the Italian education scholar Ernesto Codignola, the Scuola-Città in Florence was a "center for spontaneous social activity", and at the Giardino d'infanzia italo-svizzero in Rimini the tutors were prepared for "the tasks of the active school and the new school". Clearly the term "children's community" covered a range of experiments, even if they had a point in common: moving away from the model of "barracks schools" and "traditional schools", seen as an "artificial, obtuse and boring" environment, as Codignola put it, and unable to meet the need to retrieve "energies, intellectual capacities and the power to work", according to another education scholar.[3]

The idea of a republic, of democracy, of an autonomous government of children was only found in some communities – such as Civita Vecchia, Moulin Vieux, Marcinello, the Pestalozzi villages in Florence and Trogen, and Gaudiopolis – and seemed to be more an idea than the children's own spontaneous, autonomous behavior. It might mean at most that decisions were voted on, or councils or unions elected.

What made the model so innovative at the time was probably the combination of various forms of aid and protection: social, educational, therapeutic and based on research in the field of psychology. It was a sort of World War II version of the project of new schools in the countryside, which were intended to teach the individual within and for a collective group, and the desire for scientific research that made it essential not only to train tutors but also to accompany them with psychologists and psychiatrists within the communities themselves. Some already known methods of educational therapy were applied: individual files, life histories, tests, drawing and painting. The children's communities often had an adjoining "center for psychology and educational therapy studies and careers advice", making them, as it were, into psychology, educational and careers advice laboratories where the child became the "object" of expert reports within the field of development sciences, such as child psychiatry and development psychology.[4]

Bernard Drzewieski and educational reconstruction

Nothing in Bernard Drzewieski's past predetermined him to work for UNESCO or concern himself with children's villages. He was born in Lublin and was educated internationally: he spent his final year of secondary school in Odessa, four years at university in Geneva and at the Sorbonne in Paris, and he spoke fluent Russian and French. He gained his secondary-school teaching certificate in comparative literature at Warsaw in 1919 and taught in Poland until 1934. He then became a headmaster and, alongside his work as an activist, chair of the Polish union of secondary teachers, then vice-chair of the national teachers' union. He also worked with the network of the international New Education Fellowship, wrote a number of articles for its journal, *The New Era*, and became general secretary of its Polish branch.

However, the war disrupted his career, as it did that of so many others. In September 1939 he fled first to Romania, where he was education consultant for the Polish Refugees' Committee, and then in 1940 took refuge with his wife, Wanda Schoeneich, in England, joining Władysław Raczkiewicz's government-in-exile in London. He ran the Education Department of the Ministry of Social Affairs, although the ministry's activities were mainly strategic and military, in liaison with the resistance in Poland.

Living in London, he not only learnt English but this period was also the determining factor in Drzewieski's later UN career. He was working with the Conference of Allied Ministers of Education (CAME), set up in 1942 and already preparing a plan for the reconstruction of education in wartorn countries. The aim appealed to him as a one-time teacher, given the news he was hearing of the destruction in his homeland.

In 1945 he was appointed as cultural attaché at the Polish Embassy in London. This was in line with his political leader, Stanisław Mikołajczyk, who had been deputy prime minister since Władysław Sikorski was killed in a plane crash in 1943, and had agreed to return to Poland and join the Polish Committee of National Liberation (or Lublin Committee), which was a provisional government authority formed on 23 July 1944 at the initiative of the Soviet Union.[5]

Consequently it was as the representative of the Polish delegation that he took part in the November 1945 conference preceding the preparatory commission for UNESCO, and on 16 October 1946 he was asked by Julian Huxley, the head of the provisional commission at the time, to join the reconstruction section. Drzewieski agreed but asked for a postponement while he completed his diplomatic mission representing the Polish Government at the General Conference in 1946 that officially created UNESCO. However, he did agree to act unofficially to contact the bodies involved in reconstruction, and he was the rapporteur at the first session of the Commission for the Reconstitution of Education, Science and Culture.

In January 1947 he officially joined the UNESCO secretariat as head of the department for the reconstruction and rehabilitation of the education system. This was a strategic post which enabled him to return to his homeland from 23 August to 11 September for an initial mission, while remaining at arm's length from the pro-Soviet Polish Government, whose line was hardening after the scandalous fraud in the January 1947 general elections leading to the resignation and chosen exile of his former chief, Mikołajczyk.

Drzewieski found accommodation in Paris in the 16th arrondissement, a stone's throw from UNESCO House, but also travelled widely elsewhere, such as in the USA, Switzerland and Czechoslovakia. These trips were both diplomatic and a chance to assess the need for reconstruction, focusing initially on schools and support for teachers. Using his longstanding networks, Drzewieski contacted teachers' associations, educational publishers and various mutual aid bodies, such as the International Bureau of Education, the Commission for International Educational Reconstruction, Don Suisse, the International Children's Fund, the United Nation's Relief and Rehabilitation Agency and the International Union for Child Welfare, to coordinate fundraising and the collection of teaching equipment.[6]

To make the coordination more effective, he held three meetings that led to the creation in Paris on 23–24 September 1947 of the Temporary International Council of Educational Reconstruction, an intermediate body between the NGOs and UNESCO, and for which he acted as secretary.

It was while on one of his missions that in the summer of 1947 he came across the Trogen Children's Village, which he described as a "most amazing and inspiring venture in post-war Europe" due to the way it was run. He was even more impressed by the fact that one of the first national houses to be built in the village was the Polish one. Later he visited the very different children's village at Otwock. Here, 600 children had been dumped in a former sanatorium outside Warsaw and were funded like Trogen by the charity Don Suisse. At the suggestion of one of Trogen's founders, who wanted UNESCO to exercise some form of sponsorship, Drzewieski proposed to hold a conference, under the auspices of his department, for the leaders of the children's villages already existing in various countries so as to design a general scheme for this movement.[7]

Thérèse Brosse and child war victims

The personal career of Thérèse Brosse is just as singular and unlikely. She was French, born at La Fère, Aisne, in 1902. As a child, she and her mother moved from one garrison town to another with her father's postings until in 1920 he was appointed, with the rank of colonel, as military commander to the presidency of the Republic, a largely ceremonial role. Her schooldays were spent entirely at the Jeanne d'Arc private school in the Rue Saint-Jacques until she took her *baccalauréat* in 1920. She then studied medicine and became a doctor in 1931. Although she spent some time during her studies at the Hospital for Sick Children, her speciality was not pediatrics but cardiology at Broussais Hospital. She then practiced as a consultant cardiologist first at the La Roche Posay Spa in Vienna and later at Royat Spa in the Puy-de-Dôme in Massif Central.

Despite her Catholic upbringing, she found her path in other forms of spirituality, first within the masonic lodge "Human right", where she met university lecturer Jean Émile Marcault, who later converted her to theosophy – a system of esoteric philosophy that was strongly influenced by Hinduism and intended to create the nucleus of a universal brotherhood of humanity without distinction of race, creed, sex, caste or color.[8]

In 1936 she was appointed as head of the cardiology clinic at the Paris Faculty of Medicine and in September 1939, at the same time as Marcault, she began a political career as deputy head of the sociotechnical cabinet of the Ministry of Public Health.

During the war she also worked secretly in the resistance and became the regional head of the Clary group of agents. She was later arrested by the Gestapo and held in the German prison at Moulins Castle.

At the Liberation, Brosse was elected secretary-general of the liberation medical committee for Puy-de-Dôme and soon inspector of public health in Puy-de-Dôme, where she was in charge of combating infectious diseases, creating hospital facilities and rehabilitating the physically defective, until in March 1946 she went the USA for six months as the public health mission head for the Ministry of Foreign Affairs.

As she wrote later, this trip revealed to her "the applications for paediatrics of her work in the three areas of technical research, therapy and childcare".[9]

During her mission she was asked to carry out a thorough study of public health schools, particularly the child health departments of Yale University and Johns Hopkins University children's hospital.[10]

Shortly after her return to France, she applied for a job at UNESCO, specifying her new interests: "Any work dealing with mental hygiene of children and child guidance any study about de-socialized children during the war: appraising the needs, planning to promote new mental health, developing training facilities for physicians and teachers to get experience in child guidance, diagnosis centers, re-educational centers, and so on."[11]

On 21 May 1947, with the agreement of the Ministry of Public Health and Population, Brosse was seconded to the reconstruction section to undertake a survey of the educational, scientific and cultural needs of Poland and Czechoslovakia. Unlike Drzewieski, it was not so much school reconstruction that interested her as the psychological state of children.[12] She recommended priority support for teacher-training, child psychiatry and psychology institutions, and argued for more international exchanges between specialists in these fields.[13]

On her return she was asked by the education section of the Carnegie Endowment for International Peace to be an instructor at UNESCO's first practical summer school on education in international understanding, held for six weeks in Sèvres, France. She led a group on "educating the emotions in order to develop international relations" and gave a long talk on "the psychological foundations of international understanding", applying to the post-war period the ideas of a "new era" and a "broadening of human consciousness". She proposed a genuine progressive education of children's and adolescents' emotions so as to encourage them to transcend their own cultures and cooperate in the "development of humanity".[14]

In September 1947, as her second contract was coming to an end, Brosse resigned from her job at Public Health to continue her research and work with UNESCO, and in March 1948 she became a member of the secretariat in charge of the project in the 1948 program entitled War-Handicapped Children. She then undertook study trips to nine countries in Europe, including her visit to the Trogen Pestalozzi village, the final stage of a one-week stay in Switzerland in April 1948.

The children's village trademark

The General Conference in Mexico City in November 1947 confirmed UNESCO's official support for children's villages. Drzewieski had taken care to launch a preliminary report on "the effects of the war on children and the treatment that has been most successfully applied to children whose physical, intellectual and emotional development had been disturbed by the war".[15]

The report was entrusted to a French psychiatrist, Simone Marcus (1911–2012). Although her time as a consultant was only brief, Marcus' role significantly influenced UNESCO's choices, and her career illustrated the growing interest in children's villages. She was a doctor in child psychology and in the 1930s was given a travel scholarship to the USA to study approaches to juvenile delinquency (a trip she considered in hindsight to have been one of the reasons for her work for UNESCO) but later returned to Paris to run her private practice.[16]

In 1946 she produced an initial survey, based on a questionnaire circulated in the liberated countries by the Inter-Allied Conference, on behalf of the Ministry of Education's educational research center. It focused on the psychological effects of the war, French experience in particular, with data collected by private and public organizations, and in her own clinical practice. While she played down the direct effects of the war on children's psychology, there were so many problems at that time that new solutions had to be found.[17]

She did not mention the children's villages but was aware of these tentative experiments since she gave a talk at the European Congress for New Education, held in Paris in August 1946, alongside Walter Corti, who had come to speak about the Trogen Pestalozzi village.[18]

However, Marcus' 1947 report presented a picture of the situation in Europe, first describing the material, physical and psychological effects of the war and then concentrating on experience with reconstruction.[19] The children's villages were seen as some of the most promising attempts, but not immune to objection, because following the pre-World War II antecedents in the USA, the UK and Mandatory Palestine, there were as many versions as there were villages, making it something of a "catch-all" concept.[20]

Autonomy for children was already raising questions, challenging the place of adults, just as the continual promotion of a child's individuality rather than the collective in a model international village raised the question of children's return to their home countries. However, the novelty and experimental value of the children's village were worth trying out, leading perhaps, so Marcus thought, to greater similarity in methods. She saw the Pestalozzi village as emblematic, but it needed to be brought closer to other similar experiments.

The Mexico City conference entrusted the director-general with the role of "devising, in cooperation with the national and international organizations concerned, a study and work plan on the problems raised by the education of children who have suffered from war" and "asking experts from various countries to provide information and detailed reports and attempt to have the most conclusive experiments in this field studied in the field".[21] Although Drzewieski was aware that in these villages "only a hundred or so children were being cared for, while there are millions of orphans", he repeated in this second session the idea of a conference in Trogen to "study the ways of integrating them into the official education systems of the various countries".[22]

As a result, the conference in 1948 became a UNESCO event, bringing together the work of its various sections, and Drzewieski became one of the formulators of the invitation to the directors of children's village and other experts who would become the Trogen conference and managed to include the topic in the resolutions and action plans of the Temporary International Council of Educational Reconstruction. In June and July, Brosse was put in charge of organizing the technical side of the international conference of children's village directors, for which she later wrote the report entitled *Homeless Children,* published in a number of languages by UNESCO in 1949, supplemented by another report, *War-Handicapped Children,* published the following year.

Bringing everyone to Trogen

Early in 1948, Bernard Drzewieski set up a committee of representatives from the reconstruction and education sections. They turned to international organizations such as the International Bureau of Education, Semaines d'Études pour l'Enfance Victime de la Guerre and the International Union for Child Welfare, which had pools of experts, for their technical opinions on the children's villages and educational experiments in general, because they had field knowledge of value in inventorying the various villages, which UNESCO had to confess it lacked at that point.[23]

These bureaus and unions produced expert reports, sent out delegations to countries to report on the situation of children, brought together specialists in clinical psychology and education at conferences and regional sessions in war-stricken countries, and made training courses for teams of educational therapists, but they had not worked together, so organizing the directors' conference became the first cooperative effort. In spring 1948 the various secretariats exchanged data and made lists of the directors to be invited and the valuable experiments to be selected. Some villages were added and others removed from the original list.

As for the venue, Switzerland was chosen because of its supposed neutrality and desire to "improve" its image after the war, and there was also the Institute of Educational Sciences in Geneva, headed at that time by Jean Piaget (1896–1980), director of the International Bureau of Education, who had attended UNESCO's inaugural meeting in 1945. At the same time, the name Pestalozzi of the Trogen children's village had a symbolic value because the Swiss philosopher and educationalist's name was known worldwide. Walter Corti, the founder of the village, was an idealistic intellectual and the editor of *Du* magazine, in which he promoted the vision of harmony in the world and individual responsibility, which he shared with his co-founders: Marie Meierhofer, a pediatrician and psychiatrist, Elisabeth Rotten, a Quaker and head of the cultural relations office at Don Suisse, and Hans Fischli, an architect, painter and sculptor.

Trogen was also the model of a community of nations similar to Switzerland's own federal cantons. It embodied this international understanding in its division into houses representing national families, with tutors from the children's own countries to ensure the continuity of culture and language.

The conference held on 4–11 July 1948 marked the high point of UNESCO's support in the history of children's communities. It marked the organization's firm commitment, primarily financial, since UNESCO covered the travel and accommodation expenses of the directors, experts, translators and secretaries. Ultimately the conference was attended by 14 community delegates, 11 experts to apply their particular skills and report back on "their most typical experiences as technical experts", 4 miscellaneous participants and 12 observers from 11 UNESCO member states.[24]

As the conference talks advanced, day by day, there was a crisscrossing of concepts that revealed the diversity of practice. All agreed, however, on a discourse in praise of new education. Speakers mainly recommended a "liberal education" that did not confuse free will with an internal law shared by all the children in the community, and which was now "appropriate for their stage of development" and took account of their ambitions – particularly regarding their careers.

Self-government was also championed. On the one hand, there was practical education and an active school that argued for pupil autonomy, closely linked to the idea of a gang of children organizing themselves, and on the other, an education in international understanding, where self-government ensured the child's civic skills practiced in councils, unions or "a little democratic state", as in Civita Vecchia and the Children's Republic.[25] The idea of autonomy was linked to that of the child's life: children were not to play at being adults but rather live as children. Self-government was in no way a "Bolshevik workers' council", or even like Moulin-Vieux, which had adopted the legend of a republic, "Torchok", from an imaginary model one boy had brought back from Crimea, but a real educational instrument for developing a universalistic spirit among both children and tutors. It was in order to educate the latter to a way of being, an international awareness and a culture of understanding, that since 1944 had been run "international courses for monitors in homes for war-handicapped children".[26] The aim was to change mentalities and also provide methods for coping with violence and "anti-social manifestations".[27]

The Trogen meeting also laid the foundation for a trauma psychology that went well beyond the immediate post-war needs: war neurosis, juvenile delinquency, "abandonment complex", "victim complex" and "compensation complex", as the psychologist André Rey put it.[28] New categories of disorder were mentioned for a psychology of war victims: "troubled emotional life", "disappointed and repressed emotionality", "apathetic, indifferent or grumpy behavior", "excessive sensitivity or timidity", turbulence, anxiety, rudeness, bullying and "lack of self-control" were listed by Thérèse Brosse.

The resolutions passed on 10 July demonstrated a principled concern for past crimes against children and emphasized that education must take account of the "particular needs" of children with no distinction of faith, race or political opinion. At the same time, the creation of an international federation – a sort of "active academy" – was approved by the conference. Its aim would be to ensure these objectives were met and maintain links with UNESCO.

For UNESCO, the children's community was altogether a way of embodying its ideals and making the concept the vector of a number of values it sought to promote through children, in the hope that this would extend the influence of a culture of "international understanding".

UNESCO as a propaganda unit

A certain concept of the child was developed after the Trogen conference, making children into forces of renewal and factors for change. Child war victims became the bearers of hope and also of political projects, such as reconstruction, pacification and the growing importance of international aid policy.[29]

One of the first ways of spreading the spirit of Trogen was the practical application of education to international understanding, via international children's conferences and teachers' courses set up under UNESCO auspices. Following the Trogen conference, it was decided to hold an international camp to which the children of one of the communities would invite representatives of others. In 1949 the first camp was held at Moulin-Vieux, set up between 1944 and 1946 by a schoolteaching couple in the Isère in the south-eastern part of France. A second took place the following year at Esch-sur-Alzette in Luxembourg. Similarly, in order to exchange educational experience between children's communities, international teachers' courses began in 1949 at the Île-de-France school village at Longueil-Annel, run by the president of the International Federation of Children's Communities, Robert Préaut.

UNESCO also had major propaganda instruments in the publications it sponsored or directly produced, and these reported on the children's communities.[30] Using its networks, the organization made it possible for a special issue of *New Era at Home and School* to be brought out in September 1948, entirely devoted to children's communities and UNESCO's role, and in 1949 the journal *Impetus* featured on its front cover the photo of a boy with the caption "He governs himself" and devoted an illustrated report to Moulin-Vieux with a map of Europe showing the location of the other children's communities.

As early as March 1948 *The UNESCO Courier* gave generous space to the children's communities, with interviews with the founders of the

International Federation of Children's Communities and inspiring photographs, all intended to raise interest and demonstrate the value of UNESCO's support for these experiments, and also recount history as it happened, a sort of story of the rehabilitation of the war children. In 1950 a film called *Everybody's Child* was made by the Swiss director and mountaineer Othmar Gurtner for Swiss Cultural Films in Zurich, with funding from UNESCO.

A clear example of the genre, in its expressive force and staging, was the book of photographs published in 1949 under the title *Children of Europe*. They were taken by David Seymour, one of the founders of the international photographic cooperative, Magnum, during a journey through Austria, Germany, Greece, Italy and Poland.[31] The purpose was to illustrate the report he had been commissioned to produce by UNICEF and to be published by UNESCO.[32] It contains 52 photographs, with an introductory "Letter to a grown up" from an imaginary child seven years old when the war began. The tone is not one of resentment but rather an appeal to adults from the 13 million abandoned children in Europe. The story is told by the pictures and their captions in French, English and Spanish: from bombed streets to concentration camps, the photographs show children's emotions, "the fear of the men who kill", the instinct for survival (gangs, looting, black market, selling cigarettes, prostitution, theft) and dreams. The photographs also show the facts of the children's lives, such as police raids, concentration camps and orphanages, and figures are given for the number of children injured, burnt, disabled, amputated, blind and deaf.

After the factual report, the message becomes a call for help with these children's needs – a roof, milk, meat, teachers – sent out to the relief agencies and the vast project of UNESCO, even though it is barely mentioned by name. The aim is to rebuild the schools and to give the children instruction and training. The villages run by children are described in the letter as "glad" news. This model is presented as a children's dream, a republic where the laws are worked out by children. Gang leaders become citizen-administrators with a special currency used to reward, and also to buy things and to be exchanged. The message between the lines is to bring the children reduced to poverty and survival by crime back into the "true social community among men" to work at the jobs they have learnt.

The photographs are obviously framed to evoke anguish and compassion in turn: black-and-white, with deep shadows, such as ruins where children play; street scenes with games and crime; police clamp-downs; scenes in juvenile court and reform prison; in a cell, where a girl seen from behind sits looking out through the bars; groups of children, famished and emaciated, their eyes sunken or closed, looking dreamy or sad, holding their tin cups out to us. All these visions of abandonment and deprivation are calls for help, particularly from women – from women breastfeeding to policewomen or social workers. In a carefully staged scene, the arrival of the doctor brings children's smiles back: they will be cared for, and the disabled will be fitted with limbs and will be able to play again, learn a trade and get an education. However, the children's own force of will makes it possible too, according to the pictures, and the captions stress their youthful energy, energetic bodies and smiles of achievement and concentration in class and in workshops. By showing them usually working alone, Seymour expresses one of the great values of rehabilitative education: autonomy. The children equip their own laboratories, sew their own clothes, make their own shoes, learn to print books and build their own school, "to build the new world", in the allegory of children dancing in a circle.

Internationalist dreams and national control

In a report on war-handicapped children from 1950, Brosse highlighted the challenge the post-war period posed for UNESCO, which, according to her, was "the problem of discovering a type of education which may be able to prevent the repetition of such a social cataclysm", and she cited one of the fundamental ideas in UNESCO's constitution, which specified that "educational methods [should be] best suited to prepare the children of the world for the responsibilities of freedom".[33]

Behind the grand declarations of humanism and universalism, however, was the problem of how in practice to house and re-educate these war-handicapped children, many of whom in the post-war chaos were not only homeless but displaced persons. Before anyone could make world citizens of them, she noted, they first had to be attached to a nation, and in her report she suggested a surprising extension of the notion of orphan:

> For the purpose of education, the definition of an orphan should cover not only a child who has lost his or her father and mother or one of them, it should also cover a child who, for practical purposes, had no home and, for material or moral reasons, could not receive from his or her parents the care to which a child was entitled and who therefore had to be looked after by society.[34]

After a tour of children's communities, Brosse had realized "how much the children need a country at their own if they are to be psychologically normal and to feel 'like other people' ", and in a world in which it is still necessary, for legal purposes, to have a nationality, their youthful independence was not strong enough for them to become world citizens immediately without first being a citizen of a smaller community.[35]

And here the educational model praised as a good example and a solution was the Pestalozzi children's village at Trogen, "the village where the world is one" and "the children from the various houses come together and thus form a little 'family of the nations' ", to quote the fundraising brochure produced in 1948, and indeed the words of Elizabeth Rotten during the conference of children's community directors, that among "the three main ideas which had inspired the establishment of the village" there was "the desire to demonstrate that children and adults belonging to different nations, some of them ex-enemies, could live together in harmony and good-will, realizing the differences between them and yet united in their diversity".[36]

Brosse saw this as having been achieved and praised the virtues of living together:

> At the Pestalozzi children's village at Trogen, Polish children showed aggressive ill-feeling towards German children whose fathers had killed their own, only during the first few weeks. Then games and joint activities very soon aroused and fortified a spirit of friendship which leaves no room for antagonistic feelings.[37]

Like a scale model of UNESCO itself, Trogen, it appeared, recreated in a microcosm a world of concord and an international community where children instinctively recognized each other, just because they were children, and growing up together for a while, and it wove indissoluble ties of fraternity that would last their adult lives, when each had resumed their national home and identity. The director Arthur Bill's detailed description in *New Era* of how the village ran in fact showed that cosmopolitanism could certainly not be taken for granted and the teaching dispensed there swung ambivalently between defense of national identity and international understanding. Any "exchange of ideas and experience" there was rather "between the teachers", with "meetings every month", and every day at "four o'clock when they go to collect their letters and drink a cup of tea". Among the children this mixing was limited because all their lessons took place in their national house "in their own languages", where "careful preparations are made for the children's return to the common life in their own country". Only when they were older was it planned that they should learn the "village language", in the hope that a "school community" would "arise spontaneously". One small detail that appeared not to bother anyone, despite what some children had suffered under Nazi oppression, was that the "village language" was German, "since the village is in German-speaking Switzerland". But internationalist dreams were not to be hampered by the language barrier, and Bill was glad to see that "the children from each house join up for gymnastics, music, gardening, drawing and manual work" and "as soon as they have any facility, certain subjects will be learnt in common in that language: cultural history, general geography, natural history, mathematics, physics and chemistry".[38]

This optimism was not shared by everyone, as can be seen from Volkov's critical remarks as she addressed all the children's community directors at

Trogen in 1948, asking them what they were doing to prepare the citizen-children of these villages to see their future lives in the outside world, and what the underlying ideal was. The easiest, most dramatic and most compassable of such visions, she thought, was some kind of nationalism – service to some homeland, remembered or imagined, however vaguely. The hardest would be service to mankind, and the most sensible would seem to be service to one's immediate neighbors.[39]

This shows the Utopian nature of the project, this gap between the founders' intentions to design an international village embodying the ideal of international understanding and the difficulty in achieving it. As early as 1945, the psychiatrist Oscar Forel, president of the Semaines d'Études pour l'Enfance Victime de la Guerre, with a certain condescending skepticism, saw the plan for the Trogen village as an "outbuilding-children's home" that would give Switzerland the role of "nursery and child-minding service for the devastated children of Europe", leading, he suggested, to a "Landi-baby-Hollywood", named after the Landi 1939 Swiss national exhibition, and reduced to a cute Utopia, in contrast with more practical policies of aid for wartorn countries.[40]

Laboratories for experiments in education: A conclusion

As Zsigmond Ádám, director of the children's town at Hajdúhadház, Hungary, said at the Trogen conference, children's communities were undeniably laboratories for experiments in education, even of extreme experimentation. Because of those they sheltered, homeless and stateless children, it was possible to play down or even ignore the family or country they came from, which were denied, erased and replaced by the new framework of care.

The issue of war-handicapped children thus took on a particular resonance in the context of World War II and the highly visible destruction it had wreaked. This cause and the international mobilization it brought forth became synonyms for reconstruction, reparation and clearing society's name during a post-war period favorable to relatively Utopian universal pacifism, for which UNESCO became a mouthpiece. Then, once warfare had ceased and peace had apparently returned, and the war-handicapped children had grown up, the issue appeared to fade into the background and was replaced by other causes and displacements, some from earlier times: juvenile delinquents, gangs, street children, unhappy childhood, social cases and so on.

In addition to the divergence of views about children's communities, any extension of the model came up against the new political borders formed by the Cold War, which were reflected in tensions in the bodies with international pretensions. In 1949–1950, the International Federation of Children's Communities, for example, was caught in the middle. It was accused of sympathizing with the Eastern Bloc because one prominent member was Bernard Drzewieski, still representing the Polish Government; because Moulin-Vieux was controversially chosen for the first international children's camp, where the Juliens were suspected of being Communist Party members; and not least because of UNESCO's help given via the federation to communities taking in Greek refugee children in Bulgaria, Romania and Czechoslovakia, whereas the representatives of the new Greek Government spoke rather of children abducted by Greek communists at the end of the civil war and demanded they should be returned to Greece.

Conversely, the federation was gradually seen to be representative of the West, with its funding from Canada and the USA, which threatened to suspend it if grants were made to children's communities in Eastern Europe. The Western influence was personified particularly in the active presence of Carleton Washburne (1889–1968), creator of the Winnetka School Plan in Illinois that was commissioned by the Allied armies to reorganize education in Italy, and it became even more visible due to the "empty chair" policy of the Eastern Europeans on the federation's committees. The situation led to the despair of its president, Robert Préaut, who declared that without Eastern European engagement "this 'children's international' we want to create would not really deserve its name".[41] But the image of a Western-dominated federation only became even more obvious by the forced "repatriation" decided after 1950 of all the refugee children from Hungary and Poland who had been placed in the Trogen village.

In addition, the tensions between the USA and France concerning their views about culture and education, and then the Cold War – worsening relations between UNESCO and Czechoslovakia, Hungary and Poland – led to a reduction if not in US supremacy at least in its subsidies, to the detriment of some programs.[42] UNESCO's role as a "magnificent instrument of propaganda . . . for spreading American culture and thought throughout the world", and a base for acquiring new export markets, in the 1950s turned toward the so-called "underdeveloped" countries.[43]

Nevertheless, the memory of the children's communities as practical laboratories for experimental education would still stand as a solid cornerstone for the legitimacy of new education in the post-war world.

Notes

1. Martine Ruchat, "Entre militance et science: la cause des enfants anormaux à l'Institut J-J. Rousseau, 1912–1933", *Les Sciences de l'éducation. Pour l'ère nouvelle, CERSE-Université de Caen* 35:4 (2002): 63–84.
2. Chloé Maurel, *Histoire de l'UNESCO, les trente premières années, 1945–75* (Paris: L'Harmattan, 2010), 218.
3. Thérèse Brosse, *Homeless Children* (Paris: UNESCO, 1950), 26.
4. Samuel Boussion, "La République d'enfants à l'épreuve de la pédagogie curative: le Hameau-école de l'Ile-de-France 1945–1950", paper, AREF, Geneva, 2010.
5. A dissident Polish Government continued to exist but the USA and the UK withdrew their recognition of it on 6 July 1945 and it was forced to move out of the

Polish Embassy in Portland Place, London. Most of its members, unable to return safely to communist Poland, settled in other countries.

6. Mission reports of B. Drzewieski, X07.83, UNESCO Archives.
7. Report of journey to Switzerland from 5 July 1947 to 16 July 1947 and Report of journey to Czechoslovakia and Poland between 7 September 1947 and 11 September 1947, X07.83, UNESCO Archives.
8. Marie-José Delalande, *Le mouvement théosophique en France 1876–1921* (PhD dissertation, Le Mans, University of Maine, 2007).
9. Autobiography, Thérèse Brosse personal file, 9, Per/SNAFF/Records, 1965, Box 22, UNESCO Archives.
10. Ibid., 10.
11. Job application form, 1 March 1947, Per/SNAFF/Records, 1965, Box 22, UNESCO Archives.
12. Report on mission to Poland, 30 May 1947–20 June 1947, 361.9 A 20 (438), UNESCO Archives.
13. Ibid. and Report on Czechoslovakia, 361.9 A 20 (437), UNESCO Archives.
14. Thérèse Brosse, conférence au stage d'études pratiques sur l'éducation pour le développement de la compréhension internationale, Sem.Lec.II/1/1947, UNESDOC.
15. 1947 programme, Section A, Education, general conference, Session 1, 20 January 1946–10 December 1946, Paris, UNESCO Archives.
16. Simone Marcus, *Pour comprendre les enfants* (Paris: Bourrelier et Cie, 1947), 3 and Pascale Le Maléfan, "Rencontre avec Simone Blajan-Marcus", *Bulletin Marionnette et Thérapie* 4 (1995): 8–10.
17. Simone Marcus-Jeisler, "Réponse à l'enquête sur les effets psychologiques de la guerre", *Sauvegarde* 8 (February 1947): 4.
18. "L'enfance victime de la guerre", *Pour l'Ere nouvelle* (special conference issue, August 1946): 115–120.
19. Simone Marcus, "The problems of war-handicapped children: preliminary report," November 1947, UNESCO Archives.
20. In 1895 the industrialist William Reuben George (1866–1936) founded the George Junior Republic in Freeville, a small town near Ithaca (New York), and 20 years later the Irish-born priest Edward J. Flanagan founded Boys Town in Omaha (Nebraska). See Mathias Gardet, "Freeville et Boys Town: de la fascination à l'oubli. Les deux modèles américains précurseurs des républiques d'enfants", paper, ISCHE, 2012. The children's villages at Ben-Shemen and Kfar Yeladim in Palestine were founded in the 1920s. See Dr Siegfried Lehmann, "Le village d'enfants de Benschemen", *Pour l'ère nouvelle* 66 (March 1931); Fedi Laurent and Yaffa Wolfman, "Kfar-Yeladim ou le village des enfants: l'expérience pionnière de Pougatchev en Eretz-Israël, vue par Joseph Kessel (1926)", *Le Télémaque* 2:32 (2007): 137–148.
21. Mexico City general conference, Resolutions 1.7.1–1.7.3, UNESCO Archives.
22. Citations from Minutes of Session 1, 23.9.1947, IVO 3 Conf/Rec & Rech/SR 1.2.3.4, 7 and Minutes of Session 2, Mexico City, 17 November 1947, UNESCO Archives.
23. Report of the preliminary meeting of sections concerning children's villages, 4 February 1948, 371.95 A06 (494) "48", UNESCO Archives.
24. Invitations to the Trogen Conference, June 1948, 371.95 A06 (494) "48", UNESCO Archives. Also Brosse, *Homeless Children*, 14–15.

25. Adolphe Ferrière, *L'autonomie des écoliers dans les communautés d'enfants* (Geneva: Delachaux and Niestlé, 1950).
26. Martine Ruchat, "Le cours international de moniteurs pour home d'enfants victimes de la guerre: une formation originale pour le 'rapprochement des peuples' (1944–1956)", paper, AREF, University of Geneva, 2010.
27. Brosse, *Homeless Children*, 40.
28. Brosse, *Homeless Children*, 38.
29. Mathias Gardet and Martine Ruchat, "Le Village Pestalozzi, un modèle de communauté d'enfants pour l'Europe: entre utopie pédagogique et propagande politique, 1944–1954" in *Entre assistance et contrainte: le placement des enfants et des jeunes en Suisse 1850–1980*, ed. Markus Furrer et al. (Itinera 36, Schwab, 2014), 123–138.
30. Mathias Gardet, "Le modèle idéalisé des communautés d'enfants à l'épreuve de la réalité française 1948–1955", paper, AREF, Geneva, 2010.
31. Carole Naggar, *Chim. Children of War* (New York: Umbrage editions, 2013); Cynthia Young, *Chim: We Went Back. Photographs from Europe 1933–1956* (New York: Barnas, 2013).
32. David Niget, "Pieds nus dans les ruines: le regard de Chim sur les enfants de la Guerre", *Revue d'histoire de l'enfance "irrégulière"* 15 (2013): 143–145.
33. Thérèse Brosse, *War Handicapped Children. Report on the European Situation* (Paris: UNESCO, 1950), 12, 14.
34. Brosse, *War Handicapped Children*, 28.
35. Brosse, *War Handicapped Children*, 20–21.
36. Minutes of first session, 5 July 1948, ED./Conf.1/SR.1, UNESCO Archives.
37. Brosse, *War Handicapped Children*, 105.
38. Arthur Bill, "Self-Government and Community in the National Houses of the Pestalozzi Children's Village", *The New Era in Home and School* 29:8 (September-October 1948):182–183.
39. Peggy Volkov, "The Re-education of War-Handicapped Children: Some Questions about Re-education in Loyalty", Conference of Directors of Children's Villages, p. 7, UNESCO Archives.
40. Unpublished text by Oscar Forel, Forel Archives, Lausanne.
41. Minutes of the FICE AGM, Charleroi, October 1949, UNESCO Archives.
42. Maurel reports that in 1947 France was only contributing 7 percent of UNESCO's total budget, the UK 14 percent and the USA 44 percent. See Maurel, *Histoire de l'UNESCO*, 102, 116.
43. Foreign minister Georges Bidault, second session of executive council and France's role in UNESCO, 3 August 1947, 6–7, quoted in Maurel, *Histoire de l'UNESCO*, 96, 99.

5

Returning to the International Community: UNESCO and Post-war Japan, 1945–1951

Takashi Saikawa

Introduction

Since its formal entry in 1951, Japan has consistently been one of the most ardent advocates among the member states for the fundamental principles of UNESCO and its programs, and the country is today one of the organization's biggest financial contributors. The Japanese diplomat Matsuura Koichiro[1] served as the director-general of UNESCO from 1999 to 2009, and there are at present 270 UNESCO associations throughout the country with a view to advancing private cooperation activities in conformity to the constitution of the organization. Altogether, it is evident that UNESCO has in general been widely and positively recognized in Japan for decades.

This chapter examines why the Japanese people and government have been attracted to UNESCO and how the country's relationship with UNESCO has been envisaged, by focusing on the period from 1945 to 1951. This is not only because the relationship between UNESCO and Japan was formally and informally established during this period but also because the basic image among the Japanese people of UNESCO was formulated in line with the advancement of specific cooperation between both sides in those years. At the same time, in consideration of its characteristics as a defeated nation of World War II and one of the non-Western countries, considerable attention is given to discussions about Japan in UNESCO with a view to offering a view of the organization transcending Allied-centric and Eurocentric perspectives that could provide new insights into the history of UNESCO from a global point of view.[2]

Embracing UNESCO in Japan

Since Japan was politically, economically and culturally devastated by World War II and it subsequently came under the occupation of the Allied Forces, little attention among the Japanese people was paid to UNESCO when it opened the first session of the General Conference in November 1946.[3] Even so, some intellectuals were greatly attracted to UNESCO with limited information. For example, the Japan PEN Club – the national branch of the worldwide association of writers – showed its desire to cooperate with UNESCO as early as in February 1947. Furthermore, following the establishment of the Sendai UNESCO Cooperative Association in July as the first UNESCO association in the world, such cooperative associations were prepared to be set up in different cities in Japan including Kyoto, Nara, Kobe and Osaka. As the UNESCO movement thus increasingly grew and spread throughout the country, these associations met in Tokyo in November and held the first national convention of the UNESCO cooperative associations in Japan.

The UNESCO movement in Japan was underpinned by two strong motivations: renouncing war and becoming a cultured nation. First, apart from the question of how sincerely the Japanese people of the time responded to their responsibilities for the war, there is no doubt that their wartime experiences instilled in them a common sense of aversion to war and military power of any kind. In this context, not only Japanese intellectuals but also its public at large were fascinated by UNESCO and found one of the guiding principles for post-war Japan in the constitution of the organization. Second, based on the shared sense of "war allergy", the subject of lively discussions, particularly in the early post-surrender period, was the idea that Japan should be a "nation of culture" (*Bunka Kokka*).[4] As a nation of culture was understood as a complete break with the past militarism and a foundation of the newborn peaceful nation, UNESCO was idealized in Japan as an international symbol of democracy and peace. In this regard, the Japanese movement for UNESCO was initiated and promoted as part of a popular peace movement that connected the democratization of Japan with world peace.[5]

To this extent it is commonly stressed that the UNESCO movement in Japan was put into practice at the initiative of intellectual individuals and private organizations. Yet it is also demonstrated that the movement was in nature overtly and covertly influenced by the motives of the Japanese Government, particularly of the Education and Foreign ministries, to restructure the domestic cultural policy and facilitate Japan's reintegration into international society. In fact, the establishment of the Sendai UNESCO Cooperative Association, which is often mythicized as the birth of the private Japanese campaign for UNESCO, was initiated by officials of the Foreign Ministry engaged in liaison works with the Allied occupation authorities.[6]

On the other hand, UNESCO's headquarters also increasingly became interested in Japan. At the General Conference in November 1946, the Dutch delegates submitted a proposal, with Germany in mind, to study educational and cultural influences of democratic countries on former enemies.[7] Based on this, UNESCO started to advance negotiations with the Allied authorities and formulate its possible programs in Germany, whereas no consideration was given to Japan for most of 1947.[8]

At the General Conference in December 1947, however, the Chinese delegation submitted a general resolution in order to extend UNESCO's education for peace to Japan. The outcome was that the director-general was asked to consult with the Allied authorities in order to collect information regarding Japanese education and to ascertain the ways in which the objectives of UNESCO could be promoted in, and in relation to, Japan.[9]

The director-general accordingly sent a letter to the Supreme Commander of the Allied Powers (SCAP) in February 1948 asking whether it would be willing to initiate conversations with a view to devising the best means to implement the resolution. Soon after receiving SCAP's endorsement, UNESCO's headquarters sent Kuo Yu-shou, a Chinese diplomat and a special adviser to the director-general on Asia and the Far East, to Japan to pursue negotiations. Kuo visited Tokyo in September and had conversations with relevant section chiefs of SCAP, officials of the Japanese Government and leaders of UNESCO cooperative associations in Japan.[10]

SCAP had paid little attention to UNESCO, although it had been suggested that the organization could encourage the democratization of Japanese education and even that Japan should be admitted to it in the near future. However, the growth and spread of the UNESCO movement in Japan had made SCAP aware of the potential of cooperation and it began a study on the objectives of UNESCO and its possible effects on the occupation of Japan.[11]

As a result of the negotiations, it was agreed that a close relationship should be developed and maintained between UNESCO and SCAP, as well as between UNESCO and Japanese agencies through SCAP. In addition, various actions were suggested as possible educational and cultural programs that UNESCO could implement in Japan. UNESCO's activities were expected to encourage the Japanese reconstruction efforts under the direction of SCAP by UNESCO's continued manifestation of interest and by its extension to Japan of special assistance in certain areas and projects.[12]

On his return to Paris, Kuo gave an oral report to the executive board on his negotiations with SCAP and his impressions of the Japanese movement for UNESCO. He said that various circles in Japan had for a long time been enthusiastic about UNESCO and that SCAP and official circles were therefore very anxious for UNESCO to extend its work to the country. In response, while appreciating the result of Kuo's mission, the Chinese representative expressed a dissenting view against the SCAP- or US-led UNESCO programs in Japan and stressed the significance of the Far Eastern Commission as the highest decision-making body regarding the Allied occupation of Japan. By the same token, the Chinese representative also suggested that, after the model of Germany, it was necessary to form a small committee of experts for the re-education of Japan.[13]

As a result of the suggestions, the director-general was requested to submit to the General Conference a plan of action in Japan for 1949, to make sure to keep the Far Eastern Commission fully informed of all developments in UNESCO's activities in Japan, and to include in the plan a provision for a small committee of experts on the re-education of the Japanese population.[14]

UNESCO programs in Japan

In SCAP, the director of the Civil Information and Education Section, William K. Bunce, was in charge of UNESCO affairs. He thought positively about UNESCO programs in Japan, mentioning that "probably the time will never be better than at present for getting the Japanese people a hearing for UNESCO ideals".[15]

Bunce was thus sent to participate as an observer at the General Conference in November 1948. When relations with Japan were discussed at the conference, Bunce stated that SCAP was greatly interested in the contributions that UNESCO could make towards the re-education of Japan, also introducing the keen interest shown in UNESCO's work by the Japanese people.[16]

Accordingly, the UNESCO programs in Japan for 1949 were finally adopted. They were modeled on the programs in Germany and included a range of activities, mainly consisting of six items. It should

- distribute to interested groups in Japan, and especially to educators, the documents, publications and other materials of UNESCO, and make known the aims and achievements of the organization by all appropriate means;
- facilitate the exchange, between Japan and other countries, of publications and of scientific, educational and cultural works and information calculated to further the aims of UNESCO;
- study the question of textbooks in Japan and define, from UNESCO's point of view, the criteria that should guide the preparation and publication of such textbooks;
- survey the problems involved, and the opportunities that exist, in the exchange of persons between Japan and other countries, in accordance with UNESCO's objectives;
- encourage objective research by Japanese social scientists with a view to promoting a better understanding by the Japanese people of their own problems and of their relationships to other members of the international community;
- select experts from Japan to attend technical meetings called by UNESCO, when their attendance is deemed advantageous to the execution of UNESCO's programs.

At the same time there was established a committee of experts to advise UNESCO on matters affecting the present and future programs of the organization's work in Japan.[17]

Although Bunce thought that these projects were too ambitious to be implemented in 1949, UNESCO had for the first time presented a concrete action plan for Japan, based on which its programs were put into action until 1951.[18]

To start with, UNESCO, at the recommendation of Kuo, decided to send the Chinese educator, Lee Shi-mou, as its representative in Japan. He had so far been a commissioner of education for the Shanghai Municipality in China, and arguably the appointment of Lee as a UNESCO representative in Japan was part of a Chinese policy intention at the time to ensure its influence in the Allied occupation of Japan via UNESCO.[19]

After intensive training at UNESCO's headquarters, Lee, accompanied by Kuo, went to Tokyo in April 1949. Soon after their arrival, they opened a UNESCO office in Tokyo and called on representatives of UNESCO cooperative associations in Japan, the officials of SCAP as well as those of the Foreign Affairs and Education ministries of the Japanese Government. At the same time they took an extensive tour from Tokyo through Kyoto, Nara, Osaka, Kobe, Fukuoka and Sendai to get acquainted with the UNESCO cooperative movement in Japan. Furthermore, according to the decision of the General Conference, the expert committee on UNESCO programs in Japan was set up and held its first meeting in May.[20]

However, the UNESCO office in Tokyo faced some difficulties at the beginning of its activity. First, because it was difficult, in spite of SCAP's assistance, to find office space in the devastated city, Lee had no choice but to carry out his responsibilities from his hotel room for some months. It was not until October 1949 that he could officially open the office in the YMCA's new building in Tokyo. Second, with no staff, all the work of the office had to be undertaken by Lee in the first year. Nevertheless, with the office and its representative Lee Shi-mou, UNESCO was able to take the first step in its programs in Japan until the UNESCO's headquarters decided to increase the budget for a staff cost so that the Tokyo office could take on full-time employees.[21]

In response to the start of UNESCO activities, the Japanese Ministry of Education established a Liaison and UNESCO Section to the ministry's secretariat in June 1949. This section hosted a biweekly liaison meeting, which was attended by the officials of the Ministry of Education and the Ministry of Foreign Affairs, as well as the representatives of the National Diet Library, SCAP and UNESCO's office in Tokyo. As the domestic system cooperating with UNESCO programs was increasingly consolidated, politicians from the Diet (the Japanese Parliament) also became interested in the strengthening of the relationship with UNESCO. In fact, following the formation of the Federation of Diet Members for UNESCO (Kokkai Yunesuko Giin Renmei), both houses of the Diet adopted a resolution on the promotion of the Japanese movement for UNESCO. It is notable that both these resolutions, from November and December, expressed a strong desire to be immediately granted admission to UNESCO. In fact, being under the Allied occupation where SCAP was in control of Japanese diplomatic relations, UNESCO was understood among Japanese politicians as the only way for Japan to work directly with the outside world. For this reason, as with the motivation of the Ministry of Foreign Affairs, Japanese politicians also called for the country's admission to UNESCO as proof of its reintegration into the international community.[22]

In these circumstances it was anticipated that Japan would be allowed to participate as an observer in the General Conference in September 1949, but the executive board decided to invite neither SCAP nor Japanese observers to the conference on the grounds that the interests of the Allied Powers occupying Japan could be represented by the state members of the USA and others in UNESCO. As a result of the strong opposition of the US representative, however, the invitation was eventually extended to SCAP but not to a Japanese observer. As Lee Shi-mou reported, this came as a great disappointment to the Japanese people and organizations involved in the UNESCO cooperative movement.[23]

SCAP sent W.K. Bunce to attend the General Conference in September 1949. Here it was also discussed whether or not it was appropriate to invite qualified Japanese individuals to attend meetings and conferences held under the auspices of UNESCO. The delegations of the Philippines and Australia expressed strong opposition to the further strengthening of the relationship between UNESCO and Japan, particularly the Japanese participation in international affairs. In their view, if a Japanese observer was admitted to UNESCO conferences, the organization might run the risk of giving certain international status to Japan in the situation where the democratization and re-education of the nation were still only half done. An invitation of a Japanese observer, not to mention its admission to UNESCO, should be considered only after the conclusion of a peace treaty with Japan. The USA, on the other hand, defended the invitation of a Japanese observer on the grounds that it did not necessarily entail the granting of international status to Japan. In the end, the counterproposal from the Philippines and Australia was rejected and the resolution was adopted with the additional statement that objections were made against the invitation of Japanese nationals to UNESCO conferences. In the light of the subsequent events, however, it would be fair to say that the opposition of the Philippines and Australia certainly read the future course of the relationship between UNESCO and Japan. In fact, Japanese observers participated in the General Conference in 1950 and the Japanese Government was admitted to UNESCO months before the conclusion of the peace treaty with Japan in September 1951.[24]

The resolution also stipulated the increased budget to provide for more employees at UNESCO's office in Tokyo, and it became staffed with assistant officers and an adviser to its representative, Lee Shi-mou, in 1950. First, Ueda Koichi, one of the founders of the Sendai UNESCO Cooperative Association, was appointed full time to the Tokyo office. In April 1950, Ayusawa Iwao, the secretary-general of the Federation of UNESCO Cooperative Associations in Japan, joined the office as an adviser. In addition, an Italian, Mario Schubert, was sent by UNESCO's headquarters to assist Lee that November.

In short, however, the activities of the office were soon limited to only a few programs among the six original items, because domestic reforms in the educational, cultural and scientific fields in Japan were primarily undertaken by the Japanese Government under the direction of SCAP. Therefore what it could do was confined almost exclusively to enlightenment activities on ideals and activities of UNESCO for the Japanese people. For example, the increased budget for 1950 also aimed at additional aid for the publication of UNESCO documents in Japan, the broadening of UNESCO fellowship facilities for Japanese applicants, the provision of fees to Japanese social scientists with a view to promoting objective research relative to the relationship of Japanese with other people, and the extension to Japan of the UNESCO Book Coupon Scheme. Furthermore, in the following months a series of UNESCO publications were translated into Japanese with the assistance of UNESCO's office in Tokyo and published under the title of "UNESCO Collections". This included *Yunesuko to Shokun* (UNESCO and You), *Gunpuku wo tsukenu Hitobito no Yosai* (12 Speeches Delivered by Dr. Jaime Torres Bodet) and Julian Huxley's *Yunesuko no Mokuteki to Tetsugaku,* (UNESCO: Its Purpose and Philosophy).[25]

During this time, Lee energetically traveled all over the country and gave lectures about UNESCO at conferences and seminars held by UNESCO cooperative associations in Japan.[26] Also, in October, UNESCO's office launched a new educational project for the Japanese youth with a view to promoting wholesome youth activities, placing special emphasis on international understanding. After Japan's admission to UNESCO in June 1951 and the subsequent inauguration of the Japanese National Commission for UNESCO, however, this and other projects were taken over by the National Commission, and UNESCO's office in Tokyo was closed in July 1952.[27]

In sum, the significance of UNESCO's office in Tokyo can be found not so much in its activities but in its very existence in Japan. Serving as the only way for the Japanese people to communicate with the outside world and foster international understanding, the office enabled them to restore and maintain a sense of connection with the international community. For this very reason, the Tokyo office was easily forgotten in Japan through the course of its reintegration into UNESCO, the UN and the international community as a whole.[28]

Japan's admission to UNESCO

Since the next General Conference was scheduled for Florence in May 1950, Bunce encouraged UNESCO's headquarters to consider the possibility of inviting Japanese individuals to attend. In this way, with strong support from SCAP, Japan swiftly became integrated into the international system of UNESCO, first as an observer and then as a full member.[29]

The executive board in February 1950 thus examined differing proposals regarding the invitation of a Japanese observer: one in which SCAP was invited to send an observer accompanied by Japanese advisers and one which was an invitation to send an observer addressed directly to the Japanese Government. The USA supported the latter for the reason that Japan's statehood was unquestionable in spite of its present situation of limited sovereignty, whereas this proposal faced strong opposition from Australia and even China.[30]

The executive board decided to only invite SCAP accompanied by Japanese advisers, and with the selection of the Japanese Government, three Japanese, led by SCAP representatives, arrived in Florence in May and for the first time represented Japan on the international stage of UNESCO's General Conference.[31]

As with the previous sessions, Bunce glorified the progress of democratization in Japan since 1945, appealing for "any UNESCO action aimed at the integration of the Japanese people in the family of the democratic nations of the world".[32] Arguably, SCAP and the US Government intended to realize the admission of the Japanese Government to UNESCO on the basis of this Japanese participation in the General Conference as a first step for the Japanese Government to return to the international community altogether, not least the UN.[33]

The Japanese participants were not given a voice at the conference and the representatives of the Philippines and Australia voiced again their negative views of giving international status to Japan, but the great potential that SCAP and the Japanese Government saw in the participation in UNESCO and for Japan's reintegration into international society stimulated and accelerated the move toward its official admission.[34]

The Japanese Government readily started the groundwork with SCAP for the application for admission to UNESCO, and in December 1950, officials from the Foreign Ministry called on SCAP and showed its draft application form. SCAP added a letter of authorization to the application and the Japanese Government sent the application to UNESCO's headquarters on 12 December 1950.[35]

Since Japan was not a member of the UN, the application was forwarded to and examined first by the Economic and Social Council of the UN in March 1951. Although communist countries and the Philippines voted against Japan's admission to UNESCO, the Economic and Social Council endorsed the Japanese application by a majority vote and then forwarded it to UNESCO.[36]

Undoubtedly, with the approval of the Economic and Social Council, a mood of optimism prevailed not only in Japan but also at UNESCO's headquarters. Under the circumstances, the Japanese Government prepared for the organization of its delegation so that Japan could be represented in UNESCO as soon as the General Conference decided its admission.[37]

As the executive board adopted Japan's application without much trouble, the Japanese delegation arrived in Paris and got everything in readiness to savor the jubilant moment in the General Conference, and on 21 June, the conference at last deliberated on the application to UNESCO of Japan as well as those of another four states.[38]

As predicted, the Japanese application was in general warmly welcomed and approved by most of the member states of UNESCO. It was only the Philippines' representative who voted against Japan's admission on the grounds that Japan had not recognized its obligations toward the Philippines with regard to the damage inflicted on the country by the Japanese invasion during the war.[39]

The voice of the Philippines was, however, drowned out in welcoming cheers and applause from the conference floor. As Maeda stated in his speech, UNESCO had been "hope and light" for the defeated Japanese people as well as a guiding principle for post-war Japan in its reconstruction as a democratic and peace-loving nation. In this way, leaving the problem of its war responsibility unsolved, such a self-centered motivation bore fruit as the admission to UNESCO that accelerated the following course of Japan's reintegration to the international community.[40]

The decline of the Japanese UNESCO movement

Contrary to these steady steps toward Japan's admission to UNESCO, the UNESCO cooperative movement in Japan experienced many twists and turns. As mentioned above, there is no doubt that the private movement played a major role in establishing a constructive relationship between UNESCO and Japan at the beginning. Following the first national convention of UNESCO cooperative associations in Tokyo in 1947, the National Federation of UNESCO Cooperative Associations in Japan was formed in May 1948.[41] Thereafter, holding a national convention annually, the federation achieved significant progress in increasing UNESCO cooperative associations in cities throughout Japan. In fact, at UNESCO's General Conference in 1949, the federation on behalf of 70 UNESCO cooperative associations in Japan delivered a statement vowing its cooperation with UNESCO.[42]

In 1950, however, the Japanese grassroots movement for UNESCO confronted a serious crisis resulting from ideological conflicts in the National Federation of UNESCO cooperative associations in Japan. In particular, in the face of the outbreak of the Korean War, a clash between the ideologies of "idealism" and "realism" among members of the federation erupted over the ideals of UNESCO.[43] In addition, it was revealed that the head office of the federation, the Tokyo UNESCO Cooperative Association, was virtually an empty shell without organizational and financial basis.[44] Although efforts toward the reconstruction of the federation were made on the initiative of local UNESCO cooperative associations, particularly in the Kansai, Shikoku and Tohoku areas, it was obvious in 1950 that the private movement for UNESCO had been driven to the brink of collapse.[45] This was mainly because the primary purpose of the movement was to achieve Japan's earliest admission to UNESCO, and thereby it became less meaningful as the Japanese Government embarked on negotiations with SCAP for the same purpose.[46] In sum, the Japanese movement for UNESCO was not a product of genuine interest in the ideals of UNESCO but rather a passing fashion that most Japanese people were attracted to in the context of their isolation from the international community after the country's defeat in the war.[47]

Under these circumstances, even leading members of UNESCO cooperative associations in Japan envisaged that the private movement would come to an end in parallel with the realization of Japan's admission to UNESCO and the establishment of the Japanese National Commission for UNESCO.[48] The Japanese UNESCO movement nonetheless secured its survival through the reorganization of the federation as the National Federation of UNESCO Associations in Japan in 1951. This was because it was thought that the national commission was primarily a government agency to implement UNESCO programs while the federation was a voluntary private organization of those people who work in the interest of UNESCO.[49] In other words, it was concluded that, in light of limitations of the governmental cooperative activities with UNESCO, private initiatives such as UNESCO associations in Japan could justify their own *raison d'être* in disseminating the ideals of UNESCO and its programs all over the country. In this way the decline of the Japanese UNESCO movement paradoxically demonstrated that cooperative activities with UNESCO needed not only to be undertaken by intergovernmental relations but also to be backed up by the people in general as well as by non-governmental or private organizations.

Conclusion

Altogether, the relationship between UNESCO and Japan started and developed under domestic and international conditions.

First, in the aftermath of the devastation of World War II, the UNESCO movement was launched and expanded through the country on the initiative of intellectuals and private organizations such as UNESCO cooperative associations. Following the decline of the grassroots movement, the Japanese Government instead led the coordination of UNESCO programs in Japan and eventually accomplished its admission to UNESCO. As a result it is obvious that both private intellectuals and the government understood UNESCO to be an important point of contact with the world and a foothold for Japan's return to the international community. In this regard, UNESCO served as a benchmark for measuring Japan's position in international society.

Second, the collaboration between UNESCO and Japan, particularly Japan's admission to UNESCO, would not have been possible without particular international circumstances of the time. More specifically, it was as a result of the Cold War that the Chinese Government showed a conciliatory attitude toward Japan and SCAP, as well as the US government continuously supporting and accelerating Japan's admission to UNESCO in the discussions at meetings of the executive board and at the General Conference. It was for the same reason that the righteous oppositions from the Philippines and Australia ended in vain. In light of this, the admission of the Japanese Government to UNESCO can be seen as a product of the Cold War.

Notes

1. Japanese and Chinese names in this chapter are rendered according to local custom: the family name before the given name. Sources in Japanese are all Romanized.
2. Little historical research in the Japanese language has been conducted on the relationship between UNESCO and Japan. In fact, there are only brief descriptions or personal stories about UNESCO written by its former officials. See Noguchi Noboru, *Yunesuko: Gojunen no Ayumi to Tenbo* [UNESCO: Its 50-year History and Prospect] (Tokyo: Shingurukatto Sha, 1996); Matsuura Koichiro, *Yunesuko Jimukyokucho Funtoki* [Laborious Days of the Director-General] (Tokyo: Kodan Sha, 2004); Matsuura Koichiro, *Sekai Isan: Yunesuko Jimukyokucho wa Uttaeru* [World Heritage: An Appeal from the Director-General] (Tokyo: Kodan Sha, 2008). On the other hand, it is also arguable that very little attention has been paid to Japan in the historiography of UNESCO. For example, see Chloé Maurel, *Histoire de l'UNESCO: Les trente premières années. 1945–1974* (Paris: L'Harmattan, 2010); Fernando Valderrama, *A History of UNESCO* (Paris: UNESCO Publishing, 1995); James P. Sewell, *UNESCO and World Politics: Engaging in International Relations* (Princeton: Princeton University Press, 1975).
3. For details about the social circumstances of Japan in the aftermath of World War II, see John W. Dower, *Embracing Defeat: Japan in the Wake of World War II* (New York: W. W. Norton & Company, 1999).
4. Ibid., 63. See also Nihon Yunesuko Kokunai Iinkai ed., *Nihon Yunesuko Katsudo Junenshi* [Ten Years' History of UNESCO Activities in Japan] (Tokyo: Nihon Yunesuko Kokunai Iinkai, 1962), 3; Peter J. Katzenstein, *Cultural Norms and National Security: Police and Military in Postwar Japan* (Ithaca: Cornell University Press, 1996); Thomas U. Berger, *Cultures of Antimilitarism: National Security in Japan and Germany* (Baltimore: Johns Hopkins Press, 2003).
5. In concert with "A Statement by eight distinguished social scientists on the cause of tensions which make for war" issued by UNESCO in July 1948, leading Japanese intellectuals formed Heiwa Mondai Danwakai (Peace Problems Discussion Group)

and expressed principles for the problem of world peace, as well as the problem of the peace settlement for Japan. Advocating the overall peace, the maintenance of neutrality, the opposition to giving military bases and the objection to rearmament, this group played the central role in the debate over peace in post-war Japan. For Japan's pacifism, including this group in this period, see Rikki Kersten, *Democracy in Postwar Japan: Maruyama Masao and the Search for Autonomy* (London: Routledge, 1995).

6. Nihon Yunesuko Kyokai Renmei, *Yunesuko Minkan Katsudo 20 Nenshi* [Twenty Years' History of Private Activities for UNESCO] (Tokyo: Nihon Yunesuko Kyokai Renmei, 1966) and Liang Pan, "Senryoka no Nihon no Taigai Bunka Seisaku to Kokusai Bunka Soshiki" [Japanese International Cultural Policy under the US Occupation and International Organizations: The Case of UNESCO Cooperation Movement], *Kokusai Seiji* 127 (2001): 185–205.
7. UNESCO, "General Conference: First Session, held at UNESCO House, Paris, from 20 November to 10 December 1946", Paris 1947, 154, UNESCO Archives.
8. UNESCO, "Report of the Director General on the Activities of the Organisation in 1947: Presented to the Second Session of the General Conference at Mexico City, November–December 1947", Paris 1947, 73, UNESCO Archives.
9. UNESCO, "Records of the General Conference of the United Nations Educational, Scientific and Cultural Organization, Second Session, Mexico 1947", vol. I, "Proceedings", Paris, April 1948, 600 and vol. II, "Resolutions", Paris, April 1948, pp. 55–56. It is interesting that China, one of the countries most damaged by the Japanese military invasion, made this forgiving proposal, as contrasted with the case of Germany, in which victims of Nazi Germany, particularly Czechoslovakia and Poland, expressed their vehement opposition and asserted that UNESCO should prioritize the cultural reconstruction of the war-affected countries over the commitment to Germany.
10. "Schedule: Dr. Kuo Yu-shou, Adviser on Asia and the Far East, UNESCO, Japan visit: 9–20 September 1948", n.d., CIE (B) 7856, Kensei Shiryoshitsu [Modern Japanese Political History Materials Room of the National Diet Library of Japan].
11. Religions and Cultural Resources Division, "Press Translation and Summaries", 7 November 1948, CIE (B) 8000, Kensei Shiryoshitsu. The suggestion of a Japanese membership of UNESCO was made by the US educational mission to Japan, which was requested by SCAP and dispatched in March 1946. The head of the mission was George D. Stoddard, who later became a US representative to UNESCO and a strong supporter of UNESCO activities in Japan. For the Stoddard Mission, see Takemae Eiji, Robert Ricketts and Sebastian Swann transl., *Inside GHQ: The Allied Occupation of Japan and Its Legacy* (New York: Continuum International Publishing Group, 2002), 352–359.
12. "Suggestions concerning Extension of UNESCO's Program to Japan", 19 September 1948, CIE (B) 7854, Kensei Shiryoshitsu, and "Report to the Executive Board on Negotiations with the Supreme Commander of the Allied Powers in Japan", Paris, 13 October 1948, UNESCO Archives.
13. UNESCO, Executive Board, Eleventh Session, "Provisional Summary Record of the Third Meeting held at Unesco House, 19 Avenue Kléber, Paris 16e on Thursday, 14 October 1948 at 10 a.m.", Paris, 16 October 1948, UNESCO Archives.
14. UNESCO, Executive Board, Eleventh Session, "Resolutions and Decisions adopted by the Executive Board at its Eleventh Session from 12 to 15 October 1948", Paris, 28 October 1948, UNESCO Archives.
15. Letter from W.K. Bunce to G.D. Stoddard, 26 November 1948, CIE (B) 7853, Kensei Shiryoshitsu.

16. UNESCO, "Records of the General Conference of the United Nations Educational, Scientific and Cultural Organization, Third Session, Beirut 1948, Volume I, Proceedings", Paris, April 1949, 416, UNESCO Archives.
17. UNESCO, "Records of the General Conference of the United Nations Educational, Scientific and Cultural Organization, Third Session, Beirut 1948, Volume II, Resolutions", Paris, February 1949, 66–69, UNESCO Archives.
18. Bunce nevertheless recognized the significance of UNESCO programs in Japan for SCAP and its occupation policy, noting that "the UNESCO program will supplement American efforts to advance occupation objectives by promoting the same ideas without having the American trademark", according to "Report on Trip to the United Nations Educational, Scientific and Cultural Organization Conference", 24 December 1948, General Headquarters, Supreme Commander for the Allied Powers, Civil Information and Education Section, CIE (B) 7842, Kensei Shiryoshitsu.
19. Letters from Kuo Yu-shou to the Director-General, 10 January 1949, and from Kuo Yu-shou to W.K. Bunce, 14 February 1949, X07 (520), UNESCO Archives.
20. "Activities in Japan undertaken by Dr. Kuo and Dr. Lee", 1 July 1949, X07(520), UNESCO Archives. The committee was chaired by W.K. Bunce and composed of Chang Feng-chu (adviser to the Chinese Mission in Japan), Bernardo P. Abrera (chief, Reparation and Restitution Section, Philippine Mission in Japan), and T. W. Eckersley (first secretary of mission, political adviser, Australian Mission in Japan). Later, J. Chazelle of France and Robert van Gulik of the Netherlands joined the committee.
21. Letters from Lee Shi-mou to André de Blonay, 6 July 1949, from Kuo Yu-shou to W. Laves, 8 August 1949, and from André de Blonay to Lee Shi-mou, 3 November 1949, X07(520), UNESCO Archives.
22. Lee Shi-mou, "Report on UNESCO Programme in Japan", 24 August 1949, X07(520), UNESCO Archives.
23. Letter from Lee Shi-mou to Walter Laves and André de Blonay, 9 August 1949, X07.21(520), UNESCO Archives.
24. UNESCO, "Records of the General Conference of the United Nations Educational, Scientific and Cultural Organization, Fourth Session, Paris, 1949, Resolutions", Paris, November 1949, 79–82 and UNESCO, "Records of the General Conference of the United Nations Educational, Scientific and Cultural Organization, Fourth Session, Paris, 1949, Proceedings", Paris, December 1949, 420–421, UNESCO Archives.
25. Beikoku Yunesuko Kokunai Iinkai ed., Nanba Monkichi transl., *Yunesuko to Shokun* [UNESCO and You] (Tokyo: Nihon Kyobun Sha, 1950); Gaimusho Bunkaka transl., *Gunpuku wo tsukenu Hitobito no Yosai* [12 Speeches Delivered by Dr Jaime Torres Bodet] (Tokyo: Nihon Kyobun Sha, 1950); Julian Huxley, Ueda Koichi transl., *Yunesuko no Mokuteki to Tetsugaku* [UNESCO: Its Purpose and Philosophy] (Tokyo: Nihon Kyobun Sha, 1950).
26. Lee Shi-mou, "Heiwa Undo to Yunesuko" [Peace Movement and UNESCO], *Kaizo* 30:8 (August 1949): 38–41.
27. Letter from Lee Shi-mou to the Director-General, 5 October 1950, X07(520), UNESCO Archives.
28. "Yunesuko Daihyobu no Sosetsu to sono hatashita Yakuwari" [Establishment of the UNESCO Office in Tokyo and its Work], in Nihon Yunesuko Kyokai Renmei, *Yunesuko Minkan Katsudo 20 Nenshi*, 36–37.
29. Letter from W.K. Bunce to Kuo Yu-shou, 6 December 1949, X07(520), UNESCO Archives. Considering the fact that UNESCO was regarded among the Japanese

people as their only contact with the outside world, SCAP thought that "since the chief object of the Occupation – and now of the Japanese people – is to help Japan adapt herself to international society, UNESCO activities should be promoted by both" according to "Fact Sheet No. 14", 10 October 1949, General Headquarters, Supreme Commander for the Allied Powers, Civil Information and Education Section, CIE(B)7853, Kensei Shiryoshitsu.

30. UNESCO, Executive Board, Nineteenth Session, "Summary Record of the Nineteenth Meeting held at UNESCO House, on Thursday, 23 February 1950, at 10 a.m.", Paris, 30 June 1950, 11, UNESCO Archives.
31. UNESCO, Executive Board, Twentieth Session, "Summary Record of the Seventh Meeting held at UNESCO House, Paris, on Wednesday 29 March 1950 at 4 p.m.", Paris, 20 July 1950, 3–4, UNESCO Archives. The SCAP mission consisted of two observers and three advisers, among which one observer and two advisers were Japanese. It included as observers W.K. Bunce and Suzuki Tadakatsu (adviser of the Kanagawa UNESCO Cooperative Association), accompanied by advisers such as Bowen C. Dees (head of the Economic and Scientific Section of SCAP), Katsunuma Seizo (president of Nagoya University and president of the UNESCO Cooperative Association in Nagoya), and Otaka Tomoo (professor of Tokyo University and a member of the UNESCO Committee of the Science Council of Japan).
32. UNESCO, "Records of the General Conference of the United Nations Educational, Scientific and Cultural Organization, First Session, Florence, 1950, Proceedings", Paris, November 1950, 475–476, UNESCO Archives.
33. "Report of SCAP Observer to Fifth Session, General Conference, UNESCO", 13 July 1950, Civil Information Section, CIE(B)7841, Kensei Shiryoshitsu. See also Monbusho Kanbo Shogai Yunesuko ka ed., *Yunesuko Dai Gokai Sokai ni Shusseki shite* [Observations on the 5th General Conference of UNESCO in Florence in 1950] (Tokyo: Monbusho Daijin Kanbo Shogai Yunesuko ka, 1950), 6.
34. UNESCO "Records of the General Conference of the United Nations Educational, Scientific and Cultural Organization, First Session, Florence, 1950, Proceedings", Paris, November 1950, 468, 476, UNESCO Archives.
35. Joyakukyoku Jyoyakuka, "Yunesuko Kanyu Tetsuduki ni kansuru Gaikoukyoku tono Uchiawase" [Discussion with the Diplomatic Section on the Admission Procedure of UNESCO], 8 December 1950, and "Nihon no Yunesuko Kanyu ni kansuru Shireibu Gaikokyoku Tantoukan tono Kaidan" [Conversation with the Diplomatic Section of SCAP about Japan's Admission to UNESCO], 16 November 1950, B.2.3.4.1-4, Japanese Foreign Ministry Archives; UNESCO, Executive Board, Twenty-fifth Session, "Application for Membership of UNESCO", Paris, 15 January 1951, UNESCO Archives.
36. Johobu, "Kokusai Rengo Keizai Shakai Rijikai Nihon no Yunesuko Kanyu Shinsei wo Shoninsu" [United Nations Economic and Social Council Approved Japan's Admission to UNESCO], 15 March 1951, B.2.3.4.1-4, Japanese Foreign Ministry Archives; UNESCO, Executive Board, Twenty sixth Session, "Application of Japan for Membership of UNESCO", Paris, 11 May 1951, UNESCO Archives.
37. Letter from Hagiwara Toru to Yoshida Shigeru, 4 April 1951, B.2.3.4.1-4, Japanese Foreign Ministry Archives. The Japanese delegation was eventually headed by Maeda Tamon (former minister of education). The members included Tokugawa Yorisada (president of the Federation of Diet Members for UNESCO), Harada Ken (executive board member of the UN Association of Japan), Fujiyama Aichiro (president of the Federation of UNESCO Cooperative Associations in Japan) and Hagiwara Toru (director of the Diplomatic Representative Office of Japan in Paris).

The delegation was also accompanied by advisers: Nishimura Iwao (chief of the Liaison and UNESCO Section of the Education Ministry) and Hasumi Yukio (former chief of the Cultural Section of the Foreign Ministry).

38. UNESCO, Executive Board, Twenty-sixth Session, "Summary Records, Third Meeting held at UNESCO House, Paris, on Saturday, 9 June 1951, at 10 a.m.", Paris, 20 August 1951, and UNESCO, "Records of the General Conference, Sixth Session, Paris 1951, Resolutions", Paris, 1951, UNESCO Archives. Five countries were eventually granted admission to UNESCO at the Sixth Session of the General Conference: the Kingdom of Laos, Cambodia, Vietnam, Japan and the German Federal Republic.
39. UNESCO, "Records of the General Conference, Sixth Session, Paris, 1951, Proceedings", Paris, 1951, 112, UNESCO Archives. Objections of the same kind could be voiced by other Asian nations that suffered from Japanese militarism, such as South Korea and China. However, while the representative of South Korea was absent due to the Korean War, the Government of the Republic of China adopted an appeasement policy toward Japan with a view to competing with the People's Republic of China over the representation of China in UNESCO.
40. UNESCO, "Records of the General Conference, Sixth Session, Paris, 1951, Proceedings", 115, UNESCO Archives.
41. The federation was chaired by Nishina Yoshio, a physicist often mentioned as "the father of nuclear physics" in Japan. Although Nishina had been involved in the development of the UNESCO movement at his own expense, he died in January 1951 without witnessing Japan's formal entry.
42. Nihon Yunesuko Kyokai Renmei, *Yunesuko Minkan Katsudo 20 Nenshi*, 151–152.
43. Ibid., 84–85. While idealists claimed that UNESCO should adhere to the principle of non-intervention and absolute pacifism in accordance with its constitution, realists regarded UNESCO as one of the agencies under the UN system and accepted its involvement in the war for peaceful purposes.
44. Ibid., 23–27. For these reasons, the chairman of the federation, Nishina, announced his resignation, and the head office was transferred temporarily to the Osaka UNESCO Cooperative Association.
45. In this regard, SCAP reported that "the Federation has always been weak, has had no active program, and was not what it purported to be: an organization representing the some 70 UNESCO cooperative societies scattered throughout Japan. Its finances, dependent upon contributions, have always been shaky", according to the Civil Information and Education Section, "Inter-section Memorandum: UNESCO Federation", 13 May 1951, CIE(B)7853, Kensei Shiryoshitsu.
46. Nihon Yunesuko Kyokai Renmei, *Yunesuko Minkan Katsudo 20 Nenshi*, 16.
47. Ibid., 83.
48. Ibid., 93–94.
49. Lee Shi-mou, "Report on Unesco Programme in Japan", August 1951, UNESCO Archives.

6
UNESCO, Reconstruction, and Pursuing Peace through a "Library-Minded" World, 1945–1950

Miriam Intrator

In June 1941, US President Franklin D. Roosevelt wrote that libraries "are directly and immediately involved in the conflict which divides our world". Throughout the history of war and conflict, libraries and other cultural institutions have been purposefully or collaterally damaged or destroyed. According to Roosevelt, there were two reasons why this historic pattern was manifesting in World War II: first, because libraries "are essential to the functioning of a democratic society", and second, because "the contemporary conflict touches the integrity of scholarship, the freedom of the mind, and even the survival of culture, and libraries are the great tools of scholarship, the great repositories of culture, and the great symbols of the freedom of the mind".[1] Weighing particularly heavily in 1941 among concerned individuals was the dark cloud of May 1933, when Adolf Hitler's supporters throughout Germany flagrantly confiscated and burned books they considered *undeutsch* (un-German). Images and detailed reports of the burnings had circulated widely and been discussed and protested the world over.

The resultant anxiety about the impact of World War II on information, education, libraries and other cultural institutions, coupled with the conviction that the Allies would be the ultimate victors, inspired a broad spectrum of preparatory activities. One culmination of these was the establishment, in November 1945, of UNESCO. With culture as a key target of and essential tool in fascism's rise and spread, the founders of UNESCO envisioned the new organization as an equally powerful cultural force, dedicated to overcoming fascist and other militarist ideas and to forging an enduring world peace.

This chapter will focus on initiatives developed between 1945 and 1950 at UNESCO, in its Libraries Section specifically, to address the war's impact on libraries and the long-term need to improve and expand library services throughout the world. These initiatives included tools for obtaining books, such as the International Clearing House for Publications (ICHP), its associated *Bulletin for Libraries* and UNESCO's Book Coupon Scheme, but also means for developing and promoting free and open access to libraries and information, such as the Summer School for Librarians in 1948 and the Manifesto for Libraries from 1949. To varying degrees, these initiatives had a measurably positive impact on reconstruction, and on the development of modern public libraries and on vital, transnational interlibrary networks of exchange and cooperation in the post-World War II.

At UNESCO, reconstruction was never viewed as a self-contained undertaking but was formulated to evolve smoothly into expansion, improvement and development. In shaping UNESCO's 1946 program, the vital interconnections between immediate needs and long-term goals was emphasized: "it has become increasingly clear that many of the projects desirable for UNESCO because of their long-term significance have an immediate relationship to the tasks of rehabilitation. The development of adequate library resources for the world, with which the library program of UNESCO is inherently concerned, is intimately connected with the re-establishment of library centres in war devastated areas."[2] In other words, reconstruction was central to UNESCO's long-term ability to build a better, more peaceful future.

Publications such as UNESCO's information booklet and fundraising tool, *Libraries in Need* from 1949, illustrated what had been lost and articulated a vision for the future. "War cut off the mental food supplies of millions", then-director general Jaime Torres Bodet wrote in the foreword. He continued, "War gutted, or sacked, or demolished thousands of libraries... and we always had too few", again referencing the need for both reconstruction and expansion and development.[3]

UNESCO's Libraries Section

There are three broad categories of where and how World War II impacted libraries: physical losses due to confiscation, censorship, theft and war damage; the long isolation between countries that prevented cross-border purchasing or exchange; and the interruption of already often inadequate efforts to improve and expand public libraries and librarianship.

We know much about how books and libraries were lost and destroyed, but little about how they were replaced and revived after the war's end. How were emptied shelves replenished? How were library buildings rebuilt? What about irreplaceable losses? How were the years of missed acquisitions made up for? Who was responsible for prioritizing which texts the damaged libraries needed? Formulating solutions to questions like these was the primary undertaking of UNESCO's Libraries Section.

More broadly, the section was motivated by one burning question: What kind of a library program would make the maximum contribution toward building a lasting peace? And its response was to develop a vision of a global pattern of communications. "A diagram of the Libraries Section's work would be of a complex network of lines of communication, each line justified in the pattern by the existence at each end of it of receiver and producer bodies", a statement on the section's future work claimed, and emphasized that "UNESCO's task most simply defined is to see that these lines of communication exist, that they are well served and that they are kept in good order."[4]

In other words, the Libraries Section saw itself as an organizer and facilitator. To that end, its primary post-war goals were to enable the distribution of publications that had been unobtainable during wartime, as well as attempt to restore or replace stolen, confiscated, destroyed and otherwise lost texts, and to create forums through which librarians and their supporters worldwide could cooperate to advance culture, knowledge and understanding, despite existing or intensifying divisions between their respective countries.

A key element of this vision was to expand free and open access to modern, public libraries in order to improve education and literacy, or, as per UNESCO's constitution, "building peace in the minds of men". That required "selling" the public library idea worldwide. Exchange and cooperation were vital to this vision as much from necessity due to insufficient monetary or material resources as from ideology due to the belief that transnational, cooperative networks would help to create a more peaceful world.

ICHP

The mission of the ICHP and its associated *Bulletin of the ICHP* was "promoting and facilitating the exchange and distribution of publications throughout the world" and specifically "exchange without discrimination" so that "exactly expressed needs may be exactly met".[5]

In the early post-war years the focus was on reconstruction: "A great task before us is the replenishment, on as large a scale as possible, in each of the allied countries, of national, university, public and other libraries, which are open freely to serious readers, and whose books have been destroyed."[6] The ICHP functioned as an international transit center where donated publications were stored until being requested by and distributed to libraries in need. It began as a matter of informational catch-up made necessary by wartime isolation.

Indeed the consequences of isolation were viewed by many as potentially the greatest immediate threat to reconstruction and a long-term threat to libraries and education. Isolation as a result of blocked lines of communication and transportation impacted virtually every library worldwide, whether Allied, enemy or neutral. With little or no cross-border sharing or contact possible, countries fell behind in their knowledge of the innovations, inventions, patents, theories and various other advancements that had been achieved and documented in publications during the war in other countries. This knowledge gap threatened the ability of war-impacted countries to reconstruct in accordance with the most current technological, industrial, material and other developments, creating an urgent need for both wartime and the most up-to-date post-war information and publications.

As the French ambassador put it in January 1947, "the fact that the continent found itself deprived for six years of all communication with the Anglo-Saxon world provoked in Europe's libraries a sort of asphyxia. Every day, my European colleagues can testify that, like me, we are rebuilding in our countries from the effect of the rupture of these contacts."[7] A similar report came from Polish librarian Maria Danilewicz, who wrote: "The need for British books and periodicals is as great as ever in Poland and the reading public is waiting for the new publications on scientific and technical subjects."[8] While countries acknowledged their profound need for replacement and new materials in English, they also wanted to prevent being "simply swamped by American materials".[9] During these years, the ICHP greatly expanded the source base and made it possible for librarians themselves, initially in any member state, to receive from and contribute to an international exchange pool of available publications.

National exchange centers were viewed as an ideal means to empower individual countries to request and receive precisely what they wanted and to distribute material among their libraries. UNESCO, via the ICHP, served "as a co-ordinator of national activities".[10] Few national centers existed in the early post-war years, however, and UNESCO looked to those that did for successful models. Interestingly, many of them were in non-UNESCO member states at the time, such as Finland, Japan, Latvia, Mandatory Palestine, Portugal and the USSR. UNESCO, in order to justify turning to those countries, argued that "all our Member States wish to obtain publications from these countries; therefore, information we can obtain about their exchange establishments is directly in aid to our own Members".[11] This willingness to look to and learn from non-member states, as well as those, such as Japan, which were considered the enemy, and the USSR, an outwardly hostile non-member, provides a concrete example of UNESCO's often-stated determination to transcend physical and ideological boundaries and to look and act beyond borders, perhaps especially in a time of heightening hostility.

According to UNESCO's Reconstruction Program, the "need for a co-ordinating organisation, like UNESCO's ICHP, has also been proved over and over again. Books must be directed to the library where they are best used; centralized documentation on needs, resources and allocations made must be established and kept up-to-date."[12] *UNESCO Bulletin for Libraries* became the forum for communicating the activities of the ICHP, furthering its reach and distribution abilities.

UNESCO Bulletin for Libraries

The ICHP could only reach full functionality by advertising its contents and activities via a means of communication accessible to all interested libraries. With *Bulletin of the ICHP* first appearing in November 1946 and its successor, *UNESCO Bulletin for Libraries* following in April 1947, the title was UNESCO's first periodical publication and the first truly international library publication. It provided means for librarians all over the world to communicate with one another regarding the material that was available in or through the ICHP, making it possible to answer questions such as: "What has been published in enemy and occupied countries? What periodicals have ceased? What published material is still available? Which should be reprinted or reproduced?"[13]

Initially published in bilingual English and French editions, the brief posting format helped make it possible for librarians not necessarily fluent in either language to extract basic information regarding titles, publishers, editions and so on – details that could be "intelligible without translation", providing sufficient information for readers to decide whether or not they wanted any of the materials listed for their own library. Importantly, *Bulletin for Libraries* was "provided to the war-ravaged libraries free of charge" and sought to function "beyond political and commercial influence".[14] These features were all integral to the Libraries Section's dedication to ensuring the broadest possible access and availability to the ICHP.

Much evidence of *Bulletin for Libraries'* success exists. Almost immediately after its launch, the Libraries Section announced that "one European librarian has reported to us that announcement in a single issue enabled him to establish more than fifty exchanges".[15] As stated in the first issue, its success would "be judged by the extent to which the information is actually used".[16] Numbers tell of a statistical success, and in 1949 *Bulletin for Libraries* had, at 7,500 copies, the greatest distribution of any UNESCO periodical (except for the *Courier*, the general informational outlet of the organization), a success which the Libraries Section felt justified its continued free distribution.[17]

The sections of *Bulletin for Libraries* devoted to libraries' postings, such as "Publications Wanted", "Exchange", "Free Distribution" and "Publications for Sale", steadily expanded with each new issue, telling of an even more concrete success. *Bulletin for Libraries* was being actively used as a forum of communication and cooperation between libraries the world over, including, in early 1949, the USA, China, Switzerland, France, Belgium, the UK, Poland, Czechoslovakia, Hungary, Greece, the Netherlands and Brazil.[18]

Had the postings not been successful, the same libraries and countries would not have posted month after month, year after year. Soon, the Libraries Section reported, demand began to expand as well: "Very soon, many libraries approached us, asking us to make available not only those books and magazines published in recent years, available, but also to help them get the books of today and tomorrow."[19] In addition to replenishing what had been lost, libraries needed assistance purchasing new and forthcoming publications, demand that inspired another Libraries Section innovation the book coupon.

UNESCO book coupons

Introduced in 1948 as "a new international currency to be used exclusively for the purchase of books and publications", book coupons were a response to libraries' pressing need to purchase books, periodicals and other texts and to the "difficulty of libraries in soft currency countries to purchase material the required from hard currency countries". In other words, it was not only a problem of wealth. As a 1949 article in *The UNESCO Courier* pointed out, it was "not a question of price. For a university library in Poland it has been at least as hard to get a scientific paper priced fifty cents from the United States as the last edition of the expensive *Encyclopaedia Britannica*."[20] Currency restrictions had, of course, existed prior to the war, but they were greatly exacerbated by its circumstances and aftermath, making it virtually impossible for libraries, universities and other institutions in soft-currency countries to purchase books and periodicals from hard-currency countries. Intended to contribute "not only to the international circulation of ideas, but also, and more particularly, to the restoration of cultural institutions in the war-devastated countries", book coupons were launched on a one-year experimental basis on 6 December 1948. By that time the coupons were highly anticipated as they had been discussed since the first UNESCO General Conference in the fall of 1946. Due to the severity of the post-war currency crisis and the extent of need, most talk of the Book Coupon Scheme was met with optimism and praise.[21]

The Chinese, Czechoslovakian, Greek and Polish governments expressed their support from these earliest discussions, and during the summer of 1947, British scientist Joseph Needham visited Czechoslovakia and Poland on behalf of UNESCO. Reporting back from the University of Warsaw, he wrote: "Great interest was expressed in the Book Coupon Scheme, essential to break the stranglehold which the currency exchange regulations at present impose on the exchange of learned literature." Yet Needham found these countries, in dire financial straits after the war, to also be seeking reassurance regarding being asked to contribute monetarily to UNESCO, a new, still unknown or unproven, organization. He encountered doubts about whether they "would get any concrete benefits of any kind in exchange for the national monetary contribution". Needham encouraged those at UNESCO to consider "how it feels to be in the position of a devastated or relatively poor country, and to wonder 'what advantage shall we get for our contribution to this organisation?' " Book coupons were decidedly one of those advantages and Needham repeatedly described their success in calming such concerns.[22]

In 1948 UNESCO gifted USD44,126 from its educational reconstruction funds of "free" coupons to Austria, China, Czechoslovakia, Greece, Hungary, Indonesia, Iran, Italy, the Philippines and Poland, all considered "war devastated" by UNESCO.[23]

Normally the coupons, available in fairly small denominations such as USD0.25, USD1.00, USD3.00 and USD10.00, were prepaid for by each country's designated distribution agency, often their Education or Culture ministry, or a similar body, in their choice of US dollars, pounds sterling or French francs. The distribution agency then sold the coupons to libraries, schools, institutions and individuals in their own currency, which used the coupons like checks. The program immediately received an overwhelmingly positive response as the coupons actually provided a practical solution to the many bureaucratic obstacles and other barriers disrupting the post-war book trade between countries. In addition, like the ICHP and its *Bulletin of the ICHP*, participants in the program could freely select the texts they purchased rather than rely on what was donated or otherwise made available to them. As stated in a UNESCO report, the book coupon as a "form of distribution is greatly welcomed, not least because it enables the recipient countries to exercise a free choice in the purchase of the most urgently needed books". With book coupons, the Libraries Section created another branch in its transnational network of communication and empowerment, ensuring that libraries all over the world had various means to access, to the greatest extent possible, what they wanted, down to the precise title, author, language, edition or issue.

Originally conceived of to help libraries and institutions, coupons quickly became a vital tool for smaller groups and individuals as well. Early in 1949, for example, a medical student at the Sorbonne in Paris wrote to the Libraries Section. He and his classmates were unable to obtain the texts they needed, especially those published in other countries. As a result, they had begun to build up their own small library, with each student contributing whatever money they could and with all meeting to agree on which books and periodicals would be purchased with their pooled resources. His question to UNESCO was whether they use book coupons. These students were far from alone and book coupons, especially allocations in relatively small amounts of two to three dollars, offered an accessible solution to professors and students who urgently needed access to foreign publications.[24] In a 1950 report to UNESCO, France stated that the "most varied circles have benefited from the book coupon scheme: university professors, primary and secondary school teachers, students, aeronautical experts, engineers, artists, etc".[25] This accessibility is further illustrated by a note published in the *Christian Science Monitor* in early 1951, stating that the coupons made "it possible for the man in the street to participate directly in the work of the United Nations".[26]

In the fall of 1949, UNESCO implemented a new shipping label reading "Please hurry it through customs", to be adhered to packages containing materials purchased and shipped using book coupons. Looking to save border control officials "time and trouble", the labels were coupled with UNESCO requesting all participating governments to do whatever they could internally to ease the entry of book coupon parcels into their countries. Shipping was covered by three main sources: free shipping donated by transport companies, the use of diplomatic pouches, and funds from UNESCO's modest budget. Combined, these methods contributed to making the book coupons the "immense success" UNESCO declared them to be only a few months into the initial trial period.

In November 1949 *The UNESCO Courier* reported that France, initially allocated USD20,000, wanted USD150,000 instead, and that Czechoslovakia, allocated USD50,000, wanted USD240,000. In addition, more countries regularly joined the program, both as bookselling and as purchasing countries. Given this initial success and rapid expansion, the greatest obstacle UNESCO faced was financial. It had to find sufficient sources of "hard currency against which more coupons can be issued".[27] Many possible solutions were discussed, but ultimately the project relied primarily on donations to UNESCO's hard currency reserve and the purchase of coupons, especially by hard currency countries; money used to buy coupons went back to UNESCO, which used it to produce and sell more coupons.

US libraries could make the biggest difference. By using book coupons to purchase publications from abroad, they could directly contribute to the ability of foreign libraries to purchase publications. Book coupons generated a unique, symbiotic relationship between libraries worldwide, ensuring that money spent by libraries in one country would directly benefit libraries in another. Thus, in order for the coupons to succeed, effective publicity was key. To that end, notes and announcements about the coupons can be found in most UK and US scholarly, trade and general periodicals published at the time. Ultimately, according to a 1952 French report, the coupons were their, and UNESCO's, own best advertisement: "It should be noted that the UNESCO coupons do far more than any article, lecture or film to spread knowledge of the Organization, its program and its aims."[28]

By March 1951 the sale of UNESCO book coupons had reached USD1,000,000. Around the same time, in another signal of its success, the program expanded from book coupons to gift coupons with which purchasers could also buy films and scientific, laboratory and other educational and research equipment. Less than two years in, the success of the coupons was indisputable. By mid-1951, 21 countries were participating, with new additions including West Germany, Belgium, Burma, Ceylon, Indonesia, Pakistan, Persia, Thailand and the Union of South Africa.[29]

Even more indicative of their success is the fact that coupons, evolved in form and function, remain active today. Due to varying exchange rates and currency variations, they are furthermore one of the only internal projects to have generated some income for the agency, and in his personal history of UNESCO published in 1978, the former assistant director-general, Richard Hoggart, wrote: "quite simple arrangements can be life-lines, such as the UNESCO book coupon scheme".[30] Between 1949 and the mid-1960s, over USD50 million in coupons were used by over 35 countries, and it all began as a reconstruction tool to empower and enable war-damaged libraries to purchase much-needed materials.[31]

Coupons proved to be an ideal response to how expensive the international publications market was, how restricted post-war budgets and currencies were, the reality of existing and increasing tensions between countries, and the fact that libraries virtually always struggle to obtain sufficient funding. Simultaneously, they contributed to further expanding the network of transnational library communication and cooperation that UNESCO had established with the ICHP and *Bulletin of the ICHP*. Such networks were only possible, of course, with the dedicated participation of people. The Libraries Section therefore focused equal attention on the individual.

Summer School for Librarians

There were two driving themes behind the 1948 Summer School for Librarians: reconstruction and the right of all people to public libraries. During UNESCO's General Conference in 1947, the school was categorized as a goal "of first importance".[32] Mandated to emphasize the personnel and program needs of war-devastated countries, the theme of the summer school was "public libraries with particular emphasis on their services to popular education and the promotion of international understanding".[33] The school also maintained an integral connection to reconstruction: "It will be limited to approximately fifty participants and preference will be given to qualified applicants from European countries, and particularly from those in which post-war reconstruction of educational, cultural and scientific institutions is most urgent."[34]

Held from 2 to 28 September 1948 in London and Manchester, the school was attended by 48 librarians from 20 countries: Australia, Belgium, Brazil, Canada, China, Denmark, France, the UK, Greece, Haiti, Hungary, India, the Netherlands, Norway, Poland, South Africa, Switzerland, Turkey and the USA. This diversity reflected UNESCO's rapid membership expansion in the early post-war years and was made possible because the school, funded primarily by UNESCO, did not charge participants tuition, room, board or other fees. On the basis of international representation alone, UNESCO considered the school a success.

The presence of librarians from ideologically opposed countries was viewed as an opportunity. The school provided a forum in which librarians could discuss means of cooperating, with an eye to overcoming international ideological barriers then solidifying, rather than diminishing, between countries. It brought together individuals who could have an impact back in their home countries as well as in international cooperative networks. Each participating group of librarians submitted a report to the school outlining the situation of libraries in their country. These recount war damage and losses, as well as hopes and plans for the future. According to Edward J. Carter, first head of the Libraries Section, it was the extent of loss that opened many up to new possibilities. He wrote: "it is perhaps in those countries precisely where the effects of the war were greatest that the spirit of adventure and enterprise most vividly prevails. The will to create is at flood-tide, and this School is a craft that can ride on the flood." And speaking to the librarians, he said: "In your discussions we hope that the needs of the war-damaged areas will be borne in mind and that in framing your picture of library services positive contributions will be made to their development in these countries where ruin and destruction have both heightened the need for libraries and made it most difficult to meet the need."[35]

Indeed their reports reveal that development and improvement were in many cases considered an integral element or direct extension of reconstruction. This is illustrated by a comment made by the Polish librarians, who wrote: "The problem of attracting a large public to culture, by a strong organization of public reading, is one of the principal directives of cultural reconstruction in Poland."[36]

There were tensions over the school being held in the UK. In a reflection of increasing Cold War divisions, librarians from Hungary and Poland had their arrival delayed due to visa issues, while librarians from Czechoslovakia were unable to gain entry into the UK at all, despite all being fellow UNESCO member nations. Some non-attendees also criticized what they perceived as a UK bias in the school, specifically identifying a "somewhat chauvinistic attitude" in the UK library community.[37] However, participants, generally expressed admiration and gratitude for the opportunity to view first hand what was widely considered to be the world's first modern public library system. The French librarians admired the system's impact on UK society, writing that "the frequenting of libraries having become a habit, public reading has attained a very high degree of development".[38] The Polish librarians agreed, writing: "It was instructive above all for the librarians from war devastated countries, where we have to completely reorganize the libraries, to get to know the highly developed network of English public libraries; the choice of England as the seat of the course was therefore very fortunate."[39] Each of the most advanced public library systems of the time, the UK, US and Scandinavian, were represented within the structure of the school, which took place in the UK and was directed by Norwegian librarian Arne Kildal who had trained and worked in the USA. Additional instructors came from India, Belgium, the UK and the USA.

The school, by providing "librarians with an excellent opportunity to widen their professional horizons", exemplified UNESCO's belief that its member nations could learn from one another, and that repairing and expanding transnational networks of communication and cooperation, paralyzed during the war years, was the primary way in which the world would become more tolerant and peaceful.[40]

To that end, in his welcome address to the school's participants, Carter said: "International understanding and reconstruction are not things that can be developed in a purely mechanistic way. Mechanism and technique are important, but it is the individual men and women who are serving public libraries on whom the responsibility for applying techniques rests." He continued: "This School is, therefore, a direct contribution to the growth of personal contacts which we hope will have a lasting effort insofar as those of you who meet your colleagues and will make friendships which will have lasting value both spiritual and practical."[41]

The growth of personal contacts was viewed as essential by participants as well. As the Polish librarians explained, "because of the war and the German occupation, our relations with foreign countries lapsed. Our participation in the course will not be without influence on the renewal of these relationships. The characteristic atmosphere of goodwill and friendship, as much on the part of the direction as the participants, greatly eased our interactions with colleagues from different countries."[42] The school also provided an opportunity for participating librarians to compare and contrast their library practices with those of other countries. The Polish librarians were reassured that their country's unprecedented post-war legislation, a 1946 public library decree, was in line with "modern postulates" based on what they saw first hand in English libraries, primarily in the "laboratory" for the school, the Manchester Public Library.

Evidence of the positive influence of the school includes the fact that additional summer schools, each having a different theme, were held in future years – in Sweden in 1950 on Adult Education, in Brazil 1951 on Public Library Development in South America, and in Nigeria 1953 on Libraries in Africa. In addition, UNESCO's three-volume Public Library Manuals – *Education for Librarianship*, *Public Library Extension* and *Adult Education Activities for Public Libraries* – emerged out of the 1948 school, just as additional manuals were published after the subsequent schools. These manuals were referred to in 1953 as " 'best-sellers' among UNESCO publications – one country alone has ordered 3,000 copies of each volume – and many of the editions have been reprinted".[43] They offered a way for librarians all over the world to self-professionalize by learning, on a continuous basis, about themes and ideas discussed at the schools.

The library world avidly discussed the school, often in very positive terms. Sweden, which had librarians at the International Federation of Library Associations (IFLA) session under way at the same time in London, reportedly favorably observed enough to urge "Swedish membership of UNESCO on the strength of what Swedish librarians observed of the School".[44] In his final report to UNESCO, Kildal noted that the inhabitants of most countries were not "library-minded" at all. He thus proposed that UNESCO considered "a propaganda campaign for the purpose of calling the attention of the general public to the significance of public library work, particularly emphasizing what it can do to promote popular education and international understanding".[45] To that end, in the wake of the school – aimed at librarians – another project became a central focus for the section: the UNESCO Public Library Manifesto.

UNESCO Public Library Manifesto

The plan to draft a public library manifesto was agreed upon during UNESCO's General Conference of 1947 as part of the organization's goal to "promote the publication, translation and dissemination of manuals and leaflets to aid in the development and understanding of public libraries".[46] Expressing his strong support, the then director-general, Jaime Torres Bodet, wrote: "Public libraries, with their roots deep in local and national needs and traditions, are in close contact with the daily lives of the great masses of men, women and children all over the world, and thereby have immense power to aid educational work and to contribute to social, cultural and moral growth. For this reason UNESCO should do everything in its power to encourage the development of public library services and the wide use of such services by people everywhere."[47] Having served from 1922 to 1924 as head of the Libraries Department of Mexico's Ministry of Education, Bodet, like many high-level officials at UNESCO, had a particular interest in the topic.

The manifesto, which came out in 1949, was meant to communicate "in simple but bold terms the aims and functions of public libraries".[48] It was drafted by the Libraries Section with Richard Hart and Emerson Greenaway, two librarians from Baltimore's famous Enoch Pratt Free Library. Greenaway, UNESCO's honorary consultant for public libraries at the time, played a particularly important role. First, during the summer of 1947, he traveled to Switzerland, Austria, Czechoslovakia and Poland to survey the post-war state of public libraries in Europe. Second, based on his experience as a leading figure in the US library system, and on what he observed during his European survey, Greenaway made final revisions to the wording of the manifesto.

Greenaway was sensitive to the reality that public libraries as they existed in the USA would not necessarily be a good match for all other countries, a discussion that arose from the vocal demand for public libraries that was coming from non-Western states or countries with non-Western-style liberal ideologies. In such cases, library developers worldwide often looked West, to countries with established public library systems, as much for examples of what not to do as for what to do. Nevertheless, Western, democratic and especially US influence cannot be denied in the language of the manifesto, which goes so far as to borrow from Abraham Lincoln's Gettysburg Address, employing the words "by the people for the people". Ultimately the choices and preferences of individual countries regarding specific elements of public libraries were not, and could not, all be expressed in the manifesto, which represented UNESCO's ideal guideline for "a world program" for public libraries, based almost entirely on what librarians in Western countries viewed as successful and important.[49] Still, as Edward Sydney, UNESCO library consultant in New Delhi, wrote in a recent essay, the manifesto was "probably the most important and forward-looking short statement of the purpose and service of public libraries ever issued and certainly the only one to be published with international approval".[50]

The manifesto, rightly referred to as "a document of ideas", exemplifies UNESCO's aim to inspire the public to demand public libraries, "to stimulate awareness" about their cultural right to libraries, rather than to somehow impose or dictate public libraries, or even a certain type of public library, upon a global public, an undertaking that would have been impossible in any case. UNESCO had neither the resources nor the authority for enforcement. However, it was in a position to guide, inform and motivate, forms of action for which the Libraries Section was properly lauded.[51]

To that end the section sought, in distributing the Manifesto, to reach the greatest possible number of people, providing them with information which was intended to empower them, which in turn would encourage them to participate in the cultural life of their society or state. Words led to action insofar as the manifesto educated people not only about what public libraries could provide but also that public libraries were institutions that they quite simply had the right to have, to access and, perhaps most importantly, to demand from their local and national leaders and governments. For disseminating this message the manifesto received high praise from a 1957 issue of the American Library Association's *ALA Bulletin*: "We are proud of UNESCO's Public Library Manifesto," the editor claimed. "We are proud that UNESCO has deeds as well as these fine words to its credit in support of public libraries . . . in fact, all of UNESCO's assistance throughout the world not only to create and improve libraries but to create the conditions in which libraries can flourish."[52] The emphasis on practical action over theoretical discussion is particularly key as UNESCO has, increasingly over time, contended with accusations of being an unwieldy, ineffective bureaucratic organization, consisting mostly of some great ideas and too much talk, but very little impact or effectiveness. Coming from the American Library Association, the largest and best-known library organization in the world, representing some of the most free, open and accessible public libraries globally, such praise of UNESCO's action and influence was particularly meaningful and welcome.

The manifesto was intended primarily to encourage library development in places where there were not enough libraries, or indeed none at all, and resulted in the building of three "model" or "pilot" public libraries in New Delhi, Medellín and Nigeria. Yet, like so many of UNESCO's early library-related activities, the initial targets and beneficiaries of the manifesto campaign were primarily libraries in European countries under reconstruction. Carter specifically addressed how he hoped the manifesto would be useful even in countries with existing public library systems "as a UNESCO demonstration of its belief in the importance of public libraries and their place as an essential part of the educational and cultural organisation of the world. In this way the manifesto may help to build up the prestige of public librarians as men and women fulfilling a really important role in contemporary civilization."[53]

Published in the form of leaflets and posters, the manifesto was quickly circulated to member states, primarily via UNESCO national commissions, in tens of thousands of copies, in English, French, Spanish, Polish, Italian and Arabic. About half of the first printing was intended for European countries, with the other half shipping all over the world, reaching from Afghanistan to Venezuela, Burma to Haiti, and beyond. Only Germany, still considered an enemy nation and, as such, not yet a UNESCO member state, was excluded from most UNESCO programs or activities. German librarians, however, impressed with the manifesto, informed UNESCO that they were undertaking their own translation and distribution within Germany, an enterprise that UNESCO quietly supported.[54]

With an original planned distribution of 950 posters and 9,800 leaflets, Poland received by far the greatest number of copies of the manifesto, another reflection of the particular concern UNESCO felt for the extent of the war devastation of libraries there.

It is possible, although no evidence has been found documenting it, that the disproportionate effort to distribute the manifesto in Poland may have also reflected a hope to rally public demand for the free and open democratic institution of the public library as a way to counterbalance the increasing non-democratization of post-war Poland. Unfortunately it has thus far proved impossible to trace the actual distribution of the manifesto once it was shipped out by UNESCO.

Nevertheless, evidence of the manifesto's effectiveness exists insofar as it, like the Book Coupon Scheme, has endured. At the 1996 IFLA General Conference, one speaker reminded listeners that it was "based on confidence in the liberating role of knowledge, necessary for freedom and progress", and praising the "role which it assigns to culture, serving tolerance, understanding, mutual benefit and the respect and preservation of cultures". The manifesto, the speaker continued, "remains faithful to UNESCO's humanist principles and the ideas on which the Organization is based. Finally, there is continuity in the desire to see the library recognized as a public institution essential to justice, freedom and social cohesion."[55]

The 1949 version remained in use until UNESCO asked IFLA to update it in 1972. Additional revisions were made in 1994 and most recently in 1998. Today's manifesto, evolved to reflect advances in information and technology, remains faithful to the original version and is still actively distributed and widely functioning today. It remains one of the best articulations of best practice for public libraries.[56] As one librarian stated in 1998, after that year's revision, "Without a doubt, the previous 1949 and 1972 Manifestos helped increase understanding and insight into the importance of public libraries for democracy and education in many countries, at both central and local levels." She went on to argue that the manifesto "is kept alive through our work. It is not a 'desk drawer' product for our municipalities, but can rather be a real tool in the dialogue with decision-makers and others. We can spread this means of working to more municipalities and in this way keep the Manifesto alive."[57]

Conclusions

All the initiatives discussed in this chapter began as tools for post-war reconstruction and rehabilitation focused on the European continent. All successfully transitioned, for varying periods of time, into tools for global development. This success can be traced to the fact that these programs were carefully formulated for the immediate, but with the long term very much in mind. While today UNESCO's origins in post-war cultural reconstruction are no longer widely remembered, at the time the organization's founders recognized the necessity of having an impact by successfully addressing early post-war cultural needs. If they failed, they would face a much more difficult uphill battle in establishing UNESCO's name and reputation as an effective transnational cultural body in the long run.

Libraries are easy targets for budget cuts today but nevertheless remain heavily relied upon by a broad and diverse public seeking information, knowledge, escape or even a safe haven. It behooves us to remember how, during the early post-war years, UNESCO and supporting governments both recognized the crucial importance of libraries in creating a more tolerant, informed and peaceful world. Their role toward this end is perhaps as critical today as it was then.

Notes

1. Letter from Franklin D. Roosevelt to Luther H. Evans, 13 June 1941, Roosevelt Presidential Library, Hyde Park, New York, USA.

2. Report of the Technical Sub-Committee to the Preparatory Commission, 15 November 1946, 4, and General Conference 1. Paris 1946, vol. 4 Documents, UNESCO Archives.
3. Jaime Torres Bodet's foreword, 29 February 1949, in *Libraries in Need* (Paris: UNESCO, 1949).
4. Committee on Libraries, Museums and Special Projects, 15 May 1946, Preparatory Commission vol. V; 31/11, UNESCO 1946–1947; and The Libraries Programme, Opening Remarks by the Counsellor, 31 May 1946, Preparatory Commission vol. V, UNESCO Archives. See also Ralph R. Shaw, Department of Agriculture Library, 21 August 1946, James Marshall Papers, American Jewish Archives.
5. The three quotes from *UNESCO Bulletin for Libraries* 1:1 (April 1947); UNESCO, National Surveys of Library Needs and Resources, 1 August 1946, 2. Preparatory Commission vol. V; and 1.Supplementary Report on ICHP, 5 November 1946, Preparatory Commission, Box 41, 37/2/328, Libraries, ICHP, UNESCO Archives.
6. "Restoration of Libraries", Conference of the Allied Ministers of Education, Inter-Allied Book Centre, Polish Institute and Sikorski Museum, London, A.19.II/32, Ministry of Religious Affairs and Education, CAME, 1942–1945, UNESCO Archives.
7. The French Ambassador, "Restoration of Libraries", *The Library Association Record* 49:1 (January 1947): 4.
8. Maria Danilewicz, Memorandum, 31 December 1945, UNESCO Archives.
9. Review with William Farr, Secretariat UNESCO Preparatory Commission, 24 July 1946, Rockefeller Archives Center, RG 12.2 Officers Diaries, John Marshall, 1 July 1943–31 December 1943 to 3 June 1946–27 September 1946, UNESCO Archives.
10. Edward J. Carter to Hawkes, 13 June 1947, PRO, ED 157/197, UK National Archives.
11. Carter to Exterior Relations, 24 March 1948, 02 A 855 International Exchange of Publications, Part I, UNESCO Archives.
12. Untitled, unsigned, undated [circa 1946–1947], 361.9:02 Reconstruction Libraries, UNESCO Archives.
13. Unsigned to Julien Cain, 6 May 1946, 04:341.383 "39/45" Confiscated Books, UNESCO Archives.
14. The two quotes from Jacob Zuckerman, [unpublished] Memoirs, 63, with grateful thanks to the Zuckerman family, and *Bulletin of the ICHP*, Preparatory Commission vol. V, UNESCO Archives. See also *UNESCO Bulletin for Libraries* 1:1 (April 1947): 2.
15. Edward Carter, "UNESCO's Library Programs and Work", *The Library Quarterly* 18: 4 (October 1948): 237.
16. *UNESCO Bulletin for Libraries* 1:1 (April 1947): 1.
17. Subscription, first instigated in 1951, was USD2 per month according to a letter from Carter to Department of Cultural Activities, 19 February 1951, X07.353.321 UNESCO Publications – General Part IV, and see also 4C/PRG/6 Annex II, 1 August 1949, X07.353.321 UNESCO Publications – General Part 1(B), UNESCO Archives.
18. General Conference, 4th Session, Programme and Budget Commission, 1 August 1949, 2. X07.353.321 UNESCO Publications – General Part 1(B), UNESCO Archives.
19. Zuckerman, Memoirs, 57.
20. "Book Coupon Scheme Proving a Success", *UNESCO Courier* II:2 (March 1949): 3. See also "New Book Currency Now in Circulation", *UNESCO Courier* I:11–12 (December 1948–January 1949): 6.

21. "New Book Currency", 6.
22. Dr. Joseph Needham, Notes on a short visit to Poland and Czechoslovakia, June 1947, 2, 4, X 07.21 (438), Relations with Poland – Official, Part I, and see also UNESCO, National Co-operating Bodies in the United Kingdom, Notes on a draft proposal for UNESCO Book Coupon Scheme, 24 September 1947, TNA PRO ED157/197, UNESCO Archives.
23. UNESCO Book Coupon Scheme, 8 February 1949, 332.55:02 Book Coupons-General, Part II, UNESCO Archives.
24. Julien Cain to Director General, 30 May 1949, 332.55:02 Book Coupons-General, Part II, UNESCO Archives.
25. France, 144, General Conference, Fifth Session, 1950, Reports of Member States, UNESCO Archives.
26. "UNESCO Offers Gift Plan", *Christian Science Monitor*, 6 February 1951.
27. Jacob Zuckerman to John Jay McCloy, 11 March 1949, 332.55:02 Book Coupons-General Part II, UNESCO Archives. See also "Has UNESCO Anything to Declare?" *UNESCO Courier* 11:10 (1 November 1949): 8.
28. Extract of the Reports of Member States presented to the General Conference at its Seventh Section, Paris, November-December 1952, France, 74, 332.55:02 Book Coupons General Part III, UNESCO Archives.
29. "UNESCO's Book Coupons", *Physics Today* 4:3 (March 1951): 30 and General Conference 1951 Report of the Director-General, 77, UNESCO Archives.
30. Richard Hoggart, *An Idea and its Servants: Unesco from Within* (London: Chatto & Windus, 1978), 37. See also "UNESCO Coupons Can Buy Knowledge and Change Lives", http://www.unesco.org/new/en/unesco/partners-donors/who-are-our-funding-partners/unesco-coupons-programme (accessed 29 October 2014)
31. Chloé Maurel, *Histoire de l'UNESCO: Les trente premières années, 1945–1974* (Paris: L'Harmattan, 2010), 238.
32. "News and Information", *UNESCO Bulletin for Libraries* II:1 (January 1948): 2.
33. UNESCO International Summer School for Librarians, The Work of the Libraries Division, 1948, Paris, 17 September 1948, accessed 29 October 2014, http://unesdoc.unesco.org/images/0014/001474/147429eb.pdf. See also Twelfth Summary Report to the Department of State, 13 November 1947, 30/4, James Marshall Papers, American Jewish Archives.
34. UNESCO, International Summer School for Librarians, 3 May 1948, Summer School, UNESCO Archives.
35. Carter, Paper read to the UNESCO Summer School for Public Librarians, 27 August 1948, 4, 6. Summer Course, UENSCO Archives.
36. Evaluation from Poland, Summer Course, UNESCO Archives.
37. Carter memorandum, 26April 1948, Summer Course, UNESCO Archives.
38. Rapport final des participants français, 3. Summer Course Part I, UNESCO Archives.
39. Evaluation from Poland, undated, Summer Course Part I, UNESCO Archives.
40. UNESCO, International Summer School for Librarians, 3 May 1948, Summer Course. UNESCO Archives.
41. Carter, Paper read to the Unesco Summer School, 4, 6, Summer Course Part II, UNESCO Archives.
42. Evaluation from Poland, Summer Course Part I, UNESCO Archives.
43. Everett N. Petersen, "UNESCO and Public Libraries", *Library Trends* 1:4 (April 1953): 538.

44. Carter, Report of the UNESCO/IFLA International Summer School for Librarians, 8, Summer Course Part II, UNESCO Archives.
45. Arne Kildal, official report as UNESCO document, 25 February 1949, 12, Summer Course Part III, UNESCO Archives.
46. Resolution 6.51112, quoted in Jaime Torres Bodet to various, undated, Public Libraries Manifesto, UNESCO Archives.
47. Bodet to various, undated, Public Libraries Manifesto, UNESCO Archives.
48. Carter to Robert L. Hansen, June 1948, Public Libraries Manifesto, UNESCO Archives.
49. Emerson Greenaway to Carter, 27 May 1948. Public Libraries Manifesto, UNESCO Archives.
50. Edward Sydney, "Public Library Development in the Post-War Years: The First Decade" in *Libraries and Information Studies in Retrospect and Prospect Essays in Honor of Prof. D.R. Kalia*, ed. J.L. Sardana vol. 2 (New Delhi: Concept Publishing Company, 2002), 342.
51. Kerstin Hassner, "The Model Library Project – A Way to Implement the UNESCO Public Library Manifesto", *IFLA Journal* 1 (1999): 143–147 and Carter, "UNESCO's Library Programme and Work", 241.
52. Lucile M. Morsch, "Promoting Library Interests Throughout the World", *ALA Bulletin* 51:8 (September 1957): 582.
53. Carter to Robert L. Hansen, Statens Bibliotekstilsyn, Copenhagen, June 1948, Public Library Manifesto, UNESCO Archives.
54. "Bulk shipments of the Public Library: A Living Force for Popular Education". Public Libraries Manifesto and Guido Geyer to Peterson, 13 February 1950, Public Libraries Manifesto, UNESCO Archives.
55. Abdelaziz Abid, "Revision of the UNESCO Public Library Manifesto", *62nd IFLA General Conference – Conference Proceedings* – August: 25–31, 1996.
56. For example, in 1999 the US National Commission on Libraries and Information Science passed a resolution adopting Principles for Public Library Services based on the UNESCO Public Library Manifesto.
57. Hassner, "The Model Library Project".

Figure PIII.1 School of Technology at the National University of Engineering in Lima, Peru, 1967. The post-secondary school for the training of technicians in the fields of mechanics, electricity and chemical processing was funded by UNDP while UNESCO was responsible for the implementation of the project. (Photographer: Dominique Roger © UNESCO)

Part III
Experts on the Ground

The most extensive fieldwork carried out by international organizations has its origin in the UN's Expanded Program of Technical Assistance, established in 1949 to provide technical assistance for the economic development of underdeveloped countries. UNESCO participated, and requests for technical assistance could be sent by any country or territory in need. When approved, the organization would send mission experts and equipment. The requests came mainly from the many newly independent countries and within all fields of science, education and culture.

The implementation of UNESCO's mental engineering activities was, however, only possible if people could read and write, and one of its top priorities was therefore fundamental education. India was the first country to acquire assistance in fundamental education specifically. UNESCO's intervention was not accidental since the expression "underdeveloped" or "developing" evoked India as a prototype. As soon as Indian experts had been trained, some of them carried on to Indonesia where only 25 percent of the school-age children went to school and 80–84 percent of the adult population were illiterate.

During the decolonization process, UNESCO went from being a mainly Western organization to being a truly global institution. From 1958 to 1964 a total of 27 newly independent African states joined the organization, and as a result the focus on development and capacity-building in the newly independent countries became even stronger. One of the most noteworthy and extensive field operations was UNESCO's engagement in building up an entire educational system in the Democratic Republic of Congo.

Broadly speaking, UNESCO's aid program provided for assistance in technical, elementary and fundamental education, advice on the production and use of materials for education, the training of teachers and other specialized personnel, and the planning and organization of scientific research and training laboratories. Thousands of field officers have since been sent via Paris to every corner of the world.

7
UNESCO's Fundamental Education Program, 1946–1958: Vision, Actions and Impact

Jens Boel

Fundamental education was part of UNESCO's initial program adopted by the General Conference in 1946. This chapter explores its origins, vision, scope and activities until the General Conference in 1958 decided to abandon the concept of fundamental education.[1] No decision was taken on what should replace it and in practice no single term was chosen. "Out of school education" and "community development" were widely used as direct replacements.[2] The already well-established "adult education" was also used for some activities that would before 1958 most likely have been called "fundamental education". For the purpose of this chapter it therefore makes sense to end at that point, although the idea of fundamental education as a way of addressing the broad needs of communities rather than implementing standardized, specific solutions remained important for UNESCO's work in education in the years that followed.

Origins and vision

The program on fundamental education had its origins in the CAME, while the actual work on and discussion of the concept, its definition and scope unfolded in the Preparatory Commission for what was to become UNESCO and the work of its Education Section.[3] During the earliest discussions, the focus was on the fight to erase illiteracy, but in a working paper prepared by the Education Section the concept was broadened. In the words of the senior counsellor for education, Kuo Yu-Shou, providing elementary education for all young people of the world was just as important as the attempt to liquidate illiteracy.[4] The final paper on fundamental education presented by the Education Section to the Education Committee in May 1946 went even further in vision and scope by stating that UNESCO should envision education in its widest sense. Education involved instruction in all areas "which contribute to the development of well-rounded, responsible members of society." it said.[5] Therefore "it is proposed that the Organization should launch, upon a world scale, an attack upon ignorance by helping all Member States who desire such help to establish a minimum Fundamental Education for all their citizens".[6]

What exactly was meant by "fundamental education"? From the outset a certain embarrassment and even confusion was apparent. In May 1946, Kuo Yu-Shou declared to the members of the Education Commission: "We confess to a certain dissatisfaction with this title, but have found no better. We might have spoken of 'Illiteracy', 'Mass Education', 'Basic Education' and 'Popular Education'." He then went on to explain why those terms were not chosen: illiteracy would be too limited an approach, mass education had an "unpleasing connotation" since it paid insufficient attention to individual differences, popular education was similar and had in some languages a faintly patronizing sound and, finally, basic education could give the impression that UNESCO would be somehow connected with the concept of "basic English" in English-speaking countries and others might think that UNESCO was committed to Mahatma Gandhi's ideas, since "basic education" was sometimes used in connection with them.[7]

The initial approach chosen by UNESCO was not to define the concept of fundamental education but simply to describe it. The secretariat of the Preparatory Commission acknowledged that using the term "definition" would suggest "too great a degree of precision", and therefore the choice was made to start from practical examples, thereby showing what was meant.[8] However, what emerges clearly from the early discussions in 1946 and the subsequent work in the first couple of years of existence of the program was a holistic approach to education. "The aim of all education is to help men and women to live fuller and happier lives," it was stated.[9] Both the publication *Fundamental Education: Common Ground for All Peoples* of 1947 and the updated version from 1949, *Fundamental Education: Description and Programme*, start by paying homage to the Chinese educator Yan Yangchu, also known as James Yen, and by quoting him. "Three-fourths of the world's people today are under-housed, under-clothed, under-fed, illiterate," it began. "Now as long as this continues to be true we have a very poor foundation upon which to build the world."[10] The second publication would also emphasize that "it was against the background of James Yen's statement, and not as a result of a mere theory, that the UNESCO delegates accepted the term 'fundamental education' ".[11]

Therefore UNESCO's vision was broad, global and intended to change living conditions through education in a range of areas. The list – as, for example, established in the 1949 monograph – was long and included health education, domestic and vocational skills, knowledge and understanding of the human environment – including economic and social organization, law and government – and what was called "the development of qualities to fit men to live in the modern world, such as personal judgment and initiative, freedom from fear and superstition, sympathy and understanding for different points of view".[12] Altogether this was a long way from a simple literacy campaign.

In 1951 the General Conference adopted a strong resolution in favor of the creation of a world network of international fundamental education centers: "Believing Fundamental Education to be at the heart of the work of UNESCO and convinced that the general plan [to create this world network] constitutes a first attempt to combat through education the problems of ignorance, poverty and disease". The broad definition was maintained.

However, the doubts around the term, its meaning and scope persisted in spite of several attempts at definition. Criticism of the term came both from within the UN system and from member states. In the early years of the UN system, UNESCO shared concerns with other specialized agencies – the FAO, the ILO and the WHO – as to what was perceived as the UN's intentions to take over program-execution responsibilities in their respective fields. In particular, in 1947 UNESCO's secretariat had been worried that the UN Social Department would set up an Educational, Scientific and Cultural Division. UNESCO made its position very clear: the UN should exclusively have a coordinating interagency function and not a program-implementation role in areas that had been defined as the competence of the specialized agencies – such as education in the case of UNESCO. At a meeting between the UN secretary-general and the heads of ILO, FAO and UNESCO in February 1947, the secretary-general declared that he would take steps to bring about "the smooth operation of arrangements" concerning relations with the specialized agencies. However, areas of difficulty remained.[13]

The UN's concerns with regard to "fundamental education" related to the perceived risk of the term and approach becoming all-inclusive, going well beyond the field of education. This was, for example, expressed as an insistence of the need to see fundamental education as only one aspect of community development, which was defined as the field of competence of the UN. In 1956, UNESCO prepared a "Working Paper on the Definition of Fundamental Education" for the UN Administrative Committee on Coordination.[14] The purpose seemed to be to reassure the UN that fundamental education indeed only had a limited scope. Cautiously, however, the working paper emphasized that any new definition would have to be approved by the General Conference. The working group discussed the relationship of fundamental education with other forms of education and commented on "community development" that the term "is widely used, but in a somewhat more restricted sense".[15] Later the same year, a working group at the General Conference discussed the topic and concluded that there was no need for changes in terminology.[16] The group agreed on a definition of fundamental education, which was published as an annex to its report to the General Conference.[17]

However, internal criticism among UNESCO's member states finally led to a change. While fundamental education had always, in principle, been directed toward empowering (although this expression was not used at the time) the underprivileged all over the world, in both developed and less developed countries, all major fundamental education activities had in reality taken place in the latter category of countries.[18] Delegates from Ceylon, France, Italy, Morocco and Sweden observed at the General Conference in 1958 that "they were not happy about the distinction between adult and fundamental education. The proposed activities under out-of-school education suggested that fundamental education was intended for economically underdeveloped countries and adult education for more advanced countries."[19] In response to these critical voices, the General Conference decided to abandon the term "fundamental education" on the grounds that it had led to confusion.[20]

In relation to the UN it is interesting to observe that a note on "community development", submitted by UNESCO to the UN Administrative Committee on Co-ordination a few years later opened by stating that "Perhaps the most significant step towards community development during the period under review [1960–1962] has been the gradual elimination of the concept of 'fundamental education', which for many years has been a cause of confusion because of its similarity with 'community development' ".[21]

These discussions and developments concerning the concept and definition provide a framework for understanding the context of UNESCO's actions regarding fundamental education. They set the background for the activities at which we will look more closely in the following.

The idea in action

UNESCO's first ever publication was *Fundamental Education: Common Ground for All Peoples*, issued first as a report for the first session of the General Conference in November 1946 and then, the following year, as a publication in its own right.[22] Even more remarkable, the whole book, which the delegates had not had the opportunity to read in full, was adopted by the General Conference as the basis for the work the secretariat was asked to accomplish.[23] This resolution of 1946 is essential since it both authorized and requested the secretariat to immediately start working on fundamental education activities. The Program Commission of the General Conference had grouped all recommended projects in three categories, depending on their relevance and urgency. "A" was for most urgent projects, while "B" meant desirable projects and "C" stood for advisable projects. The fundamental education program was adopted as a category A activity.

The most high-profile fundamental education activity on the ground was the pilot project, which was launched in the Marbial Valley in 1947, a poor rural area in the southern part of Haiti. This pilot was soon followed by other projects in different parts of the world, but the Haiti project would remain the most emblematic and the one in which UNESCO was the most closely involved. The project was extensively covered by the media – including in UNESCO's own magazine, *The UNESCO Courier*, and press and radio in different parts of the world. As discussed briefly later, it was also controversial, its results and impact giving rise to debate.

During the first months of its existence, UNESCO established a Fundamental Education Division, which soon developed an important clearing-house function. The most significant publications were 12 monographs, published from 1949 to 1959. They were supplemented by other publications, in particular *Fundamental Education: A Quarterly Bulletin* which came out in 1949 and in 1952 changed its title to *Fundamental and Adult Education: A Quarterly Bulletin*. In 1961 the title was changed again, to *International Journal of Adult and Youth Education*. It published, in particular, case studies of specific experiences of fundamental and adult education.

Also in 1949, the Education Clearing House – a documentation center – began publishing *Fundamental Education Abstracts*, which continued under this title until 1952 inclusive, and then from 1953 to 1960 under the title *Education Abstracts*. These "abstracts" consisted of an annotated bibliography of technical material, which was shared across the world. They included filmstrips, since a number of films promoting fundamental education were also produced and distributed.

The Associated Projects scheme was also launched in 1949 and only two years later, UNESCO had accepted 34 such projects in 15 countries. The idea was that any project accepted as associated would receive from UNESCO the services of the clearing house and sometimes more direct technical assistance. The main obligation of the projects toward UNESCO was sharing all data and experiences applicable to other countries.[24] Experiences of these projects were published in the quarterly bulletin.

In 1950 the Regional Centre of Fundamental Education in Latin America (CREFAL)) was created by UNESCO in Patzcuaro, Mexico, and it served as a training center for teachers, trainers and civil servants involved in fundamental education activities not only in the Latin American region but worldwide. For example, in 1954, supported by UNESCO fellowships, ten Pakistani officials were sent to CREFAL for training in fundamental education.

Also in 1950 the Executive Board endorsed an ambitious fundamental education program, developed by the secretariat. At the heart of the program was a proposal to establish six regional centers in the main regions of the world for the training of leaders, the preparation of material, and the development of methods to fight "illiteracy and its attendant evils". The project was planned to last 12 years and depended on raising the enormous sum of USD20 million.[25] The General Conference invited the member states to consider providing the additional resources but also made it clear that this funding would have to come from voluntary contributions.[26] As early as the following year it became clear that it would be very difficult to raise such an amount, and the General Conference limited its ambitions to maintaining CREFAL and creating a second regional center, which was eventually established in 1952, in Sirs-el-Layyan, Egypt, for the Arab region. The total regular budget set aside for fundamental education activities in 1952 became a modest USD235,000, of which most would go to running the Patzcuaro Center.[27]

Fundamental education activities, on the other hand, became an integrated component of UNESCO's work with the UN Technical Assistance Program. The most common form of technical assistance consisted of sending an expert to a given country, where they would advise on how to carry out a campaign toward the solution of a problem. Sometimes the expert would go to carry out actual fieldwork. UNESCO experts would often work with experts from other specialized UN agencies in a venture that news correspondent and editor Ritchie Calder called "the greatest social experiment of our time", by mobilizing "the innate resources of the peoples themselves", thereby representing "a new attitude to world problems".[28]

Taking a closer look at UNESCO's activities in fundamental education during the years 1946–1958, certain characteristics are worth noting.

First, the geographical scope of activities is impressive. Just taking into account the list of field missions and files in the first registry series (1946–1956) in the UNESCO Archives, at least 62 countries had UNESCO-organized activities that were explicitly entitled "fundamental education". Compared with the number of member states by the end of 1958 – 82 altogether – this is already significant. Furthermore, a number of fundamental education activities were of a regional nature – in Africa, among Arab states and in Latin America. In addition to this, many other projects which did not actually carry the name "fundamental education" in fact focused on activities that may well in reality have been perceived as such – for example, in the fields of adult education, textbooks and audiovisual aids.

Second, the high level of attempted cooperation with other UN agencies at grassroots level is striking. In particular, it was almost a general rule that the ILO, FAO and WHO – all of them together, or at least, one – worked with UNESCO on fundamental education projects. This was logical in view of UNESCO's holistic approach, which implied that concerns such as nutrition and land erosion (FAO), hygiene and health in general (WHO) and the development of handicraft and local industries (ILO) were integrated parts of the global projects. However, cooperation was not always smooth. In several cases – for example, in Haiti – there were tensions and conflicts due to different appreciations of requirements and of the situation on the ground. To this must, for some projects, be added the issue of rivalry in leadership. Nevertheless, it stands out as an interesting characteristic that in spite of the well-known difficulties in coordinating and defining the respective roles of each organization at the global level, there were very substantial efforts to work together locally, on concrete projects.

Third, the political situation in the "host countries" and the support or lack of such from national authorities were key factors in the success or failure of projects. As a rule, UNESCO's projects were not "stand alone" but had to fit into major national campaigns and policies. For example, in Indonesia, the Ministry of Education estimated that an intense literacy campaign from 1946 to 1958 resulted in 900,000 new literates each year.[29] UNESCO supported and participated in this effort but sometimes met resistance in the involved ministries and had to negotiate its precise role. This complex situation makes it even more difficult to evaluate the impact of UNESCO's actions.

Finally, in 1951, UNESCO's secretariat presented, as mentioned, an ambitious plan to the General Conference for the implementation of its fundamental education program.[30] The ambitions had to be reduced significantly in the face of strong opposition to budget increases from the member states, in particular the largest contributor – the USA. As a consequence, UNESCO's initial high-flying dreams of providing fundamental education to children, and all men and women, in the world were reduced to more sporadic actions. A side-effect of this may well have been the resignation of UNESCO's director-general, Jaime Torres Bodet, in 1952. He had himself led a massive literacy campaign in Mexico before heading UNESCO and was very committed to the cause of fundamental education. He stepped down from his position in protest at the member states' refusal to increase the budget, and he cited explicitly education – as opposed to the willingness of certain countries to invest heavily in the arms race – as the field that more than any other required additional funding.

Making a difference

What did it all change and what was the impact, particularly in the long term? This is a complex question and there are several components that need to be discussed. Before going systematically through some of the key dimensions, it is worth observing that the challenge of measuring results, of evaluation and impact was present within the fundamental education program from the outset.

The most striking example is the emblematic pilot project in the Marbial Valley in Haiti. At the beginning of 1954 the Education Department sent an expert, Lucien Bernot, to evaluate this project and its results. When he sent a report in January, his first month out of four, Bernot expressed some skepticism: "I am a bit worried that there will not be many results to study and evaluate." About the 20 trainers and teachers working at the project he commented: "These instructors are full of good intentions but around them life has not changed since the passage of A. Métraux."[31]

Within the Education Department, Bernot's observations – which became even more critical over time, although never directly negative against UNESCO – gave rise to debate. Conrad Opper, the head of the Fundamental Education Division, wrote in an internal note to Lionel Elvin, the director of the department: "It is of course difficult to assess such intangible things as, for instance, receptivity to new ideas; but it would be incorrect to assess five years of fundamental education only by such results as can be measured statistically."[32] In his reply to Bernot, Lionel Elvin quite subtly tries to get this point across. After having mentioned the question of changes resulting from the pilot project, he notes: "These changes in the social behavior of individuals can sometimes be difficult to measure in quantitative terms but I'm certain that they will not escape you in spite of their singular subtlety."[33]

In the end, Bernot's later report remained quite critical with regard to the overall results achieved – mostly based on figures and statistical measures. André Lestage, the fundamental education specialist at UNESCO's headquarters to whom Bernot sent his first report, would later, in 1959, after an evaluation mission to the Marbial Valley, conclude in a very similar way that not much remained as lasting results of the pilot project: "We have in Haiti spent thousands of dollars for very little."[34]

A rare but essential source on impact at the individual level is personal accounts from the people directly concerned. An example is that of Mrs Vissière, a widow from the Marbial Valley in Haiti who in 1948 learned to read and write at the age of 42, thanks to a UNESCO pilot project on fundamental education. This is how five years later she described her experience of becoming literate and what it meant to her:

> When I am required to put my signature on any document, needless to say that I no longer settle for drawing a cross, and when the opportunity arises to read documents related to my possessions, I do not resort to friends to whom I pay tips when begging alone won't suffice. At the speculator's, I now know to inspect the weight of the coffee I sell; at the shopkeeper's, I am agile to calculate the currency to be paid and received.
>
> In truth with all these open horizons before me, I feel happy.
>
> I remember in fact with what shame, at the time when I could not read, I had to pay one dollar to Mister Ti-Pierre to search for a property deed among other, no less important documents. But today it is me who could be in a position to exploit my illiterate compatriots.
>
> At present I am an ardent supporter of the literacy campaign. Having become a teacher of adults, thanks to my relentless efforts, I encourage people in my area to educate themselves. I often tell them that it was at the age of 42 that I learnt to read; it is therefore five years ago that I could

not grasp how much knowing to read and write is the key that allows access to notions of a good life.[35]

In spite of the obvious methodological difficulty in determining how much weight can be given to this kind of individual testimony, it does embody a strong personal experience, which expresses a result of the literacy process that corresponds very well with what UNESCO's fundamental education program aimed to achieve: people and communities participating actively in gaining control of their own lives and obtaining "fuller and happier lives".

In order to establish an overview of the global impact of UNESCO's initiatives within the field of fundamental education, the following five paragraphs describe some of the essential elements which must be considered.

Statistics, reports and evaluations. There is a great deal of quantitative data from various fundamental education projects, which include the number of teachers and trainers trained, new literates, community centers, new crops introduced, handicraft production and reduction of victims of diseases. To some extent these developments can clearly be identified as results of the projects and sometimes even directly related to UNESCO's involvement. However, these are merely outcomes of activities, while the real evaluation of impact should relate to the overall purpose of fundamental education as a community approach enabling people to live "fuller and happier lives". Furthermore, studies in Indonesia, for example, show that a large percentage of people who may have seen an improvement in their life situation would revert to their prior state – or worse – if there were no adequate follow-up actions to the initial projects. In Indonesia, 40 percent of the newly literate would actually lose their new knowledge because they did not use it.[36]

Clearing house. Much more than specific results from a given project, UNESCO's goal was to arrive at changes by mutual inspiration across countries and regions – this was what the clearing house, the documentation center, was for. To what extent this goal was actually accomplished is much harder to document and quantify. What can be observed is that, in addition to the clearing house itself and the already mentioned publication activities, there were multiple links across borders and among the different projects. Often the experts worked for projects in a number of different countries. For example, this was the case for Spencer Hatch, who was responsible for projects in India and Ceylon but also had experience from Latin America; and for the Afro-American textbook specialist, Ella Griffin, who worked on the Marbial Valley project and then in Jamaica, India, Egypt, Congo-Brazzaville and Ghana, before being recruited to work at UNESCO's headquarters where in 1962 she became head of the Literacy Section.

Reactions among people in the societies concerned. The intention of UNESCO's fundamental education activities was to give people hope and realistic expectations of a better life. Thanks to literacy skills, combined with better hygiene practices and employment opportunities – the life situation as a whole – a brighter future should become possible at the community level. Evidence of success in this area is hard to find and even harder to value objectively. One specific example is the personal account from the Haitian woman quoted above. A collective expression of support, also from Haiti, emerged in 1949 when the peasants of Marbial heard rumors that UNESCO might abandon them. A number of them came together and demonstrated with signs saying "Kêbé l'Inesko Fò!", which in Creole means "Support UNESCO hard!"[37] Newspaper articles about the projects can also shed light on such qualitative dimensions. However, this evidence is scattered and not necessarily representative.

The political impact. UNESCO constantly argued with member states and within the UN, and in particular in ECOSOC, that education was fundamental for development. The work and experiences in fundamental education were used as evidence for this argumentation, and it strengthened the case that so many projects were carried out jointly with other UN agencies. UNESCO's work on fundamental education was probably a decisive element in convincing the UN to consider education as an essential tool for development. This was particularly important at the moment when the wave of independence of former colonial states, in particular in Africa, started. This gave UNESCO a head start when development issues became the top priority for the UN in the 1960s and 1970s – a process during which UNESCO achieved a leading international role in the field of education, receiving substantial funding, in particular from the UNDP, but also from the World Bank, the UN Population Fund and other UN agencies.

Internal struggles. Within UNESCO, and even within the Education Department itself, there was no consensus on the assessment of impact. Some reports and statistics gave rise to heated debates. Advocates of the relative success of the projects – and in particular the emblematic Marbial Valley project – argued that political and other "unexpected" difficulties should be taken into account as extenuating circumstances. For example, Conrad Opper, director of the Marbial Valley Project as of October 1949, at one point wrote:

> A man is climbing up a hill and travels only one mile in an hour. In purely statistical terms that is poor progress as a man's normal walking speed should be about three miles per hour. It is therefore insufficient to produce figures. A proper evaluation must assess the difficulties and measure success or failure in relation to these difficulties.[38]

Another, more significant, angle of debate was, as mentioned above, whether it actually made sense to base judgments of success and failure on purely quantitative data.

Nevertheless, there was strong pressure from some experienced staff in the field to improve statistics and evaluations. For example, the head of the fundamental education activity in India, Emily Hatch, in 1951 said the following in a statement that sounds like a cry of despair:

> We are busy all the time on activities we know are somewhat helpful. We could ... write reports about them as most people do, making the brightest side shine a little. Too many are content with this, with the result that the real issues are not faced at all, and the work is superficial. India so needs some honestly solid work, we feel we should make a fearless stab at trying to do some. Basic to all the superficiality of the long hours of work and half-way schemes to meet needs, is the vagueness of the how and why of the work. We have the feeling people are working blind-folded. And worse they don't realize it. How can they, when almost anyone's *opinion* is as good as another's, no one really know the *facts*. They all think they know, especially those who fly over villages in an airplane or those who hastily scan a few records from some village official.[39]

These observations, discussions and doubts also reflect a profound tension and dilemma for UNESCO's strategies, work and actions in general: how to reconcile the lofty ideals expressed as the ultimate goal of the organization – creating the defenses of peace in the minds of men and working towards intellectual and moral solidarity of mankind – with concrete, practical actions, and how to measure progress towards that goal.

In the context of the overall purpose of the organization, education is perceived as a moral rather than a technical objective. Essentially, this means that although UNESCO is a technical agency of the UN, its real *raison d'être* is to serve as a tool for changing the world toward another way of living together, a kind of "small utopia", where all human beings participate fully in societies. Fundamental education is in that sense a concept that goes a long way toward being a universal concept characterizing UNESCO's global mission, in a similar way to the broad concepts of "a culture of peace" and "a new humanism". Measuring impact and progress toward attaining these global goals remains a major challenge.

Conclusions

This initial study of UNESCO's fundamental education activities from 1946 to 1958 leads to some conclusions and a number of questions.

UNESCO started very early with fundamental education activities. Together with the reconstruction program, it was the only major area where actions were taken by the Preparatory Committee, even before the formal creation of UNESCO in November 1946, when the 20th country ratified the constitution. The fundamental education program involved a range of fields for action and was rapidly described in ambitious terms, rather than "simply" focusing on literacy. The purpose was to enable communities to help themselves in order to achieve better and happier lives. Therefore UNESCO's whole mission to change the "minds of men" was embodied in the program, and fundamental education during the early years of its existence in a sense became the epitome of UNESCO.

The debate about the scope and definition of the program led to some tensions with the UN. There were concerns from the UN that the very broad way in which the term "fundamental education" was used could lead to confusion with regard to what the UN saw as a vaster notion of "community development". By the end of the period discussed here, the problem seemed to be solved since UNESCO decided to abandon the term in 1958. At the same time, the ideas behind the concept lived on in UNESCO's activities. I also think – although this discussion is beyond the scope of this chapter – that it can be convincingly argued that at about the same time UNESCO also succeeded in mainstreaming, within the UN system, the idea of education as an essential and indispensable tool for development.

The impact of the fundamental education program was considerable in terms of acknowledgment and awareness in numerous countries around the world. At the General Conference in 1956, the strength of the "brand" was one of the arguments used by the member states not to change terminology. The statistical impact – for example, in terms of numbers of trainers trained – can be measured, but it is not really significant since the purpose of the program went far beyond activities where results can be counted directly. Much more important – but difficult to measure – was probably the support for nationwide campaigns and contributions to joint technical assistance activities with other UN agencies. This can be described as "mainstreaming" the idea of education for development. The "soft" impact, in terms of creating hope and motivation, and of enabling people to become active, participating citizens, is even more difficult to measure. I have given a few examples of such an impact happening, but this is a topic that requires much more research.

To better understand the impact on communities and individuals of fundamental education ideas and activities, in-depth case studies in different regions of the world are necessary.[40] Different types of sources must be combined to deepen our understanding – based on national, local, civil society and private archives. Finally, further studies should work with a longer timespan, moving up to the launch of Education for All in Jomtien in Thailand in 1990 and, in a later stage of research, to the present. The ideals and very practical dreams of "fundamental education" and the "fight against ignorance" are still relevant in the world of today where we are far from having reached the goal of inclusive knowledge societies.

Notes

Disclaimer: The ideas and opinions expressed in this chapter are those of the author and do not necessarily represent the view of UNESCO.

1. All the issues dealt with here will be described, analyzed and discussed in greater depth in a publication which the author is preparing for publication.
2. In the staff lists of UNESCO, the Fundamental Education Division appears for the last time in 1958. In 1959 (staff list of 15 October) the staff that formerly worked there had become part of a new, larger "Education Out of School Division". The French term for fundamental education, *éducation de base*, was also abandoned. Much later in UNESCO's history "basic education" and *éducation de base* would be widely used.
3. See *Fundamental Education: Common Ground for All Peoples* (New York: Macmillan Co., 1947), 1–12, as well as Director-General Julian Huxley's foreword on p. vii.
4. Ibid., 5.
5. Memorandum on the education program of UNESCO, Paper No. 1, prepared by the Education Staff of the Preparatory Commission, May 13, 1946. UNESCO.Prep.Com./Educ.Com., UNESCO Archives.
6. Ibid.
7. Ibid.
8. *Monographs on Fundamental Education, No. 1: Fundamental Education: Description and Programme* (Paris: UNESCO, 1949), 7–8. This publication is a revised version of a document prepared after the second session of the General Conference. See F.E./Conf.6 Unesco, mimeograph, 59, February 1948, UNESCO Archives.
9. Ibid., 9.
10. *Fundamental Education: Common Ground for All Peoples*, 19. The quote is from Pearl S. Buck, *Tell the People* (New York: John Day Co., 1945), 11.
11. *Monographs on Fundamental Education*, 1, 9.
12. Ibid., 11.
13. Confidential information note from the UNESCO Secretariat to the Executive Board, UNESCO/Cons.Exec./9/1947, Paris, 31 March 1947, UNESCO Archives.
14. Working Paper UNESCO/2, 15 June 1956; WS/066.59. United Nations Administrative Committee on Co-ordination. Working Group on Community Development, UNESCO Archives.
15. Ibid., 5
16. 9C/ PRG/27, UNESCO Archives. The working group, drawing on "the fullest and most detailed explanations" of the chief of the Division of Fundamental Education, John Bowers, also reached the conclusion that the specific characteristic of fundamental education was "that it is active". Thus "fundamental education would have little real or lasting effect unless it succeeded, in the end, in making every individual conscious of being at once a citizen and a producer, unless it inspired in him at once a feeling of his fellowship with his community and a sense of his own personal dignity".
17. Ibid., 11, annex. The entire definition proposed by the working group was the following: "Fundamental education aims to help people who have not obtained such help from established educational institutions to understand the problems of their environment and their rights and duties as citizens and individuals, to acquire a body of knowledge and skill for the progressive improvement of their living conditions and to participate more effectively in the economic and

social development of their community. Fundamental education, which must have regard for religious beliefs, seeks to develop moral values and a sense of the solidarity of mankind. While the object of the school is to educate children, and while 'further education' continues the education of persons have previously acquired basic training in schools, fundamental education is designed to supplement an incomplete school system in economically underdeveloped areas, both rural and urban."

18. As an example, the General Conference at its third session in 1948 adopted a resolution on fundamental education "giving priority to less developed regions and to underprivileged groups within industrialized countries", according to 3C/Res. 2.42, UNESCO Archives. The idea was, as pointed out by Wyn Courtney and Gerhard Kutsch in an unpublished study on educational policies produced for UNESCO in 1980, "that rich countries had a duty not only to the poorer countries but also their own poor". See Courtney and Kutsch, "Major issues in educational policy since 1946 as perceived by the Member States of UNESCO" (unpublished study), 9, UNESCO Archives.
19. 10C/Records, p. 153, UNESCO Archives.
20. 10C/Res. 1.51, UNESCO Archives: "The General Conference *Recognizing* that the term 'fundamental education' has led to confusion, *Instructs* the Director-General to take immediate steps to secure that a proper terminology which can be applied all over the world be used by UNESCO for all kinds of education of adults and young people and to discontinue as rapidly as is feasible the use of the term 'fundamental education' in all official documents of UNESCO." Slightly ironically, the following resolution (1.52) stresses the need to extend "fundamental education" – so, old habits die hard. Nevertheless, the term does cease to be used widely after 1958. For example, where it still appears in 21 occurrences in the records of the General Conference of 1958, this is only the case four times at the subsequent session in 1960.
21. ACC/WGCD/Working Paper No. 4/Add.1 UNESCO, UNESCO Archives. The observation continues: "Although the term 'fundamental education' still persists in a number of Member States, there is an increasing acceptance of 'adult education' and 'youth activities', as being sufficiently broad to cover the whole range of out-of-school education, with whatever adaptations may be necessary", according to p. 2.
22. Due to printing delays, only the first and final chapter were submitted to the delegates in 1946.
23. The activity was approved as part of the report of the Program Commission with the following wording: "A *Programme of Fundamental Education* under the direction of the Secretariat and a panel of experts, to help to establish a minimum fundamental education for all. See publication: 'Fundamental Education: Common Ground for All Peoples'. To be begun in 1947", according to Document 1/C/30, 223, UNESCO Archives.
24. "The UNESCO Literature on Fundamental Education" by Francisco José Monsant – Seminar Study at the University of California, 1958, Theses/12, UNESCO Archives. For a contemporary description of the Associated Projects scheme, see *The UNESCO Courier* (March 1951): 2.
25. UNESCO/ED/86; Paris, 16 November 1950, UNESCO Archives.
26. 5C/Res. 9.114 and 5 C/Res. 9.115, UNESCO Archives.
27. Report of the Program Commission of the joint meeting of the budget committee and the working party on the special project for the creation of a world network of

regional fundamental education centers in 6 C/PRG/29, 1951, UNESCO Archives. UNESCO's total regular budget for 1952 was USD 8,718,000 according to 6 C/5, Approved Programme and budget for 1952, UNESCO Archives.

28. *Men Against Ignorance* (Paris: UNESCO, 1953), 38.
29. Report on reading materials by the director of the Indonesian National Seminar for Reading Materials to the minister of education of Indonesia, 15 December 1958, CPX/REP.3/231: T. Krishnamurthy – Fundamental Education, UNESCO Archives.
30. Special project for the establishment of a world network of regional fundamental education centers, 1951, 6 C/PRG/3, UNESCO Archives.
31. Letter from Bernot to Lestage, Education Department, accompanying the report on January 1954, in Reg. file 375 (729.4) A61 F.E. Haiti – Pilot Project. Part XXI, UNESCO Archives. The quotes are translated from the French by the author. Alfred Métraux, outstanding anthropologist of Swiss origin, came from the UN to UNESCO in 1947, where he led the Marbial Valley anthropological survey (1948–1949), which was designed to be the basis of the pilot project. This survey was seen as a model for other fundamental education projects around the world. Métraux continued working for UNESCO after his essential contribution to the Haiti project and played an important role in the organization's work on race relations in the 1950s.
32. Opper to Elvin, 19 February 1954 in ibid.
33. Elvin to Bernot, 24 February 1954 in ibid. Translated from the French by the author.
34. Confidential mission report from André Lestage to the director-general of UNESCO, 17 April 1959, entitled "Mission en Haïti". Reg. file 375 (729.4) A61 F.E. Haiti – Pilot Project. Part XXI, UNESCO Archives. The quote is translated from the French by the author.
35. Reg. 1, Dossier 375 (729.4) A61 F.E. Haiti – Pilot Project. Part XXI from 1/XII/1953, UNESCO Archives. The first three sentences of this quote have been published on the UNESCO website in "UNESCO in the Making", written in 2014 by the author for UNESCO's 70th anniversary, http://en.unesco.org/70years/unesco_making (accessed 30 March 2015), and these lines were at that point translated from the French by UNESCO. The rest of the quote is translated from the French by the author.
36. Report on reading materials by the director of the Indonesian National Seminar for Reading Materials to the minister of education of Indonesia, 15 December 1958, CPX/REP.3/231: T. Krishnamurthy – Fundamental Education, UNESCO Archives.
37. Articles in *The UNESCO Courier* (June 1949 and September 2010).
38. Minute sheet from Conrad Opper to Jean Guiton, director of the Department of Education, 25 March 1954, Reg. file 375 (729.4) A61 F.E. Haiti – Pilot Project. Part XXI, UNESCO Archives.
39. Letter from Emily Hatch to John Bowers, head of the Fundamental Education Division, UNESCO, 19 May 1951, Reg. 1; 375 (540) A573 F.E. – India – Advisory Mission, UNESCO Archives.
40. For an interesting and critical reflection on early UNESCO fundamental education activities, see: Glenda Sluga, *Internationalism in the Age of Nationalism* (Philadelphia: Pennsylvania Press, 2013), 108–111.

8

Education for Independence: UNESCO in the Post-colonial Democratic Republic of Congo

Josué Mikobi Dikay

During the 1950s, many new states emerged and changed the entire balance of power within the international organizations, adding to both their willingness and their ability to help. Support came first and foremost via the UN and its specialized agencies' programs for technical assistance, which aimed to help with the economic development of new as well as already existing countries around the world, and through which also UNESCO slowly but surely expanded its field operations. In these early years the organization could offer technical assistance to requesting governments within the fields of elementary education, fundamental and adult education, technical education and science, and the work would be carried out by mission experts sent and employed by UNESCO.[1]

Numerous programs and projects were launched, in Africa not least after the UNESCO Conference of African States on the Development of Education in Africa in Addis Ababa, which took place in May 1961 and whose outcome was an inventory of educational needs and an action program corresponding to the needs drawn up in the Plan for African Educational Development, which would guide the work of the organization over the next 20 years. The conference is still considered and remembered by many as the real starting-point of the vast movement of development aid for education in Africa.[2]

Besides the general policy and its continuous implementation, which still has a huge impact in Africa, there were also – from time to time – special tasks that needed special attention. That was the case in post-war Korea and in the present-day Democratic Republic of Congo [hereafter Congo] after the Belgians had left the country. This chapter explores UNESCO's involvement in the latter, where the organization came to play an important role in the UN mass operation of the early 1960s, and in many ways Congo became an early testing ground for UNESCO's subsequent activities all over the Sub-Saharan continent.

The first years of training and learning of new freedoms

The first months and years after the proclamation of the independence of the new country in Léopoldville in June 1960 were not only formative years and a time for the Congolese to get used to the new freedoms, but were also marked by conflicts and instability, and of institutional and political crises – the so-called Congo Crises – which called for the intervention of the UN. The UN sent peacekeeping forces, its first peacekeeping mission with a significant military force, but the Soviet Union also made propositions for the support of Patrice Lumumba, the young Congolese prime minister, which tricked the USA. All of a sudden, the post-colonial Congo was a battlefield for the Cold War and a center of violent confrontations in the heart of Africa.

Until 1960 the majority of employees in the administration of Congo had comprised Belgians, and their hasty exit had left the country without a functioning administration and an educational system in deep crisis, leaving it all for unexperienced political leaders. In fact, various accounts of the time before independence show that the only Congolese with an education of some kind were almost entirely to be found in the field of teaching, and not on the higher levels of society, so there were only a few intellectuals and Congolese qualified for administrative jobs that demanded background knowledge of some kind.

It was under these circumstances that the US National Commission for UNESCO elaborated a program for the promotion of education in Africa and made provisions for urgent scholarships for Congolese, as well a plan for long-term support and exchange in the field of education.[3] In fact the Americans offered 300 scholarships, and the initiative was given high priority in the State Department in the shadow of the Cold War; and also in the UN, where Dag Hammarskjöld, secretary-general of the UN, immediately gave UNESCO's director-general, Vittorino Veronese, instructions regarding how the organization could take part in an emergency operation.[4]

There was obviously a direct connection between US foreign politics and the actions of the international organizations, which gave certain – not least financial – prospects for future action on African soil. That bond was only strengthened by a speech of the US president, Dwight D. Eisenhower, in September in front of General Assembly of the UN. "I suggest that the United Nations undertake an immense effort for helping African countries to develop education according to the methods of their choice," he proclaimed. "It is not enough to have loudspeakers on public places for exhorting people to freedom. It is also essential that population have necessary intellectual fitting for protecting and developing their freedom."[5]

After consulting with the Congolese Government, it was decided that the UN engagement should go further than just sending peacekeeping forces, and that Sture Linnér, chief of the UN's civil operations in Congo, should be helped by a consultative group of experts in a range of fields, including public administration, agriculture, foreign trade, communication and press, finances and budget, instruction of national security's forces, justice, work market, national resources and industries, public health and – not least – national education.

At the same time Congo went through the admission procedure very quickly, and became a member of UNESCO in November 1960, only five months after its independence and in recognition of the urgent situation in, and the educational needs of, the country. That year the number of member states increased from 83 to 101, and of the 18 new member states, 17 were African – completely changing the power balance and future priorities of the organization. The decolonization process was all of a sudden becoming a center point of attention in the debates and a main priority for UNESCO.[6]

At that time the assistant director-general, René Maheu, had already been to Leopoldville – today's Kinshasa – to initiate a precise and exact inventory of the needs of the Congolese authorities, and he presented them all in a report to UNESCO's executive committee. In December 1960, UNESCO's action plan for Congo was accepted.[7]

Despite – and because of – the complex situation, UNESCO immediately intervened in order to take action and to help stabilize the country in a difficult period in its short history. In this chapter I will trace the history of UNESCO's role in the UN's technical assistance program to understand the nature of the partnership between UNESCO and Congo in these early, formative post-colonial years, when it was undoubtedly an important player in stabilizing and reshaping the country. I will therefore look at how the organization helped the leaders of the first republic to reform the educational system inherited from the former colonizers and helped to solve the Congolese crisis during years in which the school system went through a profound transformation.

The relationship between Congo and the UN and its specialized agencies has never been the subject of a scholarly study. The absence of such works is in a way justified by the confidentiality and secrecy still surrounding some of the official holdings of files on these operations, and to illustrate the poverty or non-existence of academic research on UNESCO's intervention in Congo, a search for doctoral theses between 1960 and 2005 reveals only one, besides my own, written by a scholar at the University of Geneva in 1983 and never published.[8] Also, books and articles about the topic are almost non-existent, so it is in any sense a new subject in Congolese historiography, and to take it up will bring light to a topic not only important to the history of UNESCO and Congo but appreciable to the history of Sub-Saharan Africa in general, and to the history of basic education in all its dimensions.

Planning independence

The policy of development in the developing countries sometimes intensifies already existing political troubles and contributes to maintaining social injustice, such as when aid comes in the shape of loans from developed countries with political ambitions and expectations, which widen the gap between the rich and the poor due to subsequent debts that sometimes reach sums that are impossible to pay back.

It was different with UNESCO when it intervened in Congo. It came as part of a larger, international undertaking, the UN technical assistance program, which was put into action in 1949, merged with UN's special funds in 1958 and eventually became UNDP in 1965. In the case of Congo there was even made a special UN fund for Congo which lasted from 1960 to 1968 and was earmarked only for helping the country to become truly independent. However, the help did not come in the shape of financial loans or donations; it came in the form of physical and moral support for educational activities, because UNESCO's approach to aid of any kind was that it would be effective only if it was accompanied by parallel developments within the field of primary and secondary education in the first place but in fact in education on all levels. Education was a vital key to lasting economic development, and vice versa, because economic factors would also have direct consequences for people's ability to get an education, especially considering the costs of education for the poor populations of Africa.

Director-general René Maheu's plan consisted of several projects to be carried out right away. The mission was led by Maurice Dartigue of the UNESCO secretariat, and he and the other UNESCO experts went to Leopoldville to take up posts in the various ministries and departments as *ad hoc* specialists, and until Congolese with expertise could be recruited in large numbers, UNESCO's civil servants would often in practice take the lead in the fields of primary, secondary and vocational education, higher and university education, communication, literacy and adult education, scientific research and many other activities.

On the agenda was the immediate reopening of schools, where most of the teachers had been Belgian, accelerating the training of staff for primary and secondary teaching, recruiting teachers, training inspectors, organizing the inspection of schools, and an entire reform of the educational system, including the outline of national laws within the field of education and the planning of future educational development.[9]

UNESCO's contributions in Congo

The world's attention focused almost entirely on the military aspects of the situation in Congo at that time. The new civilian operations of the UN and its specialized agencies received much less attention, but the UN believed that the civilian program would in the long run prove to be of greater significance than the military operations, because it would make the foundation for a new efficient administration that could stabilize the situation in the country.

A range of agencies under the UN umbrella took part in this effort. The WHO sent doctors, nurses and medical technicians to staff the hospitals left by the Belgians. Experts from the FAO came to set up a training school for agricultural mechanics and to help control epidemics of cattle disease. Technicians of the International Telecommunications Unit came to help re-establish and service telephone, telegraph and radio communications. The International Civil Aviation Organization sent specialists to man the airport control towers while training Congolese replacements. Also the ILO, UNICEF, the Universal Postal Union and the World Meteorological Organization sent expertise to help and guide the country in the direction its political leaders wanted.[10]

The team sent by UNESCO consisted of three experts in the field of general services, seven experts in the field of general education, primary as well as secondary, six experts in the field of technical and special education, three experts in the field of teacher training and five in the field of pedagogical studies and planning. However, the team faced difficult conditions on the ground right from the beginning: difficulties of communication, isolation of some provinces and their administrations, general insecurity, tribal struggles, famine and many educational establishments that had closed around the country, such as in North Katanga, South Kasai and Kivu. The whole situation led the chief of the Department of Education in Paris to describe Congo as the country in Africa "most difficult to develop".[11] Within the field of public administration, the UNESCO experts nevertheless helped the Congolese Government to formulate a development plan and a constitution, and to reinforce the role of the state in the creation of future education policies, and trained civil servants who could carry them out.

Also the experts in primary, secondary and vocational education faced the harsh realities of early post-colonialism. They found a country with 1.5 million pupils in the primary schools, but only 3,500 of the country's 16,000 schools providing teaching beyond the second grade, and some 70 percent of pupils left school before completing four years and were thus destined almost inevitably to return to illiteracy. In fact only 152 Congolese had completed their secondary education that year.[12]

At the same time, the first Congolese politicians, especially at the local level, proved to be slightly difficult to cooperate with, mainly due to their lack of experience, but post-colonialism was also something new to the UNESCO representatives, and Maurice Dartigue expressed his worries about to what degree he could interfere in political affairs, when it sometimes seemed necessary in order to effectively revive the educational system.

Of course it would have been easiest to overtake the entire leadership of the country and reform system, and then slowly reinforce the role of the Congolese in the formulation of their own educational policies and give them a chance to take over the administration and all practicalities.[13] In fact, it was discussed seriously whether it would be a good idea simply to "let action go before all" and practice "direct management" without any interference even though that meant, for example, that the local governments would for a while be "doomed to powerlessness". In fact, some of the Congolese also from time to time expressed a similar opinion in recognition of the work being done by UNESCO, and the minister of education at one point told a representative of the organization: "Your experts give advice to us and good advice, but I would rather that they took decisions."[14] It never came to the point of a temporary recolonization, even if it was felt necessary in order to most efficiently educate the Congolese to help themselves. To do everything in coordination was after all the best way to respect the sovereignty of the new country and make sure it would be stable in the long run.

As a way to improve communication, a Congolese Reform Commission was appointed in February 1961. It was headed by Henri Takizala, the secretary-general of the Ministry of National Education, and included representatives of the administration and of the various branches of the educational institutions in the country, as well as UNESCO advisors who could help the commission to formulate the new constitution's articles on education. It also proved to be an advantage with the formation of a new government with Prime Minister Cyrille Adoula, and the young and vigorous Joseph N'Galula as minister of education, in August 1961 – the same time as P.C. Terenzio took over as the head of UNESCO's mission.

One of the reasons for the new government's strong collaborative commitment was that it refused to depend on any Belgian advisers and therefore turned to the UN for help. New requests were therefore sent to UNESCO, while the Reform Commission worked at top speed to set up a strategy to train enough Congolese to replace as many of the foreign technicians and specialists as possible. All together, the development plan on education suddenly seemed to have a reasonable chance of success, thanks to the good relationship between the government and UNESCO, and due to wide national support.

However, there was a problem that had to be solved right away. Before independence virtually no Congolese had been trained as secondary schoolteachers and the small handful that had were needed desperately for positions in the administrations, so there was no basis for secondary education at all, and the first teachers trained by the UNESCO experts would not graduate before 1964. The government therefore had to accept a temporary solution that allowed all Belgian teachers to stay. A deal was settled with the Belgian Government that these teachers could continue in their positions, and in fact an increased number of Belgian teachers in Congo helped to stabilize the situation in the years that followed. In 1961–1962 the Congolese authorities estimated that there were 1,100 of them. To this figure one can add around 1,000 Catholic and Protestant missionaries or religious teachers. At the same time, UNESCO helped the Congolese Government to recruit French-speaking teachers, and their number rose from 66 in 1960–1961 to nearly 800 three years later. These teachers were people from all parts of the world, representing 25 different nationalities, and those who wanted to help the country and to have the experience of living in Africa. The teachers altogether met the basic needs of Congo's system of secondary education. A few Congolese and about 30 foreign inspectors traveled round the country to check the quality of teaching and to offer advice.

The teachers taught in very different conditions. In the larger cities there were well-equipped schools, whereas in some of the provinces the students had to bring their own chairs from home every day. At the same time, the existing system was designed primarily for European children and used European curricula, teaching methods and textbooks, and the children were not taught in a much different way because of a tremendous shortage of books, laboratory equipment, visual aids and school supplies in general. However, the immediate effects of UNESCO's presence were massive. Enrolment in secondary schools rose tremendously – from barely anybody attending in 1960 to more than 90,000 Congolese children attending four years later.[15]

To train new Congolese teachers at the secondary level, the government had given top priority to the creation of a National Institute of Education, which opened its doors in December 1961. It was headed by a Congolese director and staff consisting of UNESCO experts from 12 different countries, and in 1963 there were 200 students at the institution. Its program of studies, in addition to the normal teacher's college subjects, placed emphasis on African linguistics, African and Congolese societies and institutions, cultural anthropology and sociology, and by stressing Congo's own heritage it would become what the minister of education, Joseph Ngalula, called "an instrument of mental decolonization".[16] The only problem was that there was a shortage of well-educated people everywhere, and the new teachers often left the teaching profession again, being strongly attracted by the administrative services and by private undertakings where they found better working conditions and higher salaries.

UNESCO also helped the Congolese Government to improve primary education through curriculum reform and by organizing mobile educational units that travelled around the country – an area the size of Western Europe – to give four-week refresher courses to selected primary schoolteachers, principals and inspectors in remote provinces on topic such as hygiene and nutrition, French, arithmetic, and education methods in history, geography, science and manual training. By April 1963 more than 1,702 teachers had taken part in courses at 14 different sites across the country. Other experts formed the School Reconstruction Office and moved around in jeeps to supervise the young nation's ambitious school-building program.

Educational radio was produced and audiovisual aids developed, but even more significant was the opening of a Purchasing and Distribution Centre for Scholastic Materials in February 1963, led by UNESCO expert Gaston Lambot. This aimed to bring some order into the haphazard system of buying school supplies which had previously prevailed, and standardize the price of school equipment, books and supplies throughout Congo. Later a textbook office and school publishing office were created with the purpose of creating textbooks for and about the Congolese.

Measures were also taken to secure technical training, and the National Institute of Building and Public Works was created by the Congolese Government with the help of UNESCO. The institute opened in Leopoldville with 73 students, and it soon proved important to overcome the shortage of technicians which faced the country after independence. Another key creation was the National Institute of Mines, located in the green hills of Bukavu above Lake Kivu on the eastern border of the country. The site was motivated by the fact that Bukavu town is next to the pewter and gold mines of Kilomoto and close to the copper, cobalt, pewter and zinc mines of Katanga, and from a geological point of view Kivu offered enormous potential in the search for new metals and energy resources.

Also the pre-existing research institutes, such as the Institute for Scientific Research in Central Africa founded by Belgian scholars at Lwiro in Congo's Kivu Province, as well as the Institute of National Parks and the Institute for Agronomic Studies, received expert help and equipment. An expert team of four, headed by Antoine DesRoches, was attached to the Office of the Minister of Information to help train editors and newscasters, and to improve the scope and effectiveness of the educational programs for journalists.

Aid was also given to the University of Kinshasa and the University of Lubumbashi, and UNESCO helped with the foundation of a law school, a medical education institute, an institute of meteorology and geography, and several other institutions. At the same time new structure for administrative services was created. Most noteworthy was that at the central level a single directorate for education was formed to ensure coordination of education. In 1965 more than 2,300 Congolese attended higher education.[17]

"With UNESCO's help, the Congolese government has adopted a global approach to education," the organization reported back to its headquarters in 1964.[18] The biggest problem by 1965 was in fact the primary schools, not so much in their number but with regard to their quality. During the previous five years it was made possible for many more pupils to attend primary school and new syllabuses were introduced. However, the new teaching regime required competent staff, and the proportion of qualified staff did not show any significant increase, at least not until the pupils who entered secondary schools had completed their studies. In that sense the development of the Congolese educational system was going to be a slow transformation.[19]

Development of a new partnership between Congo and UNESCO

As a way to better involve people on the ground and make the transmission of information easier between UNESCO and the member states, a membership of the organization involved the creation of a national commission. The national commissions could provide UNESCO with precious analyses of the situation in their respective countries and help it to implement some of the numerous initiatives. They could also develop new partnerships and reach out to the private sector with its sometimes precious technical expertise and financial resources. Members of the national commissions were not only competent governmental administrators but also representatives and distinguished personalities from every corner of the intellectual environment, whether it was teachers, scientists, technicians, artists, writers, journalists or lawyers.

A preliminary meeting that had the aim of creating a Congolese national commission for UNESCO took place in Leopoldville in April 1963 under the leadership of P.C. Terenzio, chief of the organization's mission in the country and former manager of its bureau for relations with member states. He was not successful – at least not before the country was suddenly exposed to a big change in Congolese politics. This took place when the UN's operation in Congo ended in June 1964.

Although the military phase of the UN operation in Congo had ended, civilian aid continued in the largest single program of assistance undertaken by the UN and its agencies by that time, with some 2,000 experts at work in the country at the peak of the program in 1963–1964. However, as the troops withdrew, political upheaval and conflict arose again to an even greater extent, causing thousands of deaths, with the Soviet Union and the USA supporting opposing factions.

The Congo Crisis only ended with a *coup d'état*, and suddenly the entire country was under the rule of Joseph-Désiré Mobutu. He came into power in November 1965 with the support of lobbies of the great powers, and by proving to be skilled in the exercise of political power he became an "undisputable guide of [the] Zairian revolution" and the transformation from political sovereignty to include cultural sovereignty, which in June 1967 involved the creation of a Congolese National Commission for UNESCO, the aim of which has since been to enable the Congolese to take initiatives to realize the goals of UNESCO's programs and actions, and develop regional and subregional cooperation in the fields of education, science, culture and information.

Not only did this put an end to the time when Congo could be divided into exact spheres of influence, where outside powers could consider a part of the country as their own hunting reserve, but also it marked a change in the relationship between UNESCO and the national actors, which would from then on take charge themselves and have responsibility for Congolese affairs. A new type of diplomacy began, not only through the National Commission for UNESCO in Congo but also via the Permanent Delegation of Congo to UNESCO in Paris and the Congolese membership of the executive board. Added to that was the feeling of empowerment on a local level via the establishment of UNESCO chairs and UNESCO Associated Schools or via UNESCO networks and UNESCO clubs, which promoted the visibility of UNESCO but on Congolese terms and which at the same time fostered dialogue between government actors and NGOs. At that time, 2.5 million pupils attended primary school, a million more than seven years before, while 152,000 attended secondary education and more than 5,000 higher education, compared with almost no one when UNESCO first entered.[20]

The organization would never again play such a visible role because most of its work now went through a national counterpart, but its presence would still make a big difference. In particular, the literacy and education campaigns of the late 1960s, which had as their aim to teach the greatest possible number of people to read and write, have had a lasting impact, not only on the stability of the country but also on people's lives. In 1968 alone Congo had 50,089 illiterates enrolled with 33 full-time and 690 part-time instructors in its UNESCO-sponsored national literacy program, and the same year the National Literacy Service brought out the first issue of a newspaper published for new literates.[21] These campaigns brought vital social changes and an adult population whose active participation is essential to the country, and made it possible to transfer fundamental educational management to local communities and to reinforce associations of parents of pupils. Just learning to read and write has been crucial to a range of people, making it possible for someone to fill in an enrolment form for one's child, read a medicine prescription and deal with a range of other issues common to the daily life of the Congolese people.[22]

Not much has been written about UNESCO's role in Congo, so this chapter offers a first overview of the relationship in the early 1960s and shows that UNESCO at that time came to play a vital role in laying the foundation for a fully independent country in the wake of Belgian rule. Basically the entire educational system was built up or renewed by the organization in the early 1960s due to an emergency program for teacher recruitment, via adult education and school-building programs, and by developing institutions of higher education and coordinating scientific research. Later, the Congolese took over and continued the relationship with UNESCO on their own terms, not least with the establishment of a National Commission for UNESCO in 1967.

The first initial impact of UNESCO's technical assistance program is in many ways relatively easy to trace in the sense of numbers and that some of the earliest buildings still exist and are in use. It is also worthwhile determining the exact and overall long-term mental effects of the school's existence. I have tried to give an estimate but a lot of data will have to be analyzed to deliver a more accurate account of the wider impact.

However, there is no doubt that UNESCO has played, and still plays, a role due to its good reputation, help and positive actions, and has remained a much favored partner. It is, for example, a fact that higher education in Congo has developed significantly since the organization's first involvement and that the universities have remained the centers of excellence which train the country's teachers, learners, politicians, officials and developers, and that their numbers are still growing. The country now has more than 20 universities and many senior technical and educational institutes, many of which have been established with the continued cooperation of UNESCO. For example, in Kisangani, the African project of the Regional Postgraduate School of Integrated Management of Tropical Forests is a direct outcome of cooperation with UNESCO. I should also mention the effect of UNESCO's policy of training university professors by sending young assistants abroad, and the role of the so-called BEPUZA (Organisation des études post-universitaires, Zaïre) program, a special initiative from 1987 the aim of which has been to raise the standards of instruction, promote the collective use of trained personnel and coordinate higher-education policies in the African UNESCO member states. At the same time, the organization has introduced several international conferences, seminars, workshops and working groups attended by African leaders in cooperation with national, regional and international bodies.

Having said that, it is also important to stress that UNESCO only initiated the process and that today there are still many people who have not been taught how to read and write. Social indicators relating to Congo show that the illiteracy rate is still way over 50 percent of the population. Most of those affected are women, which gives an extra dimension to the problem since they are not only subject to discrimination and exclusion but also have an important place in the household and production, so they are in many ways the key to further development.

The relationship between UNESCO and the Congolese Government has over the years been passionate, sometimes complex but always close. It has been passionate because it was UNESCO that brought hope to Congolese children by making it possible for them to go to school again and because most Congolese recognize that without UNESCO's technical assistance program there would not been reliable secondary education in Congo today. It has been complex because conflicts between UNESCO and the Congolese

Government arose from time to time. Finally, it has been close because UNESCO has often proved to be the only answer to the specific needs of the Congolese people.

Again I should like to highlight the fact that this chapter is only a first overview of the history of this relationship, and that many scholarly questions remain unanswered. For example: To what degree was UNESCO's action politically motivated and implemented to create a Western version of Africa? What was the role of Belgian professor Jef Van Bilsen and his 30-year plan of 1955 in creating a self-sufficient independent state out of the Belgian Congo for the policy implemented by the UN and UNESCO in Congo, and what was the role of his own students at the University of Kinshasa in this process? What was the local impact of the specific missions and structures created by UNESCO and to what degree did they help the country to solve the problems of underdevelopment, debt, wars and malnutrition? What role did UNESCO play with regard to the country's and many Africans' self-perception that the power balance and focus of UNESCO changed in favor of Africa in the 1960s, and in Senegalese Amadou-Mahtar M'Bow later becoming the first African director-general of UNESCO? And what was the impact of the Congolese experience in the development of subsequent projects in Africa and elsewhere? There are plenty of questions for future historians to take a closer look at.

Conclusion

It is therefore not a surprise that Africa is still a continent with crises of all kind: epidemics, political instability, genocide and corruption. It is also a place where the international community can pretend to reduce poverty without facing illiteracy, even though it seems to be the cause of many of the country's problem and a key reason for poverty in Africa. In that sense, the former colonies across the continent, including Congo – (even with the help of UNESCO) still have an enormous task ahead of them.

Notes

1. UNESCO: Tentative Proposals for Technical Assistance for Economic Development, 20 April 1950, UNESDOC.
2. UNESCO's Assistance in Educational Planning in Africa 4 September 1963, UNESDOC, and *International Symposium on 60 Years of History of UNESCO,* November 2005 (Paris: UNESCO, 2005), 42, 447, 477–478.
3. "Scholarships and Statesmanship", *Africa Today* 7:5 (1960): 4.
4. Letter from Dag Hammarskjöld to Vittorino Veronese 11 August 1960, 482.DG/214; UNOC/UNESCO/482, UNESCO Archives.
5. Newsletter 7:22, 31 October 1960, cited in Gail Archibald, *Les Etats-Unis et l'UNESCO, 1944–1963: Le rêves peuvent-ils résister à la réalité des relations internationales?: Série Internationale,* 44 (Paris: Sorbonne, 1993), 269.

6. 11/C, Rec. 8.2.1(c), UNESCO Archives.
7. 57Ex/22; 57EX/SR.14 1960; 59EX/8 December 1960, UNESCO Archives.
8. Mbuyu Mujinga Kimpesa, *L'Opération de l'UNESCO au Congo-Léopoldville et le diagnostic des réalités éducatives congolaises, 1960–1964* (unpublished dissertation) (Genève: Université de Genève, 1983) and Josué Mikobi Dikay, *La Politique de l'UNESCO pour le développement de l'éducation de base en République démocratique du Congo, 1960–1980* (unpublished dissertation) (Orléans: Université d'Orléans, 2008).
9. "UNESCO Aid to the Congo (Leopoldville) within the Framework of the Civilian Operations of the United Nations", 14 December 1960, 11 C/34, UNESCO Archives and "UNESCO's actions in the Congo", *UNESCO Chronicle* 4:12 (1960): 439–446.
10. Gary Fullerton, *UNESCO in the Congo* (Paris: UNESCO, 1964), Preface.
11. Report by R. Bergaud, Department of Education, UNESCO, on his mission to Leopoldville 21 May 1963–4 June 1963, UNESCO Archives.
12. Fullerton, *UNESCO in the Congo*, 20.
13. Memorandum no.176, "Suggestions for the elaboration of the outline law on the education", 9 November 1961, UNESCO Archives.
14. Report by R. Bergaud, Department of Education, UNESCO, on his mission to Leopoldville 21 May 1963–4 June 1963, UNESCO Archives.
15. *International Yearbook of Education,* vol. XXVII, 1965 (Geneva: International Bureau of Education, 1966), 98; Fullerton, *UNESCO in the Congo*, 11, 14, 42–43.
16. Fullerton, *UNESCO in the Congo*, 15.
17. *International Yearbook of Education,* vol. XXVI, 1964 (Geneva: International Bureau of Education, 1965), 87–89. *International Yearbook of Education*, 1965, 97–99; Fullerton, *UNESCO in the Congo*, 19, 21, 25–40 and Paul-Henry Gendebien, *L'intervention des Nations Unies au Congo, 1960–64* (Paris: Mouton et Cie, 1967), 185–191.
18. Fullerton, *UNESCO in the Congo*, preface.
19. *International Yearbook of Education*, 1965, 99.
20. *International Yearbook of Education,* vol. XXX, 1968 (Geneva: International Bureau of Education, 1969), 111.
21. *Literacy 1967–1969: Progress Achieved in Literacy Throughout the World* (Paris: UNESCO, 1970), 33.
22. This chapter is mainly based on Josué Mikobi Dikay, *La Politique de l'UNESCO pour le développement de l'éducation de base en République démocratique du Congo, 1960–1980* (unpublished dissertation) (Orléans: Université d'Orléans, 2008).

9
The Flow of UNESCO Experts toward Latin America: On the Asymmetrical Impact of the Missions, 1947–1984

Anabella Abarzúa Cutroni

Introduction

This chapter identifies some of the asymmetries in the impact of UNESCO's expert missions in Latin America. Not all the states in this region had access to the same number of experts for technical assistance, the thematic orientation of the missions was not homogenous in all the countries, and at the scientific level, certain disciplines received more attention than others. These asymmetries provide an insight into the difficulty of efficiently implementing UNESCO's program at the national level during its first 20 years of activity and reveal the significance of the support of the member states.

The first section analyzes how technical assistance for development was implemented by UNESCO's officers. It describes the UN Expanded Program of Technical Assistance in order to then analyze the distribution of experts sent on missions by states worldwide during the period studied, which is 1947–1984. The second section focuses on the impact of UNESCO's missions in Latin America to establish which member states of the region received the greatest number of experts, and the thematic orientations of the different missions and the scientific disciplines that benefited the most, first and foremost in the period 1947–1973. The third and final section briefly describes the case of mathematicians sent on a mission to Argentina to give classes at UNESCO's Regional Centre for Mathematics in Buenos Aires from 1959 to 1964.

Implementing the UNESCO program worldwide

UNESCO began work on its program in 1947 with a relatively limited budget. One of the fundamental aspects of the program involved sending experts to other countries across the world. Clarence E. Beeby, former assistant director-general and head of the educational program area of UNESCO from April

1948 to October 1949, offered details in personal testimonies about what he referred to as the "first mission of international expertise in the area of education". The former UNESCO officer argued that, based on his work in Western Samoa, a new type of international cooperation had been invented: jointly organized trips by specialists in education, and of different nationalities, so that they could advise local authorities about a particular theme by drafting a common report.[1]

In general terms, one could describe UNESCO's initiatives as activities aimed at technical assistance for development. The two fundamental goals were to train technicians capable of planning development and to implement the measures necessary to achieve development through advances in science and education. The technical assistance consisted in transferring technical knowhow that would allow the destination countries to modernize. These countries had to be willing to actively participate in the projects and to contribute to them financially, and thus the technical assistance from the developed countries was considered a way to help the destination countries to self-help.

The concept of technical assistance has a precedent in the Point Four Program announced by US President Harry Truman in his inaugural address in January 1949. Of his four points, the first reaffirmed US support for the UN, while the fourth sustained that the USA should undertake a program that should make the benefits of US scientific and industrial advances available for the growth of underdeveloped nations. Truman noted that although material resources were limited, the technological superiority of the USA was "inexhaustible".[2] According to Norwegian political scientist Olav Stokke, this discourse gave American delegates the green light to support the creation of the Expanded Programme of Technical Assistance at the UN General Assembly.[3]

A series of resolutions passed before the General Assembly were indicative of the need to improve the standard of living of the underdeveloped states. One resolution even declares that the lack of expert personnel was a factor hindering economic development. The UN, continuing the resolution, would appropriate the funds necessary in order for the secretary-general to organize international teams of experts for the purpose of advising governments in their economic development programs and arrange for the training abroad of local experts, among its other measures.[4]

The proposals from the previous resolutions led ECOSOC, "*impressed* with the significant contribution to economic development that can be made by an expansion of the international interchange of technical knowledge through international co-operation among countries", to recommend implementing the Expanded Program in August 1949.[5] The resolution on creating the Expanded Program established its guiding principles, and in an annex the technical assistance for economic development of underdeveloped countries "shall be rendered by the participating organizations", which would be

UN agencies, include UNESCO, "only in agreement with the Governments concerned and on the basis of requests received from them".[6]

The model of technical assistance that the UN adopted in 1949 partially structured the incipient UNESCO program. From 1945 until 1984, the organization's initiatives in Latin America and in other regions of the world had myriad facets. UNESCO sent the international experts; provided grants for training abroad; gathered statistics; trained or ordered training for local "experts"; organized regional and international scientific forums; published technical reports and other kinds of publication; and occasionally financed educational and scientific infrastructure, such as libraries, laboratories and schools, among its other activities. The scholars, diplomats and government officials involved colloquially referred to all of them as "technical assistance". However, the UNESCO officers took great care in the budget allocation corresponding to the program that financed the initiative based on the request from each state and did not use the term indiscriminately. The responsibility of these officers was to carry out part of the UN Expanded Program along with the organization's normal program and it Participation Program of 1956. One of their most pressing concerns during the first years of the organization's existence was how to successfully – that is, efficiently – implement this set of programs at the national level but on a worldwide scale.

UNESCO's intergovernmental nature makes the implementation of its initiatives particularly difficult at the diplomatic and bureaucratic level. In keeping with the order of "non-interference in internal affairs", UNESCO's officers had to exercise extreme care in order not to exclude any state due to its political system and to take into account the specific interests of each state. This diplomatic precaution created a pressing need to universalize the grounds of its activities, separating it from any kind of national or cultural favoritism with the aim of gaining legitimacy among its member states. However, this did not keep certain states from receiving more technical assistance than others, when these states could guarantee the implementation of the missions within their national territories.

In relation to the diverse activities and initiatives that technical assistance involved, UNESCO's network of institutional relations was broad and complex. Although the organization was sustained financially and politically by its member states and the UN, a series of NGOs also indirectly participated in its activities. Besides the institutional network, we must also consider the interpersonal networks which, though less evident, played a significant role in the success of UNESCO's initiatives, especially during the first years of its activities. Therefore the creation and consolidation of networks, and of institutional and personal contacts, represented a continuous and vital task. UNESCO's officers of the middle and upper ranks often went on tours to meet with different political and academic figures in each of the member states. The other areas of the organization were later informed of the outcome of these meetings, depending on the level of confidentiality of the information, through mission memorandums and reports.[7] The connections made in these meetings were consolidated at the national level by the national commissions, the experts sent on missions and former UNESCO fellowship recipients.

According to the information in the "Index of Field Mission Reports", from 1947 to 1984, UNESCO sent approximately 7,770 experts[8] on missions to 147 countries around the world as part of its Expanded Program and its normal program – including the Participation Program.[9] Among these experts we see certain names that appear several times due to the fact that some specialists were sent on several missions. The organization did not establish missions in the religious sense of the term, although it did "preach" certain ideas and knowledge. This is because the site where the experts worked at the institutional level was determined by the host state. There the specialists from UNESCO's different areas of expertise carried out a range of tasks, such as organizing conferences; teaching classes and seminars; coordinating the creation or reform of teaching and research institutions; launching science laboratories and research projects; and advising governments on a specific aspect of public policy. In most cases this was not a task they carried out alone since they worked in conjunction with scholars from the local universities and/or governmental planning organizations. Before a mission could be sent, UNESCO would sign a contract with the state interested in receiving technical assistance to establish the terms and conditions of the international cooperation.

The 1947–1968 index registers 3,300 missions during the period although it does not provide a count of the number of missions per year. Figure 9.1 shows the variation in the number of experts sent by UNESCO each year between 1968 and 1984. The number of experts sent on field missions worldwide peaked in the 1960s, while in 1974 it plummeted abruptly. This was probably due to financial problems (inflation, devaluations, etc.) that UNESCO and the UN were facing, and to the change in the strategies used by the superpowers to contain communism in the Third World. We are specifically referring to US support for *coups d'état* in Latin America and the resulting shrinkage of local academic fields.[10]

Considering all of UNESCO's member states between 1947 and 1984, the ten countries that received the greatest number of experts were India with 390, Iran with 208, Brazil with 192, Afghanistan with 187, Pakistan with 185, Egypt with 184, Thailand with 182, Indonesia with 167, Nigeria with 162 and Chile and Mexico each with 161. If we expand the list a bit further, we can see that 26 states received 50 percent of the experts sent on missions while 54 states received 75 percent. In general there are no major differences between the number of experts received if we compare any two states, with the exception of India and Iran. This reveals a relatively homogenous dispersion between the destination states of the UNESCO experts.

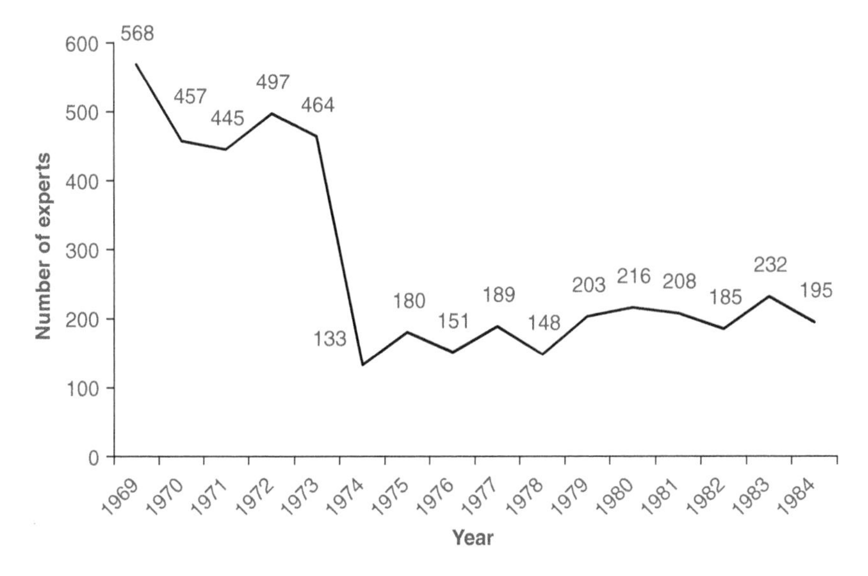

Figure 9.1 Number of UNESCO experts sent on missions worldwide, 1969–1984.
Source: Author's own elaboration based on the "Index of Field Mission Reports", 1969–1984, UNESCO Archives.

In terms of Latin America, the destination states that received the greatest number of experts are Brazil with 192, Chile with 160, Mexico with 161, Ecuador with 121, Venezuela with 117, Peru with 109, Colombia with 97 and Argentina with 80.[11]

Not all of UNESCO's member states received technical assistance. In fact, 55 states received scarce assistance or none at all during the period analyzed here. That was due to two principal reasons. The first was that they did not require assistance, for diverse academic and political reasons. This group included countries considered global education centers such as France, the USA and the UK, along with other "semiperipheral" academic centers such as Canada, Sweden and Belgium,[12] as well as the states belonging to the Eastern Bloc, such as the USSR, East Germany, Belarus and Hungary. The second was that they could not guarantee the successful development of the missions in their territories because of a lack of relatively established academic fields. A great number of underdeveloped states are included in this group, countries that generally did not attain independence until the 1970s and 1980s, such as Namibia.[13]

The UNESCO's main responsibility was to recruit experts, such as professors, administrators and engineers, capable of providing technical assistance. Unfortunately, the index does not list their nationality or institution, which makes it impossible to offer a quantitative analysis of the directional flow of experts for technical assistance worldwide during the period in question. However, we have some data on the Expanded Program that allow us to fill in the gaps and assume that the flow of experts was mainly North–South, although in some cases there was also an intraregional flow.

In 1952, given the rising number of requests from states, UNESCO re-evaluated its mechanisms for recruiting experts for the Expanded Program. In this process, during the Seventh Session of the General Conference, a meeting was organized with representatives from the states that "were providing the greatest quantity of experts" at the time. The participating states included West Germany, Australia, Austria, Belgium, Canada, Denmark, Egypt, the USA, France, Haiti, India, Italy, Japan, Mexico, New Zealand, the Netherlands, the UK, Switzerland and Yugoslavia. Uruguay participated as an observer.[14] This list provides information on which countries the UNESCO regularly requested experts from, or at least the countries with which the organization hoped to expand its recruiting sources.

Already known figures regarding the origin of the experts in the UN's Expanded Program includes the missions carried out by the secretary-general and by the UN's subsidiary agencies, including UNESCO's missions. Although the data we provide below are therefore not limited to the experts recruited by UNESCO, the origin of the experts generally coincides with the list provided in the previous paragraph. Olav Stokke, who has provided these figures, states that during the first 18 months of the organization – that is, from 1950 to 1951 – the recruited experts represented 61 countries, while in 1964 experts hailed from 87 different countries, including 13 from Africa and 22 from Latin America. Between 1949 and 1965, the year in which the UNDP began, 55 percent of the experts were recruited in Western Europe. The colonial powers – the UK, France and the Netherlands – were the main sources of experts. The Nordic countries were another important source, while the contribution of the USA was modest with 17 percent in 1953, 12 in 1960 and less than 10 percent in 1964. In terms of the peripheral countries, Stokke claims that certain countries of Latin America and Asia stood out, especially India.[15]

The impact of UNESCO's missions in Latin America

When the Expanded Program was approved, the main demands of the states of Latin America were associated with regionalizing UNESCO's activities. During the term of Jaime Torres Bodet, director-general from 1948 to 1952, UNESCO initiatives were mainly directed at operating activities that produced tangible results. Starting in 1954, Luther Evans, director-general from 1953 to 1958, reiterated UNESCO's presence in the region through personal visits. Toward the end of the decade in 1950, and during the 1960s, the training of "experts for development" was taken to the peripheral academic centers such as Chile, Mexico and Brazil. One of the most outstanding examples of this process was the creation of the Latin American School of Social Sciences (Facultad Latinoamericana de Ciencias Sociales (FLACSO)) in Santiago, Chile, in 1957.[16]

We will now analyze the number of experts sent on UNESCO missions to Latin America from 1947 to 1984 and their areas of expertise. From the total number of experts on worldwide missions during this period, altogether 1,700 or almost 22 percent were sent to Latin America. Although we do not have the annual numbers for 1947–1968, we do know that 137 experts were sent to the region during these years. From 1947 on, the number sent to Latin America increased progressively; certain indicators show that the rise was more pronounced after 1954 and that the 1960s marked the peak of missions requested by the region. Figure 9.2 shows the historic series from 1969 to 1984. The number of experts remained relatively stable from 1969 until 1973, and then plummeted in 1974. From that year on, the number again remained stable, never recovering the levels of the previous years. It is also important to note the abrupt drop in the requests from countries under military dictatorships. Chile went from nine experts sent in 1973 to two in

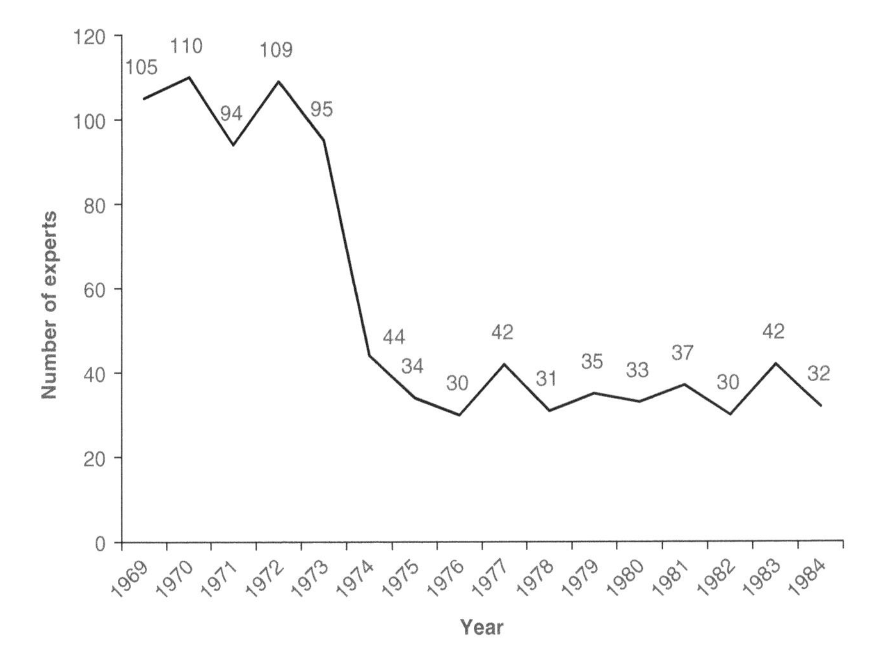

Figure 9.2 Number of experts received by Latin America, 1969–1984.
Source: Author's own elaboration based on the "Index of Field Mission Reports", 1969–1984, UNESCO Archives.

Table 9.1 Experts sent to Latin America by country, 1947–1984.

	Country	*n* experts	%	Accumulated %
1	Brazil	192	11.29	11.29
2	Chile	160	9.47	20.76
3	Mexico	161	9.47	30.23
4	Ecuador	121	7.12	37.35
5	Venezuela	117	6.88	44.23
6	Peru	109	6.41	50.64
7	Colombia	97	5.76	56.41
8	Bolivia	88	5.18	61.58
9	Argentina	80	4.71	66.29
10	Cuba	68	4.00	70.29
11	Costa Rica	64	3.76	74.05

Source: Author's own elaboration based on the "Index of Field Mission Reports", 1948–1984, UNESCO Archives.

1974, while Argentine requests for experts either remained stable or fell by one each year after 1966.[17]

Table 9.1 shows the distribution by country of the experts sent to Latin America. We can see that 11 of the 24 countries received 74 percent of the experts sent to the region, which indicates a certain degree of concentration of the missions in certain states. Brazil leads the table with 11 percent, followed by Chile and Mexico, both with more than 9 percent. Argentina is ninth on the ranking with less than 5 percent.

We will now focus on the thematic orientation of the UNESCO specialists sent to Latin America and of the request sent by the host countries. From 1947 to 1983, the indexes offered a uniform classification system for the thematic and disciplinary orientation of the results. Starting in 1974, the documents no longer included this classification and the comparison thus becomes impossible for that year on. Due to this methodological obstacle, we are only analyzing the thematic and disciplinary orientations for the period 1947–1973. Table 9.2 shows the orientation of the missions according to the topics of expertise of UNESCO. As can be seen, missions in the education sector, a category in which we include all initiatives aimed at primary and secondary education as well as school administration, have always been predominant in our region. Missions in education exceeded 55 percent every year, and for the entire period they represent 60 percent of the total. Under science missions we group a great range of missions in the natural and exact sciences, the applied sciences and engineering, the social sciences and missions providing consulting services on matters of higher education and science policy. However, the sciences never exceeded 30 percent of the UNESCO programs carried out in Latin America, and in the entire period they represent just less than 29 percent. The culture and

Table 9.2 Orientation of the missions carried out in Latin America, 1947–1973.

Orientation	**1947–1968**		**1969**		**1970**		**1971**		**1972**		**1973**		**Total**	
	N	%	*n*	%	*n*	%	*n*	%	*n*	%	*n*	%	*n*	%
Science	239	29.91	23	22.55	34	29.06	25	26.04	32	28.32	28	28.57	381	28.75
Education	494	61.83	64	62.75	65	55.56	56	58.33	65	57.52	55	56.12	799	60.30
Culture and information	66	8.26	15	14.71	18	15.38	15	15.63	16	14.16	15	15.31	145	10.94
Total	**799**	**100.00**	**102**	**100.00**	**117**	**100.00**	**96**	**100.00**	**113**	**100.00**	**98**	**100.00**	**1325**	**100.00**

Source: Author's own elaboration based on the "Index of Field Mission Reports", 1948–1973, UNESCO Archives.

information missions, which include those related to archives, publications, libraries, museums and certain cultural missions, such as those dedicated to preserving archeological sites represent just less than 11 percent of the total.

Now we will see whether this regional tendency changes if we analyze the orientation of the missions in the ten Latin American states that requested the most UNESCO experts. Table 9.3 shows us the percentages corresponding to each orientation category. Ecuador, Venezuela, Bolivia and Colombia – marked with asterisks – all exceeded the average percentage of missions in education in the region, which is about 60 percent – by at least 10 percent. Brazil, Chile, Mexico, Argentina and Cuba, in contrast, all exceeded the almost 29 percent average of science missions in the region, in amounts varying between 12 and 19 percent.

In terms of culture and information missions, Mexico, Peru, Colombia and Cuba exceeded the average for the region. Brazil and Uruguay were the only cases where science missions exceeded missions in education. Uruguay hosted relatively few missions – only 31 – but more than 77 percent of them were oriented toward science, possibly due to the Science Cooperation Office. Argentina and Cuba had relatively similar percentages for science and education missions, as did Brazil and Chile, though Brazil had slightly more science missions and Chile slightly more in education. In the cases of Ecuador, Peru, Venezuela, Bolivia and Colombia, there were clearly more education missions than science missions. These data indicate certain thematic areas of expertise in each of the countries as well as the concentration of science missions in the leading peripheral academic centers in the region at that time, centers with consolidated or rapidly developing science

Table 9.3 Missions by thematic orientation for the ten Latin American states that requested the greatest number of experts from 1947 to 1973.

State	**Science missions**		**Education missions**		**Culture and information missions**		**Total**	
	n	%	*n*	%	*n*	%	*n*	%
Brazil	79	47.02	72	42.86	17	10.12	168	100.00
Chile	60	42.25	70	49.30	12	8.45	142	100.00
Mexico	51	36.96	65	47.10	22	15.94	138	100.00
Ecuador	21	24.14	63	72.41	3	3.45	87	100.00
Peru	21	24.71	47	55.29	17	20.00	85	100.00
Venezuela	12	16.00	58	77.33	5	6.67	75	100.00
Argentina	30	46.15	32	49.23	3	4.62	65	100.00
Bolivia	12	19.35	46	74.19	4	6.45	62	100.00
Colombia	8	14.55	41	74.55	6	10.91	55	100.00
Cuba	19	40.43	21	44.68	7	14.89	47	100.00

Source: Author's own elaboration based on the "Index of Field Mission Reports", 1948–1973, UNESCO Archives.

institutes and universities. The UNESCO experts were an important resource in this process which started in the 1950s, but their participation would not have been possible without a certain level of local academic resources, since host professors and researchers were the ones who made each mission feasible.

Now we will more closely examine the distribution of the science missions by specific discipline in Latin America during the period 1947–1973 as it appears according to Figure 9.3.[18] Sociology is in first place with 16 percent, followed by earth sciences, mainly geology and hydrology according to observations from the mission reports, with 14 percent, and by marine sciences, mainly oceanography and marine biology. The percentages of other disciplines range between 1 and 10 percent and have been grouped in the categories "other social sciences" and "other exact and natural sciences", as appropriate. Some 11 percent of the missions cannot be defined as strictly scientific; these provided consultancy services for science and educational policy, general professor training at the university level, and scientific and technical documentation. The rest correspond to missions that we cannot classify by discipline because of their interdisciplinary nature, as in the case of the missions that were part of the Arid Zone Project. We can distinguish two major disciplinary areas: the social sciences, which represented 31 percent of the missions, and exact and natural sciences, engineering and applied sciences, which represented 56 percent.

If we observe each state's contribution to the different disciplines, we can identify a concentration of certain disciplines in certain countries and the relatively homogenous distribution of others among different states in the region. In relation to the social sciences as a thematic area – that being social sciences in general, sociology, economy, anthropology and other social sciences – Chile is responsible for the greatest number of missions with 31 percent, followed by Brazil with 21 percent. Chile was the main contributor to the category which would set a trend in the area of science, namely in sociology with 44 percent, followed by Brazil with almost 28 percent. This can probably be attributed to FLACSO and to the Latin American Center of Social Science Research in Río de Janeiro. Both institutions were co-sponsored by UNESCO. The number of economy missions stands out in Mexico, while anthropology appears as the prominent discipline in Peru and Bolivia.[19]

In terms of the area of exact and natural sciences, engineering and the applied sciences, the diversity is much greater than in the previous one. Based on the numbers, the country that led this area is Brazil with 20 percent, followed by Mexico with almost 13 percent. Chile had a relatively low incidence in this area, with a focus on engineering. However, earth sciences stand out especially in the requests from Brazil, which is closely followed by Peru and Ecuador, both countries that specialized almost exclusively in these disciplines. Argentina's area of expertise can also be seen in its requests for missions oriented towards math and physics, and, as a result, the

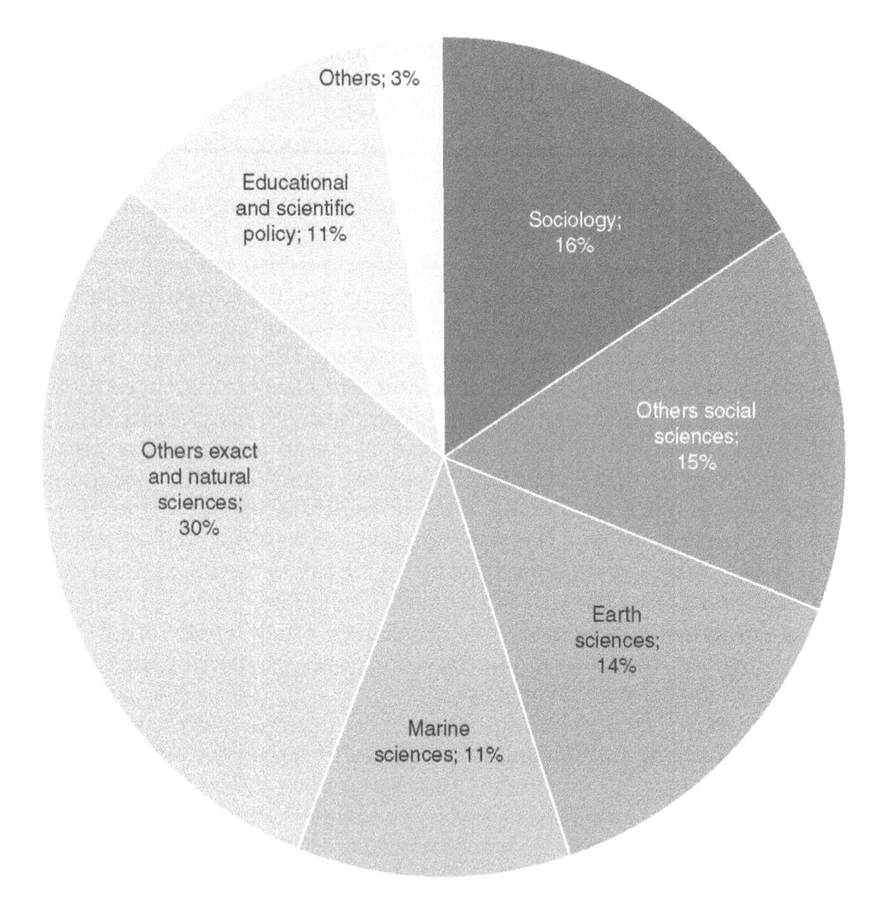

Figure 9.3 UNESCO science missions in Latin America by discipline, 1947–1973.
Source: Author's own elaboration based on the "Index of Field Mission Reports", 1948–1973, UNESCO Archives.

concentration of these missions in the country. Mexico had the highest percentage in marine sciences, but no requests in physics, math or chemistry. This last discipline had the strongest showing in Brazil.[20]

UNESCO's mathematicians in Argentina

UNESCO's science missions sent to Argentina were the result of intense academic activity in the country. In terms of mathematics, the scholarly level in Argentina at that time was far from rudimentary.[21] In addition, without the professional, institutional and personal connections of the local scholars – and without state support – UNESCO's missions would never have been successful or even feasible. In this section we will briefly describe the missions of UNESCO's mathematicians in Argentina as an example of the trends we have been discussing.

At the General Conference in 1958, the Latin American Regional Centre for Mathematics was created based on a proposal by the Argentine Government.[22] As we saw in the previous sections, Argentina's requests for experts were limited almost exclusively to science missions. A significant portion of these took place between 1959 and 1964 and involved mathematicians sent to work at this center. It was located in the Mathematics Department of the School of Exact and Natural Sciences (Facultad de Ciencias Exactas y Naturales) at the University of Buenos Aires (FCEyN-UBA). In legal terms, the situation of the center was murky from its creation until the support of UNESCO ceased in 1966. In March 1960, UNESCO's Montevideo office called a meeting of a group of the center's advisors from Argentina, Brazil, Chile, Peru, Uruguay and Venezuela. Those who attended drew up a statute that was never approved by UNESCO due to the fact that it stated that the organization would provide continuous financing for the center. In fact the statue was never signed by any of the states of the region, not even by anyone from the Argentine Government.[23]

In spite of this, the center functioned within the FCEyN-UBA Mathematics Department and it represented an important training hub for Latin American mathematicians in the 1960s. This success was owed to the joint support of both the UBA School of Science and UNESCO in terms of both financing and infrastructure. The "seal" of the regional center – no organizational structure outside the Mathematics Department was ever established – allowed the Argentine Government to include the center as a precedent in its requests for UNESCO technical assistance. In addition, the prestige of UNESCO facilitated the international recruitment of experts and fellowship recipients. The recruitment and visit of these mathematicians were not exclusively the result of the efforts and "willingness" of the UNESCO officers, however. The international connections of Argentine mathematicians – and foreign mathematicians residing in Argentina – who were professors at the Universidad Nacional de Cuyo and the UBA were decisive in the decision of the Argentine authorities to request technical assistance from UNESCO's officers.

In 1960, the university authorities appointed Alberto González Domínguez, professor at the FCEyN-UBA and director of the Mathematics Department. He was part of the Argentine delegation at the General Conference when the Argentines proposed creating the center. González Domínguez had also participated in the two Mathematical Colloquiums on Certain Math Problems Studied in Latin America that took place in Montevideo, Uruguay, in 1951, and in Mendoza, Argentina, in 1954, jointly organized by UNESCO's Science Cooperation Office for Latin America and local scholars.[24]

The missions assigned to the center were all short – from one to three months. In general, the experts sent to there had two tasks: to give seminars or classes to the Latin American fellowship recipients at the center and the university students, and to be in touch with other mathematicians. The experts offered special courses, since the regular curriculum classes were given by the university professors, and the content of these special courses was agreed by the members of the Mathematics Department according to the expertise of the visiting professor.[25]

Based on the instructions they received from UNESCO, all the experts did a particular evaluation of the center, especially with regard to the fellowship program, and commented on the level of the mathematicians in Argentina and within the region. In general, the experts found that Argentine students, usually the ones at the School of Exact Sciences, were highly knowledgeable within their fields, while the fellowship recipients from other countries were at a more rudimentary level. To rectify this situation, some proposed implementing a system involving two fellowships – one "junior" and the other advanced – though this recommendation was never implemented by either university authorities or by UNESCO's officers. One of the experts, Alexander Marcowich Ostrowski, who was there in September 1961, exemplified the general feeling among most of the experts upon arriving in Buenos Aires when he commented: "I was quite surprised by the scientific level in Buenos Aires – not only that of the staff [professors] but also that of the advanced math students. I discovered they have a modern and even ultramodern attitude towards mathematics."[26]

Final thoughts

In the case of Latin America, we noted that the "peripheral academic centers" of the region – that is, Brazil, Chile and Mexico – were the ones to receive the greatest quantity of UNESCO experts between 1947 and 1984. At the regional level, this meant that UNESCO contributed to evening out some of the pre-existing educational and scientific asymmetries, at least between 1947 and 1973. Some of the states in the region did not receive any technical assistance, while the assistance received by some others was quite homogenous – for example, with regard to education and science missions. Another group of states received mainly missions targeting primary and secondary education.

When we focused on the science missions, we first observed that most of those sent to the region focused on exact and natural sciences, engineering and the applied sciences. There were fewer Latin American missions in the social sciences, which were generally limited to specific initiatives such as that of FLACSO in Chile. The case of the mathematicians in Argentina illustrates how missions were carried out in Latin America. State sponsorship of the missions and the existing connections between the visiting mathematicians and the Argentine mathematicians were essential to the technical assistance provided by UNESCO's officers. Once the experts arrived in Argentina, it is worth noting the support of the FCEyN-UBA and of the professors at the university. This was fundamental to the successful execution of the tasks entrusted to the mathematicians.

Although in theory UNESCO's technical assistance was available for all of the member states, not all were in the same position to take advantage of the expert missions. This was due to the fact that the requests for assistance had to be drafted by the states themselves in order to guarantee the principle of technical assistance as a 'help to self-help'. The local counterparts who were made available to the experts sent on international missions were fundamental to guaranteeing the success of the UNESCO initiatives at the local level. The result was that a significant number of the underdeveloped member states both globally and in Latin America did not benefit, or benefited only minimally, from the technical assistance for development. However, the technical assistance that was provided granted recognition and prestige to other states, helping them to reaffirm their academic leadership. They did so either within the region of Latin America – as in the cases of Brazil, Chile and Mexico, which welcomed experts at their universities or research institutes – or worldwide, in the case of the central states that sent their experts to the underdeveloped countries.

Notes

1. Clarence E. Beeby, "Aux sources des programmes d'éducation" in *L'UNESCO rancontrée par ses anciens* (Paris: AAFU-UNESCO, 2006), 73.
2. Harry S. Truman, "Inaugural Address of the President", *The Department of State Bulletin* 10:500, publication 3413 (January 1949): 125. See also Anabella Abarzúa Cutroni and Natalia Rizzo, "Sin expertos no hay desarrollo: la cooperación internacional y la formación de administradores públicos y cientistas políticos en Chile" in *El desafío de construir ciencias sociales autónomas en el sur*, ed. Fernanda Beigel and Hanna Sabea (Mendoza: EDIUNC-CLACSO, 2014), and Olav Stokke, *The UN and the Development: From Aid to Cooperation* (Bloomington: Indiana University Press, 2009), 50.
3. Stokke, *The UN and the Development*, 51.
4. General Assembly, Res. 200, III, 2b and 3a-d, 4 December 1948, UN Archives. See also Res. 198, III.
5. ECOSOC Res. 222, IX, 14 August 1949, UN Archives.
6. ECOSOC Res. 222, IX, Annex I-2a, 14 August 1949, UN Archives.
7. The drafting of mission reports, with their varying levels of confidentiality, was a common and mandatory practice for the officers and/or experts carrying out a site mission or participating in a meeting or conference. The result was that the organization's officers had a large amount of information available about the status of the academic fields and about politics at the local level. In some cases the reports by the experts were released to the public as a kind of consulting service for the involved governments. In 1968 there were nearly 20,000 reports on site missions covering the 21 years of UNESCO activities offsite filed with the Division

of Relations with Member States ("Index of Field Mission Reports", 1948–1968, ii, Bureau of relations with member states, Report Division, UNESCO Archives).

8. Although we could refer to all of the agents involved in the implementation of the missions as "experts", we reserve the term for scholars from different fields who performed a mission at a specific destination. We use the term "officer" to refer to agents carrying out administrative tasks at the main headquarters in Paris or in another office outside headquarters but who only occasionally carried out tasks on site.
9. To calculate the number of experts sent to each state, we used the "Index of Field Mission Reports" edited by UNESCO since 1968. The first index included a list of reports filed by the Division of Relations with Member States and organized according to three criteria: author, country and subject. In the cases in which experts did not draft a report, the mission was still listed and a note of "no reports" was added. Later, UNESCO edited a supplementary index annually with the reports for each year. In order to calculate the number of experts sent to each country, we simply counted the authors assigned to each report. Only a few of these are repeated as they performed more than one mission or because their mission lasted for more than a year. We took care to compare the indexes with other sources such as the director-general reports, correspondence, memorandums, the mission reports themselves and the available literature (especially in relation to FLACSO), and we confirmed that the lists of the indexes are not entirely complete. We believe that perhaps certain reports were not included on the lists because of the level of confidentiality at the time when they were drafted or perhaps simply due to administrative error. In addition, in the indexes from 1947 to 1973, there is a list of projects authorized in the framework of the "Special Fund". These include 116 projects authorized for implementation as of 31 December 1968, 17 for 1969, 21 for 1971, 53 for 1972 and 43 for 1973.
10. S. Romano, *Integración, desarrollo y dependencia: la Asociación Latinoamericana de Libre Comercio (ALALC) en el contexto de las relaciones con Estados Unidos (1960–1970)* (PhD dissertation, Universidad Nacional de Córdoba, Córdoba, Argentina, 2009); Fernanda Beigel, *Autonomía y Dependencia Académica. Universidad e investigación científica en un circuito periférico: Chile y Argentina (1950–1980)* (Buenos Aires: Biblos, 2010); and Fernanda Beigel, *The Politics of Academic Autonomy in Latin America* (London: Ashgate, 2013).
11. "Index of Field Mission Reports", 1948–1968, 1969, 1970, 1971, 1972, 1973, 1974, 1975, 1976, 1977, 1978, 1979, 1980, 1981, 1982, 1983, 1984, UNESCO Archives.
12. Philip G. Altbach, *Educación superior comparada. El conocimiento, la Universidad y el desarrollo* (Buenos Aires: Universidad de Palermo, 2009).
13. "Index of Field Mission Reports", 1948–1968, 1969, 1970, 1971, 1972, 1973, 1974, 1975, 1976, 1977, 1978, 1979, 1980, 1981, 1982, 1983, 1984, UNESCO Archives.
14. Conseil Executive, Rapport sur la participation de l'UNESCO au Programme Elargi, 25 March 1953, 11, 33/EX 18, UNESCO Archives.
15. Stokke, *The UN and the Development*, 557.
16. Chloé Maurel, *Histoire de l'UNESCO. Les trente premières annés, 1945–1974* (Paris: L'Harmattan, 2010); Rapport du Directeur Général sur l'activité de l'organisation en 1954 [1955], p. 20, UNESCO Archives; and Fernanda Beigel, "La FLACSO chilena y la regionalización de las ciencias sociales en América Latina (1957–1973)", *Revista Mexicana de Sociología* 71:2 (2009): 319–349.
17. "Index of Field Mission Reports", 1947–1968, 1973, 1974, UNESCO Archives.

18. Utilizing the original classification system of UNESCO's indexes implies certain ambiguities. Whenever possible we have tried to avoid using them, especially in relation to science. However, certain disciplines currently considered part of the social sciences (e.g. archeology, linguistics and pedagogy) have been classified in the other two categories because we were not always able to distinguish them clearly as disciplines (e.g. archeology was often grouped with museums or monuments). The category of the "social sciences" was also used generically by UNESCO, so we also encountered several missions classified in this way.
19. "Index of Field Mission Reports", 1948–1968, 1969, 1970, 1971, 1972 and 1973, UNESCO Archives.
20. "Index of Field Mission Reports", 1948–1968, 1969, 1970, 1971, 1972, 1973, UNESCO Archives.
21. Pablo Pacheco, "El Instituto de Matemática de la Universidad Nacional de Cuyo durante el primer peronismo ¿Periferia de la periferia o excelencia científica en la periferia?" Ponencia presentada en el II Workshop: Producción y circulación de conocimientos desde la periferia. Distintos abordajes. Mendoza, Universidad Nacional de Cuyo, 3 y 4 de noviembre de 2011.
22. Resolutions of the General Conference, tenth session, 1958, resolution 2.33, [1959], UNESCO Archives.
23. "Rapport du Directeur Général sur l'activité de l'organisation en 1960" [1961], p. 91, and "An account of the origin and beginnings of the Latin American Centres, Informe de la Oficina de Cooperación Científica de la UNESCO en Montevideo", 1966, p. 3 in Carpeta: 5 A 102 (8) 06 (82) "66" – Meeting of Directors of National Research Councils and others Organizations resp. for Science Policy in Latin American Member States – Argentina, 1966. UNESCO Archives.
24. "An account of the origin and beginnings of the Latin American Centres – Montevideo Office Report", 1966, 4; "Tenth session of the General Conference, list of delegates", 9, 1958 and "Program of the first colloquium, 1951; 2nd Symposium sur mathematiques, Mendoza", 1954, and Report by Ángel Establier, August 1954, 3, in Folder: 51 A 064 (82) "54" LASCO – Symposium on Mathematics (Argentina: Mendoza, 1954) org. by Lasco, UNESCO Archives.
25. Mathematicians who visited Argentina were F. Gaeta (28 August–27 December 1958), Jean Pierre Kahane (20 July–19 October 1959), Charles Ehresman (1 August–30 November 1959), Antoni Zygmund (mid-September–mid-December 1960), Alexander Marcowich Ostrowski (September 1961), Jean Dieudonné (10 July–9 October 1961), Jan Mikusinski (March–May 1962), Henri Morel (1963), Philippe Tondeur (March–June 1964), Stanislaw Lojasiewicz (2 May–1 August 1964) and Warren Ambrose (7 June–31 December 1964) according to "Index of Field Mission Reports", 1947–1968, 194–195 and mission reports by Jean Pierre Kahane (1959), Charles Ehresman (1959), Antoni Zygmund (1960), Alexander Marcowich Ostrowski (1961) Jean Dieudonné (1961), Jan Mikusinski (1962), Philippe Tondeur (1964), Stanislaw Lojasiewicz (1964) and Warren Ambrose (1964), UNESCO Archives.
26. Mission report by Ostrowski 1961, 1–2, UNESCO Archives.

Figure PIV.1 Lessons about the UN and its specialized agencies were being introduced into schools all over the world after World War II as the best means of bridging the gap between nations. The teaching methods had been worked out by UNESCO and aimed to promote international understanding. (© United Nations)

Part IV
Implementing Peace in the Mind

Over the years, UNESCO has studied the nature of tensions and dealt with some of the many concepts, stereotypes and viewpoints that have a tendency to split rather than unite people. One of the first and most noteworthy examples is a statement made by experts about race and issued by UNESCO in 1950. It claimed that "race" was a social myth, that it made more sense to describe human differences in cultural terms and it highlighted the unity of humankind as a species. The statement gave authority to a new way of thinking that could not be used to legitimize discrimination on the background of biology, but also resulted in massive critique. For decades, UNESCO was therefore the core of a dispute in international scientific circles about the correct definition of race.

For UNESCO, just as important as deconstructing concepts and widespread thought patterns was to offer new and meaningful ways of thinking. "Unity in diversity" was formulated early on by the organization as a goal to achieve. To support the solidarity among different groups of people, a range of projects were launched, with the Major Project for Mutual Appreciation of Cultural Values of East and West from 1957 to 1966 being among the more prominent.

However, psychological studies showed that the most entrenched forms of prejudice are established in the formative years of childhood, often reflecting nationalistic attitudes rather than a spirit of mutual understanding. UNESCO therefore saw a special task in changing the minds of the youth and it launched a strategy of providing education for international understanding by revising textbooks for primary and secondary schools. Geography and history books attracted special attention. History, because the books often reflected the positive self-image of the nation and negative counterimages of the neighboring countries, and geography, because geography books delivered an even longer series of negative, ethnocentric counterimages. Instead, new UNESCO-approved textbooks should emphasize humanity's shared past, aim to be truly global, build on the values and ethical standards of the UN, and be submitted to foreign historians for critical review.

10
UNESCO Teaches History: Implementing International Understanding in Sweden[1]

Thomas Nygren

Introduction

International organizations aiming to promote peace and development throughout the world rose from the ashes of World War II. In a world troubled by conflicts and large gaps between rich and poor, UNESCO faced many challenges and used different strategies in its struggle to shape a better world, many of them highlighted in this book.[2]

This chapter shows how history education became an important part of UNESCO's mission to promote international understanding and unity in diversity, and to safeguard world heritage. It also highlights how UNESCO was more successful than the League of Nations and the Council of Europe in reforming history education in Sweden, and how its efforts impacted the way students after World War II started to write a more peaceful, global and multicultural history.

In a global climate crisscrossed by tensions and conflicts, education was seen – and still is – as crucial for positive growth, and relationships within and between nations: In the 19th century, history education was held to be an important part of fostering good, obedient citizens and patriots, but it was also criticized for promoting harmful nationalistic and militaristic sentiments.[3] In the shadow of the world wars, the teaching of history was increasingly criticized for creating oppositions between countries and peoples. A new philosophy arose, advancing the notion that guidelines and reviews of textbooks could promote a history education that would contribute to the development of the world instead of causing splits between countries.[4]

International organizations, of which Sweden has been and remains a member, claimed that history teaching should contribute to peace and tolerance, train critical thinking, show a variety of perspectives, demonstrate the value of cultural heritage and build up identities – history education in the service of humanity. The question is, did such efforts to create a better world through teaching history actually work, and did the international efforts to reform history education change the way students wrote about history?

To investigate the relationship between the international guidelines for teaching history and how history was understood and construed by teachers and students, I will start with the first guidelines formulated by the League of Nations in 1927 and 1937, and the early recommendations for history teaching as formulated by UNESCO in 1947–1949 and by the Council of Europe in 1952–1953, when international guidelines for history education from international organizations began to be formulated and implemented.[5] I will then focus on the three main orientations that were formulated by UNESCO – namely, international understanding, unity in diversity and safeguarding local heritage. International guidelines, national curricula, syllabi, debates, teachers' views and students' work with history are placed in relation to each other to make it possible to analyze the processes of implementation. The investigation of various levels of curricula proceeds from the following questions: What goals and means in history education have been recommended internationally in international organizations of which Sweden has been a member? How did the international guidelines relate to Swedish history education with regard to implementation? In comparison with the international guidelines, how was history formulated in Swedish formal curricula, teachers' perceptions and students' work in history?

Inspired by the notions of curricula and implementation of the Canadian-American educational researcher and theorist John I. Goodlad, I see the process of implementation as one that includes direct transactions of ideas and interpretations, as well as the more independent creation of value in a complex interplay with the world at large.[6] Each curricular level can contain several different perspectives. What is formulated in recommendations and national guidelines does not automatically seep down to the students.

The levels of curricula studied here are: (1) the ideological curricula, analyzed via the international guidelines directed toward Swedish history teaching; (2) the formal curricula, examined in national guidelines, and also how history is formulated in final examinations and inspectors' reports; (3) the perceived curricula, studied in teachers' debates and interviews with experienced teachers; and (4) the experiential curricula, examined by looking at students' choices of topics in final exams, 1,680 titles of students' individual projects in history, from 1930–2002, and an in-depth analysis of 145 individual projects, written between 1969 and 2002 (see Table 10.1).[7]

The development of international guidelines and their implementation into history teaching was, and is still, complex, but in this chapter I show how especially UNESCO's guidelines were implemented in history education in Sweden. I also highlight how the implementation of international understanding was more than a top-down process.

Table 10.1 Sources reviewed at different curricular levels.

Levels of curricula	First guidelines	Post-World War II reforms	Post-colonial reforms	End of Cold War and new century reforms
League of Nations and UNESCO guidelines *Ideological curricula*	1925 Cesare's resolution 1927 How to develop the spirit of international cooperation 1937 Declaration regarding the teaching of history	1947 International understanding 1949 Aid to international understanding	1960 Post-colonial criticism 1965 UN elimination of racial discrimination 1972 World heritage 1976 Safeguarding cultural heritage 1974 International understanding, peace, human rights and fundamental freedoms 1978 Against racism	1983 International understanding, peace, human rights and disarmament 1985 The status of women 1993 UN human rights education 1995 Declaration for peace, human rights and democracy 2001 Universal declaration on cultural diversity
Council of Europe guidelines *Ideological curricula*		1952 Revision of history and geography textbooks 1953 The European idea in the teaching of history 1954 European cultural convention 1954–1958 A history of Europe (printed 1960)	1964 Civics and European education 1965 History teaching in secondary education. 1971 History teaching in upper secondary education. 1972 Religion in history textbooks 1978 The teaching of human rights 1979 Cooperation in Europe 1981 Intolerance a threat to democracy 1981 Environmental education	1983 Promotion of awareness of Europe in secondary schools 1983 Preparing young people for life 1983 Cultural and educational means of reducing violence 1984 Teacher training for intercultural understanding 1984 Against bias and prejudice 1985 Teaching and learning about human rights 1985 European cultural identity 1989 The European dimension of education 1991 The contribution of the Islamic civilization to Europe 1993 Religious tolerance in a democratic society 1993 Local history in a European perspective 1995 Protection of national minorities 1996 History and the learning of history in Europe 1999 On secondary education 2001 History teaching in twenty-first-century Europe

Table 10.1 (Continued)

Levels of curricula	First guidelines	Post-World War II reforms	Post-colonial reforms	End of Cold War and new century reforms
Sweden *Formal curricula*	1928 National curriculum 1928 Syllabus for history teaching 1933 National curriculum 1935 Syllabus for history teaching	1950 National curriculum 1956 Syllabus for history teaching 1961 Syllabus for history teaching	1965 National curriculum 1965 Syllabus for history teaching 1970 National curriculum 1976 Translation UNESCO International understanding 1969–1982 Inspector reports on history teaching 1981 Syllabus for history teaching	1993 Translation UNESCO International understanding (second print) 1994 National curriculum 1994 Syllabus for history teaching 1994 Agreed! [Överenskommet!] (Reprint of *Human rights and International understanding*)
Teachers *Perceived curricula*	Tidning för Sveriges läroverk [TfSL] *Swedish Upper Secondary Teachers' Journal* 1926–1942 Historielärarnas förenings årsskrift [HLFÅ] *Association of History Teachers' Annual Report*. 1942	TfSL 1945–1956 HLFÅ 1945–1959 Interviews experienced teachers of their school experiences	HLFÅ 1960–1982 AFHL 1968–1977 Interviews Supervision of individual projects	HLFÅ 1983–2002 Interviews Supervision of individual projects
Students *Experential curricula*	Titles of individual projects 1930–1931 and 1938–1939. Topics for exam essays, statistics from 1938	Titles of individual projects 1949–1950 Topics for exam essays and statistics until 1963	Titles of individual projects 1969 and 1982 Students' individual project papers	Titles of individual projects 1992 and 2002 Students' individual project papers

However, to understand the impact of UNESCO's work in history teaching in Sweden it is necessary to start in the interwar period and the work of the League of Nations, and also present the works of UNESCO in contrast, and in parallel, with the work of the Council of Europe, but also to present how international curricula, as formulated primarily by UNESCO, relate to the formulations of curricula on a national level, teachers' perspectives and students' writing about history.

League of Nations guidelines

The League of Nations was formed after World War I in an attempt to avoid war in the future and to enhance peace and cooperation. Failing to stop World War II, it still became a formal arena, containing institutions and networks for peace, development and education.[8] The League of Nations and its suborganization, the International Committee on Intellectual Co-operation, composed guidelines for textbook revision and teaching, where patriotism – good national self-awareness – was meant to encourage international understanding.[9] Peaceful encounters across borders would also be furthered by knowledge about cooperation between and among international organizations and by studies of "foreign masterpieces and folktales", and the history and handicrafts of other civilizations should be studied to "develop the spirit of international co-operation among children, young people and their teachers".[10]

During the interwar period, Sweden actively participated in the League of Nations' efforts to revise history teaching. However, support for the league and its work was weaker in other countries, not least in the major member powers such as France and England, which did not sign the Declaration Regarding the Teaching of History.[11] Disputes between countries both within and outside the League of Nations ended with World War II, which proved that the ambition to create peace and international understanding had failed.

Peace and understanding

In October 1945 the UN made new efforts for peace, development and cooperation among the countries of the world. The UN and also UNESCO became a meeting place for member nations. Sweden, which joined the UN in 1946, was initially an interested observer in UNESCO and became a full and active member in 1950. There was great consensus within UNESCO around the proposition that "since wars begin in the minds of men, it is in the minds of men that the defenses of peace must be constructed".[12] Nazism, fascism and ultranationalism should be banished and replaced by the implementation of international understanding.[13] UNESCO strongly propounded the value of education for the "UNESCO idea" of peace through international understanding.[14] The content of this educational program should include history and geography, cleansed of nationalism and militarism, and actively promote understanding for other countries. The focus was to be on peaceful coexistence, human labor, cultural history and scientific development, and in 1949 UNESCO published *A Handbook for the Improvement of Textbooks and Teaching Materials as Aids to International Understanding* – a volume aimed at improving textbooks and teaching, particularly in history, by, for instance, including more world history and critical thinking.[15]

Out of the ruins of Europe there also arose a West European attempt at cooperation. The Council of Europe was formed in 1949, with the purpose of augmenting a feeling of solidarity and shared affinities between and among the countries of Western Europe and safeguarding democracy and human rights. Aware of the previous work if the League of Nations and UNESCO, the council deemed it possible to build up an awareness of Europe's historical heritage, and to that end a series of conferences on history teaching in Europe were held between 1953 and 1958.[16]

The first conference in Calw in West Germany in 1953 had the straightforward title The European Idea in the Teaching of History. It addressed how Europe and the idea of Europe could be dealt with in history education. Studying European history was motivated by the importance of avoiding traditional mistakes and prejudices and guaranteeing, or confirming, facts. In its recommendations to teachers and textbook authors, the conference underlined the importance of describing Europe's contributions to the world and various regions' contributions to European development. Teachers were advised to begin by making local and regional history accessible to students and then to guide them into a greater understanding of a "European conception of history".[17] This was to be done in a part of the world severely damaged by war, placed between the two new great powers – the USA and the USSR.

Safeguarding peace was central to both UNESCO and the Council of Europe, and the guidelines of these organizations stated that through true and neutral objectivity, history education could counteract discreditable nationalism. Peace, tolerance and human rights were to be advanced through the criticism of sources and a history teaching based more upon the common creation of civilization. Social, economic, cultural and scientific history should be highlighted, while political history and the history of conflicts should be toned down.[18] International guidelines were given for shaping a better present and future through a more tolerant, objective and contemporary history teaching.[19] In contrast with UNESCO, members of the Council of Europe ideologically were in accord about the value of democracy – the council contained only Western democracies. The power struggles between East and West during the Cold War had a greater impact on the work within UNESCO. Not least, UNESCO's attempt to write a "History of Mankind" was complicated by difficulties in encompassing both western European and communist history writing.[20] Sweden's active participation in UNESCO to promote international understanding was evident not least in Nordic collaborations aiming to support education and produce material presenting the work of "UNESCO beyond all borders".[21] Education was, and is, pivotal in Scandinavian activities in UNESCO.[22] In the Cold War era, various kinds of war on different continents and the nuclear balance of terror colored UNESCO's guidelines, which during the 1980s came to recommend disarmament.[23]

Even if the Cold War ended in 1989, there remained several wars and conflicts in the world, and others started up after the fall of the Berlin Wall. UNESCO and the Council of Europe worked in parallel and together to support peaceful development in the "new" Europe.[24] During the 1990s, former Eastern European states became members of the Council of Europe, and the importance of having peaceful multifarious perspectives in history education was underlined.[25] There were strong renewed efforts within UNESCO for international understanding through evaluating previous work and producing updated and clearer guidelines.[26] Many developments in the world – ethnic enmities, terrorism, and conflicts between groups and countries due to deepening schisms – were alarming. According to UNESCO, a more peaceful world should come about through education in which universal values such as democracy and human rights were safeguarded and underscored.[27] Both UNESCO and the Council of Europe emphasized the importance of teaching students to be critical and to gather information, to see different interpretations and to compromise. The content of instruction should be comprehensive and should pay due regard to people's creativity and their capacities for coming to peaceful agreements.

Critical post-colonial history

UNESCO, burdened during the Cold War by the gap between East and West, became an arena where new and formerly marginalized countries could make their voices heard. The liberation of colonies became extremely important for UNESCO's work within several areas, not least regarding education. The early post-war efforts for peace and neutral objectivity in history teaching were followed by guidelines for a more critical post-colonial approach, with a greater focus on previously marginalized groups. Colonialism and neocolonialism were condemned.

Within the Council of Europe, however, post-colonial criticism was milder. Countries with colonies and with a colonial past were not especially keen to condemn colonialism. Both UNESCO and the Council of Europe later on focused more on an inclusive perspective on history that could promote unity in diversity.[28]

Unity in diversity

Early on both in UNESCO and the Council of Europe, unity in diversity was formulated as a goal to achieve – namely, harmony and solidarity among different groups of people.[29] Within UNESCO there were representatives of formerly marginalized groups who wanted to highlight "their" history. UNESCO also worked actively to survey global history and to call attention especially to neglected areas such as Africa, Latin America, Oceania, Asia, and the Slavic and Arabic cultures.[30]

The Council of Europe contended that instruction in history that was unbiased and unprejudiced could create better understanding between the different peoples of Europe. In the 1960s, for example, history teaching should counteract the image of Arabs as merely fanatic warriors and Scandinavians as "bloodthirsty pirates".[31] In what was described as an increasingly multicultural Europe, it was claimed in the 1980s and onwards that unity in diversity is what creates wealth in the European culture.[32] After the fall of the Berlin Wall in 1989, a "new Europe" was described – a Europe with a pan-European character. The Council of Europe declared a responsibility for bringing the people of Europe closer together in a cultural community infused by democracy, human rights, fundamental freedoms, security and diversity.[33] The European cultural identity and its "unity in diversity" was considered during the 1990s and after 2000 to provide openings for dealing with the past and other cultures – offering a precondition for peace-generating cultural encounters within and across borders.[34]

In November 2001, UNESCO adopted the Universal Declaration on Cultural Diversity, which underlined the value of identity, diversity and pluralism, and pointed to cultural diversity as constituting humankind's common heritage. "Culture takes diverse forms across time and space," it claims. "This diversity is embodied in the uniqueness and plurality of the identities of the groups and societies making up humankind. As a source of exchange, innovation and creativity, cultural diversity is as necessary for humankind as biodiversity is for nature. In this sense, it is the common heritage of humanity and should be recognized and affirmed for the benefit of present and future generations."[35]

UNESCO declared the importance of learning about humankind's cultural heritage, which should be focused upon not least in education. Member nations promised in the declaration to support "an awareness of the positive value of cultural diversity and improving to this end both curriculum design and teacher education".[36]

The neglect of women's history

During the 20th century, at least in the West, women obtained more and more rights. Not least, experiences from working life during World War II brought changes in women's socioeconomic position. In 1958 the Council of Europe included women's emancipation as an area that had been disregarded in history teaching. However, women's situation in history education did not continue to be an important issue in the Council of Europe's guidelines, even though in 1985 it was stated that teaching should aim to counteract discrimination and sexism.[37] When UNESCO in the late 1970s and 1980s was focusing more and more attention on problems with racism, intolerance and apartheid, women's subordinate position in the world and in higher education was included.[38] Within the UN, addressing the position of women was long considered secondary to working against nationalism and racism, but in 1985, UNESCO brought up "the need for men and women everywhere to work for the triumph of peace, freedom, equality and justice, for respect of human rights and self-determination, the elimination of all forms of inequality between men and women, of racism, apartheid and any form of foreign domination and for the promotion of mutual understanding and tolerance, so that peace and security may prevail".[39]

These problems, according to UNESCO, must be confronted within the various sectors of society, not least within schools. Women were conspicuous by their absence – at least as a clearly defined group – in the Council of Europe's recommendations until 1996, when specific recommendations were made to pay particular attention to women in the teaching of history.[40]

Safeguarding local heritage

In parallel with peace efforts and working for the weak groups of people in the world, the importance of increasing and spreading knowledge of cultural heritage was registered. UNESCO stressed non-European cultural heritage, while the Council of Europe was more concerned with the European. Under the auspices of UNESCO, the Nubian temple of Abu Simbel was moved, thereby avoiding being sunk under the waters of the Aswan Dam. This was an enormous project carried out between 1960 and 1980, which directed attention towards the need to preserve cultural heritage, and it also demonstrated the possibilities of technology. In an era with great technical progress, evidenced by Sputnik and the moon landing, huge sums of money were invested in education and technical development.

However, ensuring cultural heritage was not overlooked, UNESCO's Convention Concerning the Protection of the World Cultural and Natural Heritage was approved in 1972. In accord with this convention, in 1976 the organization encouraged member countries to work on behalf of the preservation of historical and culturally valuable environments and their present-day role.[41] In the recommendation, the subject of history is said to be paramount for the preservation of local cultural heritage. "Awareness of the need for safeguarding work should be encouraged by education in schools," the recommendation says. "The study of historic areas should be included in education at all levels, especially in history teaching, so as to inculcate in young minds an understanding and respect for the works of the past and to demonstrate the role of this heritage in modern lives. Education of this kind should make wide use of audio-visual media and of visits to groups of historic buildings."[42]

According to UNESCO in 1985, greater awareness of cultural heritage could contribute to upgrading the national identity, and improving the possibilities for cultural exchange and for the preservation of customs and relics and historical remains.[43] The same year the Council of Europe followed UNESCO's suit, declaring the importance of preserving cultural heritage. The school, using various means, should underline "the unity of the cultural heritage and the links that exist between architecture, the arts, popular traditions and ways of life at European, national and regional levels alike".[44] As a means for heritage education, experience-based learning, especially cross-disciplinary and using audiovisual and internet-based aids, was recommended.[45]

Central ideas in international guidelines

The guidelines from UNESCO and from the Council of Europe had many similarities. The main difference was between UNESCO's world perspective and the council's more Eurocentric view. UNESCO had world history as a "History of Mankind", while the Council of Europe pursued the "European Idea". This dividing line went through the ideological curricula with regard to both international understanding and the view of a many-faceted culture and cultural heritage.[46]

UNESCO and the Council of Europe formulated comprehensive guidelines for history education. Three clear orientations found in the international guidelines included the desire to generate greater international understanding through a more international and peace-oriented education in history; the desire to increase understanding for marginalized groups and for cultural diversity through a perspective that highlights minorities and women and that opposes racism; and the desire to safeguard world heritage and cultural heritage by studying local historical heritage with the help of experientially based pedagogy. Links to the present were considered important since it was the present and the future that were to be shaped in order to achieve mutual understanding and a feeling of unity in diversity, and to value the history that surrounds us. The international guidelines on international understanding, which largely also contained the other guidelines, were the most comprehensive and most clearly stated.

Implementation in the formal curricula

The results of the present study indicate that the League of Nations' recommendations had little impact on interwar Sweden. Even if Sweden endorsed the league's guidelines and participated in their follow-up, they remained marginal in the formal curricula. After World War II, Sweden subscribed to the recommendations, resolutions and conventions that were to be implemented in Swedish history education. The international guidelines, particularly from UNESCO but also from the Council of Europe, were implemented in the formal curricula. "International understanding" and the "Idea of Europe" as concepts were incorporated into the syllabus in 1961 and 1965, respectively. During the post-war period, Skolöverstyrelsen – the Swedish National School Board – formulated increasingly global topics for the final exams as an extension of the formal curriculum and a more direct implementation of UNESCO's recommendations for more world history.[47]

Even after 1960, international formulations and recommendations were taken into account on the formal curricular level. Study of at least one non-European culture was first discussed at an international conference arranged by the Council of Europe in 1969.[48] Later on, in 1981, most likely through the precepts of teachers, this was included in the Swedish history syllabus.[49] When Sweden carried out reforms in accordance with recommendations for international understanding, it was reported back to UNESCO.[50] On the level of formal curricula, Ulf P. Lundgren, director of Skolverket, the National Agency for Education, referred in the 1990s to UNESCO's recommendation on international understanding from 1974.[51] The recommendation was also printed and published in conjunction with a new national curriculum and guidelines in 1994, and in 2005 it was published online on the national website of the Swedish National Agency for Education.[52]

None of the Council of Europe's guidelines was noted in this way, but European history was accentuated more in the syllabus from 1994. The ideological curricular guidelines seem generally to have been incorporated into the formal curricula, which were comprehensive and also emphasized means and goals outside the extensive international guidelines.

Implementation in the teachers' debate

An examination of how the new guidelines were received in Sweden shows how teachers and students were at least co-creators in the transformation of history education up to 1961. This transformation led to more global and contemporary-oriented teaching than a top-down approach. The League of Nations' work was not included in the debate, but international guidelines from both UNESCO and the Council of Europe were addressed in the national debates and in the syllabuses. Teachers took an active part in formulating the guidelines, and even students' interests were noted nationally and internationally in discussions about the means and goals of history education. The Association of History Teachers seems to have played a significant role in emphasizing the importance of the subject of history for international understanding in the syllabus for 1961.[53]

Also apparent is how the formulations did not need to pervade practice. The cultural orientation of the subject of history, which was advocated internationally during the 1950s and 1960s, was disregarded nationally, possibly because of the national political situation and because economic and social history was becoming more emphasized. UNESCO's declaration from 1995, "Education for Peace, Human Rights and Democracy", was taken up neither on the level of formal curricula nor in the debates.

Sweden as a country of immigrants was addressed in the debate, with direct references to the stress on diversity in the international guidelines. When local history was taken up in the debates, it had national overtones – references to international guidelines were not explicitly made. Interest in local history was in place earlier than any interest in the international.[54]

Even if the debate raised questions outside the international guidelines, conferences on these guidelines were noted and teachers were active in the forming of new syllabuses in history, at least prior to 1994. During the entire period under scrutiny there were active teachers, headmasters, historians and pedagogues who participated in international conferences arranged by UNESCO and the Council of Europe, and who took part in debates on the means and goals of history education.[55]

Participating students

Direct contact between the ideological and the experiential curricula came about through teaching projects on human rights. One such contact was a trial teaching project in Arvika on peace and international understanding. The trial, which was linked to UNESCO, received a great deal of attention in 1954 in the *Tidning för Sveriges läroverk* (the Swedish Upper Secondary Teachers' Journal) where the project was praised for aiming to affect students' attitudes and knowledge of human rights through objective and factual teaching. Having the students study the position of women, historically and at the present time, was meant to show how teaching could give more space and time to the UN, human rights and international understanding.[56]

A reverse relation existed in the international organizations' formulation of their recommendations, when they took into account students' interest in contemporary history.[57] Students' experiences of history were also noted by the state authorities, which, through inspections and evaluations, tried to discover what the students preferred as well as what they needed.[58] Thus teachers and the state authorities with contacts with students interpreted and transmitted international guidelines, and also rereported the connected developments to UNESCO and the Council of Europe at conferences on history education.[59]

A clear link between UNESCO and teachers and students began in 1953 when the organization initiated the Associated School Project (ASP). This was to nurture international understanding through direct support to participating teachers and students in schools in different parts of the world.[60] Sweden was one of 15 countries to participate in ASP from its commencement in 1953.[61] The project did not have any direct counterpart within the Council of Europe. Admittedly in 1954 in the European Cultural Convention the council had proclaimed the value of exchanges over borders, but it did not initiate any student-oriented undertakings. Activities around the implementation of the Idea of Europe were mainly directed toward the formal and teacher levels, but the Council of Europe also encouraged ASP and other exchanges between schools.[62]

UNESCO's own evaluations underlined the importance of ASP for international understanding, while at the same time they also stressed the lack of significant impact on the formal curricula and academia.[63] In 2003, ASP's importance in Sweden was described as small in light of the fact that schools there, even without ASP, had good opportunities for internationalization and that the formal Swedish curricula were in accord with UNESCO's values.[64] Although the present study has not examined ASP in any depth, it may be said that there were links between the international guidelines and Swedish history education.

Students writing history

The line of demarcation between UNESCO and the Council of Europe was between a global and a Eurocentric perspective.[65] The orientation toward world history advocated by UNESCO was the one that was also most prominent at the student level. Dating back to the 1950s, students showed increasing interest in a global approach. A previous clearly Eurocentric point of departure in the national final exams was replaced in the 1950s by themes acknowledging a more non-European perspective and the students chose to write about world history rather than national history.[66] The choice of a topic such as "Egypt in world politics from Bonaparte to Naguib" by the majority of students in 1954 clearly indicates a preference and ability to write about world history, as well as in the following year when international relations were more interesting than the national "Historical Problems Around Charles XII" (see Figure 10.1). This was a shift in the experiential curricula, students' interest and writing about history, even before the national, officially binding, formal guidelines were issued in 1956.

The essay subjects in 1957, on African contemporary history and Sweden in the UN Security Council, were addressed in the debates in both teachers' journals, *Tidning för Sveriges läroverk* (Swedish Upper Secondary Teachers'

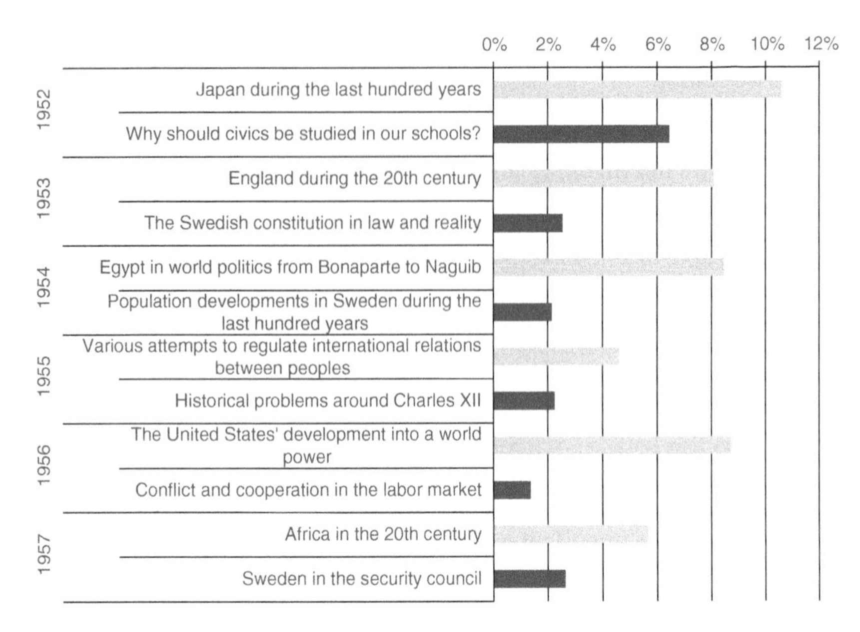

Figure 10.1 Number of students, as percentage of total, writing their essays about various historical topics, on the basis of existing statistics, 1952–1957.[67]

Journal) and *Historielärarnas förenings årsskrift* (Association of History Teachers' Annual Report). The subjects were criticized by the history teachers' association for being too contemporary and the teachers wanted an alternative topic in Swedish history.[68] In the journal we find articles complaining of a vulgarization of exam subjects, which did not reflect the schools' teaching of history but instead lent themselves to "chat around the breakfast table, but scarcely more than that".[69] However, the opposite view was also expressed – that the topics were seen by teachers as good examples of how the subject of history could include both the history of Africa and that of the UN. Thus changes were met with mixed feelings by teachers, but the students preferred African history, and world history continued to dominate the final exams into the 1960s. After World War II, students obviously cultivated an interest in international non-European history.

Students' interest in world history thereafter continued into the 21st century, in contrast with the national syllabus and the teachers' debates during the 1990s, primarily stressing European history. Statistics indicate an orientation toward the goals declared by UNESCO – international global understanding – despite the fact that during the period studied, the organization had a limited budget which was to cover many other educational and development programs.[70] The Council of Europe's propagation of more

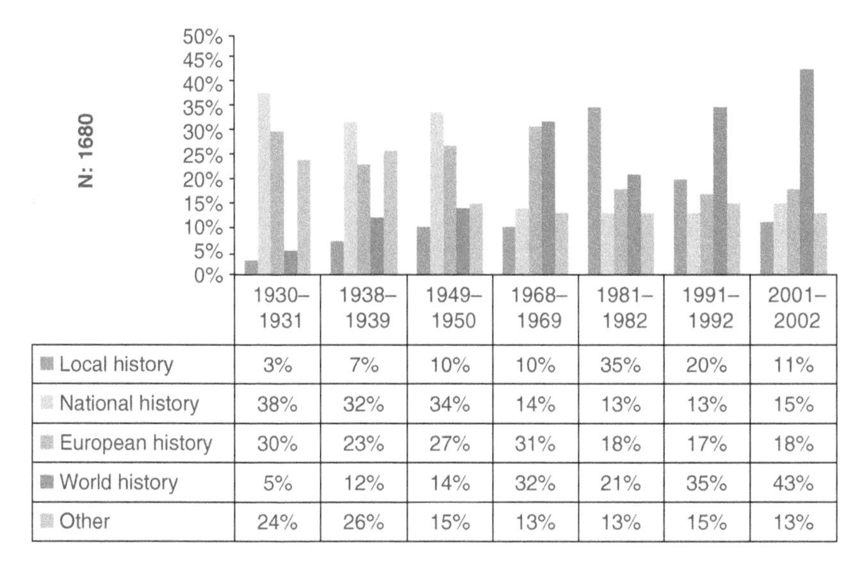

	1930–1931	1938–1939	1949–1950	1968–1969	1981–1982	1991–1992	2001–2002
Local history	3%	7%	10%	10%	35%	20%	11%
National history	38%	32%	34%	14%	13%	13%	15%
European history	30%	23%	27%	31%	18%	17%	18%
World history	5%	12%	14%	32%	21%	35%	43%
Other	24%	26%	15%	13%	13%	15%	13%

Figure 10.2 Individual projects and essays in history with clear geographical orientations.
Sources: School Archives, City and Municipal Archives, National Archives of Sweden.

European history had little direct impact on the students writing about history.[71]

Another indicator of how students' interest in and knowledge of non-European history made history teaching increasingly internationally oriented is a clear shift in choice of subject for students' individual projects since 1950, when 34 percent concerned Swedish history, 27 percent European history and 14 percent world history (see Figure 10.2). By 1969, interest in Swedish history had dropped to 14 percent, work dealing with European history had increased to 31 percent and project subjects relating to world history had risen dramatically – to 32 percent.

The statistics, presented in Figure 10.3, also indicate that interest in marginalized groups increased. The titles of these, and also my reading of other individual project papers not clearly declaring a focus on marginalized groups, show how an orientation toward more unity in diversity was possible, which was also confirmed by the statistics. Another orientation was that toward contemporary history, and in the diverse subjects treated by students, contemporary problems and entertainment subjects – for example, world politics and popular culture – became more and more dominant.[72]

It was primarily contemporary world history that attracted both male and female students. Their individual projects from 1969, 1982, 1992 and 2002 showed different orientations to a certain extent. Interestingly, two of four projects in 1969 that had a woman's name in their title were written by boys,

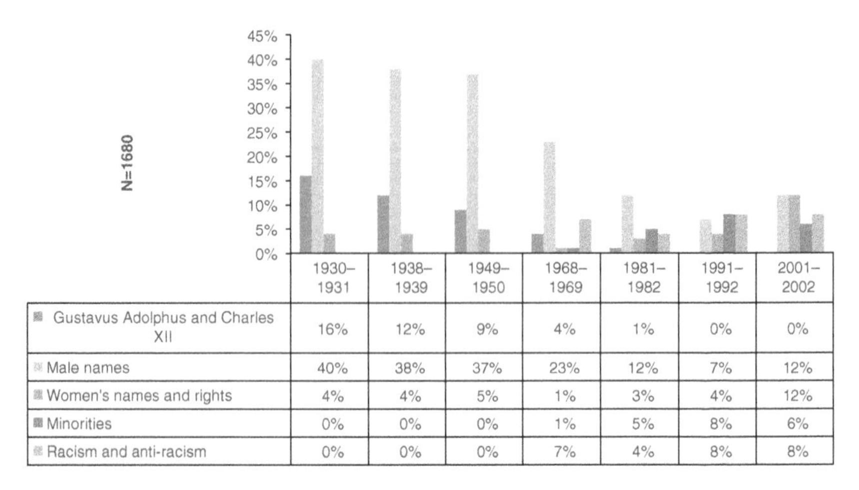

	1930–1931	1938–1939	1949–1950	1968–1969	1981–1982	1991–1992	2001–2002
Gustavus Adolphus and Charles XII	16%	12%	9%	4%	1%	0%	0%
Male names	40%	38%	37%	23%	12%	7%	12%
Women's names and rights	4%	4%	5%	1%	3%	4%	12%
Minorities	0%	0%	0%	1%	5%	8%	6%
Racism and anti-racism	0%	0%	0%	7%	4%	8%	8%

Figure 10.3 Orientations in titles toward more marginalized groups and phenomena.
Sources: School Archives, City and Municipal Archives, National Archives of Sweden.

while most of the girls' essays had a man's name.[73] In other years it was mostly boys who wrote individual projects about men, and girls who wrote about women and women's emancipation. The clearest change, however, was that over the time period, the names of women and men came to have equal distribution in the titles (see Figure 10.3). Part of the results of an orientation toward previously marginalized groups was that women were given more attention in both male and female students' work in history. Up until 2002, more girls than boys expressed an interest in minorities, but in 2002, interest was equal among both sexes.[74]

The titles of students' individual history projects also reveal that an increased regard for non-European history was followed by greater interest in marginalized groups. In comparison with titles about men and prominent kings from Sweden's era of "great power" – Gustavus Adolphus and Charles XII – racism, marginalized ethnic groups and women came more into focus from the 1960s (see Figure 10.3). For example, from 1969 onwards, more attention was paid to racism and its problems, not least in the USA, South Africa and Nazi Germany. The history of the US Civil Rights Movement was discussed in a number of students' work that I have scrutinized. The racism in the USA was criticized in 1969, when one student wrote: "Dominating whites in the Southern states today show attitudes and behaviour towards negroes that are despicable from many points of view."[75]

This more detailed study of students' essays also shows how world history in several cases brought up minorities and problems with colonialism and racism. For example, essays on Central and South American countries dealt with various Indian cultures and their encounters with conquistadors, and essays on African countries looked at similar experiences of different tribes. A stance calling for more power for the black majority and against European involvement in African countries was clear in several essays. Students criticized racism in South Africa, referring to both the UN and human rights. Nelson Mandela was depicted as a "born leader" and the African National Congress offered hope for "a free Africa in the future".[76]

However, there were also examples of currents in the opposite direction, against the international guidelines. There were students who in their essays romanticized men's use of violence – for example, the Swedish King Gustav Vasa, Commander C.G. Mannerheim, Alexander the Great and Napoleon – wholly counter to the international guidelines on peace and understanding. In their essays, some students also vindicated the use of violence in struggles against racism, and in essays on local history they portrayed immigration as a problem.[77] During the early 1970s, some students were supportive of communist dictatorships in countries such as China and the Soviet Union. Other students were very critical toward communist regimes. And after 1989, China was sharply criticized when "the government opened fire against its own people".[78] After the turmoil on Tiananmen Square, particularism was abandoned in favor of an approach emphasizing development in line with Western values.

However, in the essays analyzed, the majority of students denounced war and tyranny, both at home and abroad. For example, it was described how Hitler's "fixed ideas, which filled his confused brain during his wretched youth in Vienna, changed the world in a horrible way".[79] The history of power politics came into play in essays about Vietnam, Cuba, China, Palestine and Afghanistan. Students expressed revolutionary ideas, concern for world peace and condemnation of the horrors of war. Referring to the suffering in Vietnam, they criticized the involvement of France and the USA in Indochina, and the absence of "moral courage" among decision-makers.[80] The Soviet Union was characterized as "an inhuman barbarian state".[81] Based on interviews with relatives, war history could be an emotional family narrative, revealing ordinary people's memories and actions, with women as providers and workers when the men were at the front.[82]

Students' work that increasingly took up minorities could express strong solidarity with oppressed groups in the world, which harmonized well with the international guidelines on unity in diversity. By studying the history and traditions of the Romany people, one student claimed a greater understanding of their culture: "It feels as if I have another attitude towards the Romany now, a more positive one. When I see them in town now, I feel a sort of solidarity."[83] Students pronounced feelings of guilt and shame about what white people did in the past, indicating a clear repudiation of racism. Also, the importance of preserving cultural heritage was expressed in students' essays.

Nationalism and militarism became marginalized by the students. Men in power were often bypassed in favor of active women and more social and critical perspectives. In the scrutinized individual projects, romanticizing narratives were few and far between, and it was Raoul Wallenberg, Nelson Mandela, Mother Teresa and Mahatma Gandhi, not national kings, who were described as heroes.[84] In 1969 a student studied "Dag Hammarskjöld's contribution to world peace",[85] and Gandhi was seen as a contemporary and future model by a student in 1993 stating that "In the universal debate, his struggle against racism, colonialism, violence and the exploitation of nature and humankind is still relevant." Wholly in line with the guidelines of UNESCO, students focused on contemporary world history, criticized war and racism, and directed their attention to previously marginalized groups.[86]

The process of implementing international understanding

The organizations studied articulated means and goals for history teaching in relation to their contemporary worlds. The League of Nations and the International Committee on Intellectual Co-operation were clearly affected by world politics when working with their declaration regarding the teaching of history. The great powers – the USA, the UK and France – were keen to participate in the formation of the declaration, but then for various reasons did not sign the final document.[87] UNESCO also interacted with the contemporary context in its work with a "History of Mankind", which became plagued with difficulties when Western powers, small states and communist Eastern Europe were supposed to write a common history.[88] UNESCO was influenced by world politics, world-development questions and the desire to protect world heritage. The Council of Europe, with a base in human rights but also obviously anchored in Europe, had to take into account both powerful and less powerful countries' notions about history while at the same time protecting small ethnic groups. The formulation of international understanding, or "mutual understanding" as the Council of Europe first termed it, was a product of its time and the ideological curricula were influenced by the surrounding world.

In line with Goodlad's previous notions, the ideological curricula were in contact with other curricular levels through transactions and interpretations. In addition there were direct and mutual exchanges between the different levels.

In Figure 10.4, the two-way arrows denote how influences have been mutual, and they symbolize transactions as well as interpretations. The different curricular realities have interacted, directly and indirectly. The guidelines were formed under the influence of the surrounding society, civil servants, historians, teachers and ideas about students' interest in history. In the implementation, there have been not only direct transactions regarding the formal curricula but also indirect ones through, for example, history

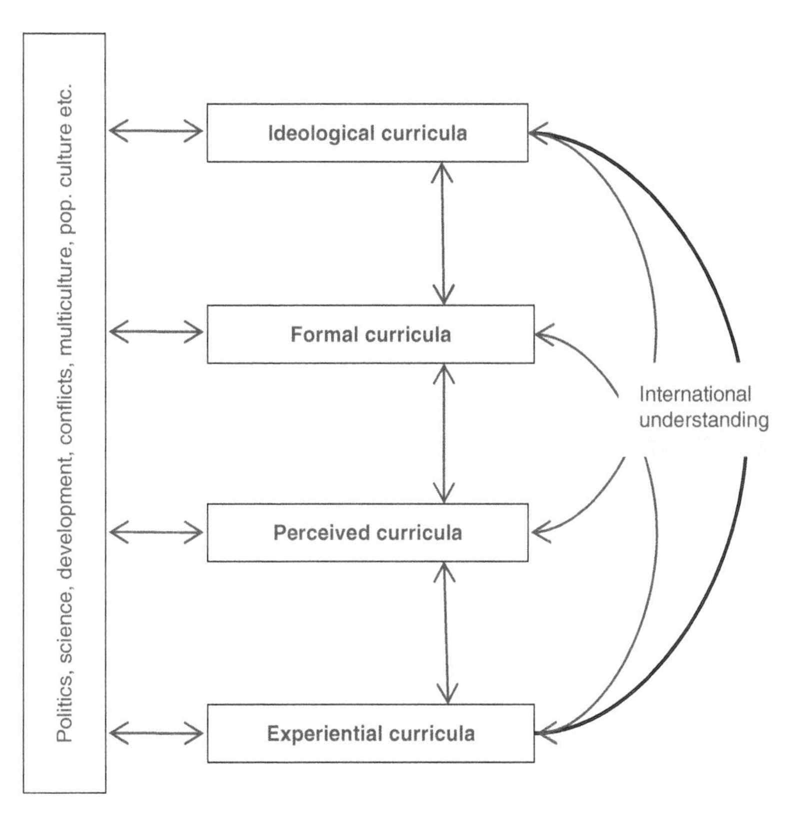

Figure 10.4 The relationship between the international guidelines for international understanding and other curricular levels and the world at large.

teachers acting as messengers and, influenced by international guidelines, debating the value of history teaching for international understanding. The formal curricula bore the stamp of international guidelines, and dispensed guidelines for teachers and students, but were also attentive to teachers' and students' opinions and their work in history. Teachers participated in international conferences, in debates and not least in the practice of teaching when meeting their students. Some teachers and students also took part in UNESCO's school project on international understanding.

Implementation may be described as (1) a sideways process where curricular levels contained different views of the means and goals of history teaching, which were affected by their own times and developments in society; (2) a top-down process where the international guidelines were incorporated into the formal, perceived and experiential curricula through transactions and interpretations via direct and indirect contacts; (3) a bottom-up process where teachers, and in an indirect way students, participated in how the subject of history was formulated and reformulated, nationally and internationally; (4) an independent creation of history, where teachers and students oriented themselves according to the international guidelines without direct links to these guidelines; (5) a disregard of guidelines in favor of other priorities – for example, cultural history, which during the interwar years was ignored because of lack of time in history lessons, and students who wrote essays on less or unacceptable subjects according to the international guidelines.[89]

Even if implementation seems to have been mainly a top-down process, the present study shows that other processes were also significant. The bottom-up process appears to have been weak, but it is nevertheless clear that students themselves were able to formulate history in manifold ways, within and beyond the guidelines.

World over European history

The present study shows that each curricular level can encompass different ideas. International understanding can be seen as spanning a number of different means and goals which tend in different directions. The orientations toward internationalism, diversity, cultural heritage and contemporary history reveal the breadth of the ideological curricular levels. The demarcation between UNESCO's global orientation and the Council of Europe's Eurocentrism exposes ideological differences in the ideological curricula. Students' disregard of more European history shows how implementation in both the formal curricula and in the teachers' debates does not necessarily bring about any changes at the student level.

One conceivable explanation for Swedish students' global orientations may be that world history was emphasized at all curricular levels up to the 1990s and was important even later. It is reasonable to assume that the teachers who focused on world history before 1994 did not fundamentally change their teaching when a new syllabus was introduced. UNESCO's ASP could also have been important for the global perspective of the experiential curricula. The Council of Europe lacked this direct link between the ideological curricula and the experiential because it was mainly concerned with the formal and perceived curricula.

Another contributing factor for Swedish students preferring world history – despite greater stress on European history – may be that Sweden is a small country in a, relatively-speaking, small part of the world. It is a small and officially neutral, alliance-free country between East and West, with an often prominent international profile regarding questions of peace, anti-racism and development in poor countries. For instance, students' growing interest in world history was contemporary to Dag Hammarskjöld being general secretary of the UN and active in the Suez Crisis, Lebanon and Congo;

and the prominent interest among students to study "the Negro problems in the US"[90] followed the international attention to Gunnar Myrdals' research in the USA, labelled "An American Dilemma: The Negro Problem and Modern Democracy".[91] Sweden's orientation toward more international understanding could be a result of more than just the international guidelines, but the efforts of UNESCO seem to have played a direct and indirect role in the shaping of a more peaceful and international focus among Swedish students' writing of history. Even if students have expressed an interest in war and dictators, this study shows that this interest was less about pro-militarism than the opposite. In the students' work that I reviewed, militarism was marginalized and dictators and war were treated, with a few exceptions, extremely critically.

Contemporary world history

The transformation of history teaching towards a more contemporary world history, and giving more and more attention to marginalized groups, can be seen as an international current affected by transnational, moral actors and part of the development of historical scholarship.[92] In this international development it is interesting to see how teachers could act as agents of change in their encounters with international guidelines, national reforms, debates and students. History teachers were co-creators of the prerequisites and content of history education.

It is also worth bearing in mind that the change in views of history education can also be explained by other factors. Teachers with different teaching strategies described how they participated in developing the subject, not on the basis of international guidelines but from personal interests, and developments in the rest of the world and in historical scholarship.

The contemporary world history orientation also encompasses a gender aspect since this orientation, in a comparison between boys and girls, was stronger among boys. Both girls and boys were mainly interested in contemporary world history but, relatively speaking, the girls set greater store by local and older history.

No impact

While the international guidelines seem to have impacted on the global orientation in history, local history was an orientation in Sweden before it was taken up internationally. The focus on local history could have been reinforced by international interest in it, but as a part of Swedish history education it was already established when UNESCO and later the Council of Europe began to note its value in preserving cultural heritage. Safeguarding the local can, of course, inflate local patriotic tendencies, but nationalism and a skeptical attitude toward immigration stood in sharp contrast to international aims. As previous research on democracy has affirmed, all good guidelines do not necessarily have good effects in practice.

The relationship between international guidelines and the study of local history shows that some of these guidelines did not need to be implemented and that one means does not always achieve its stated goal. The international understanding of Leninism and Maoism in the early 1970s exemplifies how this could also encompass dictators. Students criticized totalitarian systems but could also excuse dictators, arguing special cultural characteristics – an inbuilt problem in international understanding which teachers criticized.[93]

Towards unity in diversity

The change in the concept of culture from being defined as literature and art to leaning more toward the anthropological was clear in the Swedish debate. In the 1930s, cultural history was taken up in art classes in order to avoid taking time from hours in history, but at the beginning of the 1980s the value of history for cultural studies was asserted in arguments against ethnologists.[94] The view of the subject of history had obviously changed, and this change entailed that culture also concerned people's daily lives, traditions and patterns of behaviour, but it also meant that in the 1980s the value of history as a school subject had to be motivated. It has been claimed that a corresponding change occurred within UNESCO, where culture over time got a more anthropological definition.[95]

Back in 1949, UNESCO highlighted the importance of studying ways of life in other countries, even if these were not called "culture".[96] At the end of the 1940s, affinities with people in other cultures and countries were called "unity in diversity" in both UNESCO and the Council of Europe. Yet this positive view of diversity did not really penetrate until the 1980s, when diversity was prioritized in the Council of Europe's work with education. This was also when the council's guidelines concerning diversity were taken up in the Swedish history teachers' discussions. Possibly most clearly expressed was the desire for unity in diversity within UNESCO 2001, when a Universal Declaration on Cultural Diversity was delivered.[97] Swedish students' work in history showed that from the 1960s onwards, minorities and their negative treatment were dealt with in line with international guidelines. Students condemned colonialism before the Council of Europe did and repudiated racism and anti-Semitism. One group defined as marginalized by the Council of Europe and UNESCO was women. In the perceived curricula and experiential curricula, women's history and women's right's were noted and seemed to have gained a stronger position – even more recognized by Swedish teachers and students than in the international guidelines.[98]

Encounters with other peoples and cultures were part of students' historical consciousness, but in their work, artists like Leonardo da Vinci and Andy

Warhol were also discussed. Classical culture and popular culture cropped up in students' history essays, which could focus on Bruce Lee as well as on minorities. With empathy and interest, students addressed issues relating to the present and future in the experiential curricula, which contained more than what was in the guidelines and debates. Students' history seems to be more comprehensive than that of other curricular levels.

Students' historical consciousness seems to stretch outside the international guidelines and be palpably affected by the present. Teachers and students broadened the content of history education and oriented themselves in both world politics and local history, with perspectives that focused on minorities and popular culture in a subject clearly influenced by the present.

Shaping a better world?

Students' history seems more and more often to be filled with internationalism, and interest in contemporary history and diversity. Even if several of the students' future expectations contained fears of future conflicts, what was expressed were anxieties, not militarism. Entry ways into the past were often found in the international and local present day. Students studied the past from a variety of sources – primarily secondary sources, but also interviews and other primary sources were used to critically describe, but also closely, even empathetically, analyze the past, the present and the possible future.

When studying the recent and distant past, students articulated their understanding of minorities, the rejection of totalitarianism and war, and a desire to preserve cultural heritage – an inclusive, non-discriminatory history in the service of humankind. They take a clear position to shape a better world, wholly in line with UNESCO's guidelines.

Notes

1. This chapter presents results from the research project History Beyond Borders, directed by Daniel Lindmark and financed by the Swedish Research Council. The chapter is mainly based on Thomas Nygren, *History in the Service of Mankind: International Guidelines and History Education in Upper Secondary Schools in Sweden,* 1927–2002, PhD dissertation Umeå: Umeå University, 2011, containing reprints with permission from *Journal of World History, Journal of Educational Media, Memory, and Society,* and *Education Inquiry*. For a summary of the project's findings, see Henrik Åström Elmersjö, "History beyond Borders: Peace Education, History Textbook Revision, and the Internationalization of History Teaching in the Twentieth Century", *Historical Encounters* 1:1 (2014): 62–74. Special thanks to Christine Larson and Jan Teeland for valuable linguistic support.
2. Akira Iriye, *Global Community: The Role of International Organizations in the Making of the Contemporary World* (Berkeley: University of California Press, 2002) and Eric

J. Hobsbawm, *Age of Extremes: The Short Twentieth Century 1914–1991* (London: Joseph, 1994).

3. *A Handbook for the Improvement of Textbooks and Teaching Materials as Aids to International Understanding* (Paris: UNESCO, 1949), 10–15; Göran Andolf, *Historien på gymnasiet: Undervisning och läroböcker 1820–1965* (PhD dissertation, Uppsala: Uppsala Universitet, 1972), 8–23; Tomas Englund, *Curriculum as Political Problem: Changing Educational Conceptions, with Special Reference to Citizenship Education* (PhD dissertation, Lund: Studentlitteratur, 1986), 300f; Arie H.J. Wilschut, "History at the Mercy of Politicians and Ideologies: Germany, England and the Netherlands in the 19th and 20th Centuries", *Journal of Curriculum Studies* 42: 5 (2010): 693–723; Henrik Åström Elmersjö and Daniel Lindmark, "Nationalism, Peace Education and History Textbook Revision in Scandinavia, 1886–1940", *Journal of Educational Media, Memory and Society* 2 (2010): 63–74.
4. From the 1930s on, in Scandinavia, the Norden Associations were reviewing history textbooks with the intention of establishing a common ground of history in the Nordic countries, an effort described by UNESCO as "the most outstanding example so far of regional collaboration on textbook revision" according to *A Handbook for the Improvement of Textbooks*, 34. The efforts primarily to achieve peace through textbook research have become an academic focus. The role of history teaching and textbooks in conflict and post-conflict societies has been studied in, for example, East Asia, Rwanda, Cyprus, the Balkans and Northern Ireland. See Steffi Richter, ed., *Contested Views of a Common Past: Revisions of History in Contemporary East Asia* (Frankfurt: Campus Verlag, 2008); Sarah Warshauer Freedman, Harvey M. Weinstein, Karen Murphy and Timothy Logan, "Teaching History after Identity-Based Conflicts: The Rwanda Experience," *Comparative Education Review* 52:4 (2008); Yannis Papadakis, "Narrative, Memory and History Education in Divided Cyprus: A Comparison of Schoolbooks on the 'History of Cyprus' ", *History and Memory* 2 (2008); Jean-Damascène Gasanabo, *Fostering Peaceful Co-existence through Analysis and Revision of History Curricula and Textbooks in Southeast Europe* (Paris: UNESCO, 2006); Keith C. Barton and Alan W. McCully, "History, Identity and the School Curriculum in Northern Ireland: An Empirical Study of Secondary Students' Ideas and Perspectives", *Journal of Curriculum Studies* 37:1 (2005).
5. "League of Nations, International Committee on Intellectual Co-operation, Subcommittee of Experts for the Instruction of Children and Youth in the Existence and Aims of the League of Nations, 4–6 July 1927", C.I.C.I./E.J./24.(1), 1927, UN Archives, Geneva; *League of Nations Official Journal, 98th and 99th Council Sessions* 1937 appendix 1667 (Geneva: League of Nations, 1937); "General Conference, first session, Paris 1946, 1C, 1947, 149–153"; "Records of the General Conference of UNESCO, second session, Mexico, 1947, v. 2: Resolutions, 2C, 16, 20–21"; "Records of the General Conference of UNESCO, third session, Beirut, 1948, v. 2: Resolutions, 3C, 1949, 24", UNESCO Archives; *A Handbook for the Improvement of Textbooks*; *Resolution (52) 17. History and Geography Textbooks* (Strasbourg: Council of Europe, 1952); "Report of the Working Group No. 2", EXP/Cult (53) 33, Box 2747, 1953, Council of Europe Archives, printed in Edouard Bruley and E.H. Dance, *A History of Europe* (Leyden: A.W. Sythoff, 1960), 71–72.
6. John I. Goodlad, ed., *Curriculum Inquiry* (New York: Mc Graw-Hill Book Company, 1979), 22, 348ff.
7. For more thorough theoretical and methodological considerations, see Thomas Nygren, *History in the Service of Mankind: International Guidelines and History*

Education in Upper Secondary Schools in Sweden, 1927–2002 (PhD dissertation, Umeå: Umeå University, 2011).

8. Eckhardt Fuchs, "The Creation of New International Networks in Education: The League of Nations and Educational Organizations in the 1920s", *Paedagogica Historica* 43:2 (2007): 208.
9. "Final Text of the Resolutions on the Revision of School Text-books, Adopted by the International Committee on Intellectual Co-operation at its XIVth Plenary Session, July 1932", 189–192, Box 1750, Council of Europe Archives; Thomas Nygren, "International Reformation of Swedish History Education 1927–1961: The Complexity of Implementing International Understanding", *Journal of World History* 22:2 (2011): 329–354.
10. "League of Nations, International Committee", 8.
11. Jan Kolasa, *International Intellectual Cooperation: The League Experience and the Beginnings of UNESCO* (Wroclaw: Zakład Narodowy im. Ossolińskich, 1962), 77f; *A Handbook for the Improvement of Textbooks*, 107.
12. Conference for the Establishment of the United Nations Educational, Scientific and Cultural Organisation, held at the Institute of Civil Engineers, London, 1–16 November 1945 (Paris: UNESCO, 1946).
13. *A Handbook for the Improvement of Textbooks and Teaching Materials*; Nygren, "International Reformation of Swedish History Education".
14. UNESCO's first director-general, Julian Huxley, pointed out clearly in 1946 how peace through international understanding was central to the work for peace in a far too divided world, separated by ideologies and religions, according to Julian Huxley, *UNESCO its Purpose and its Philosophy: Preparatory Commission of the United Nations Educational Scientific and Cultural Organisation, 1946* (Paris: UNESCO, 1947), 6–7, 13, 61–62. Glenda Sluga has noted a lack of diversity in the organization and its thinking in the early years, but also how UNESCO and Huxley, in spite of the imperialist heritage, did highlight the vision of international understanding. See Glenda Sluga, "UNESCO and the (One) World of Julian Huxley", *Journal of World History* 21:3 (2010): 393–418.
15. General Conference, first session, Paris 1946, 1C, 1947, 149–153; Records of the General Conference of UNESCO, second session, Mexico, 1947, v. 2: Resolutions, 2C, 16, 20–21; Records of the General Conference of UNESCO, third session, Beirut, 1948, v. 2: Resolutions, 3C, 1949, 24, UNESCO Archives; *A Handbook for the Improvement of Textbooks and Teaching Materials*.
16. Edouard Bruley and E.H. Dance, *A History of Europe* (Leyden, 1960), 14.
17. "Report of the Working Group No. 2", EXP/Cult (53) 33, Box 2747, 1953, Council of Europe Archives, printed in Bruley and Dance, *A History of Europe*, 71. The recommendations are a literal reprint of the meeting notes from Calw.
18. Nygren, "International Reformation of Swedish History Education" as well as General Conference, first session, Paris 1946, IC, 1947, 149–153; Records of the General Conference of UNESCO, second session, Mexico, 1947, v. 2: Resolutions 2C, 16, 20–21; Records of the General Conference of UNESCO, third session, Beirut, 1948, v. 2: Resolutions 3C, 1949, 24, UNESCO Archives; *A Handbook for the Improvement of Textbooks and Teaching Materials*; "Report of the Working Group No. 2", EXP/Cult (53) 33, Box 2747, 1953, Council of Europe Archives, printed in Bruley and Dance, *A History of Europe*, 71, the recommendations are a literal print of the meeting notes from Calw.
19. Nygren, "International Reformation of Swedish History Education"; Thomas Nygren, "UNESCO and Council of Europe Guidelines, and History Education in

Sweden, c. 1960–2002", *Educational Inquiry* 2:1 (2011): 37–60; Thomas Nygren, "The Contemporary Turn: Debate, Curricula and Swedish Students' History", *Journal of Educational Media, Memory, and Society* 4:1 (2012): 40–60.

20. Poul Duedahl, "Selling Mankind: UNESCO and the Invention of Global History, 1945–76", *Journal of World History* 22: 1 (2011): 101–133.
21. Erland Sundström, ed., *Unesco: över alla gränser* [UNESCO: Beyond all borders], (Stockholm, 1962).
22. Heidi Haggrén, "The 'Nordic Group' in UNESCO" in *Regional Cooperation and International Organisations: The Nordic Model in Transnational Alignment*, ed. Norbert Götz and Heidi Haggrén (London: Routledge, 2009): 88–111.
23. Intergovernmental Conference on Education for International Understanding, Co-operation and Peace and Education Relating to Human Rights and Fundamental Freedoms, with a View to Developing a Climate of Opinion Favourable to the Strengthening of Security and Disarmament, Paris 12–20 April, 1983, ED-83/Conf.214/2 (Paris: UNESCO, 1983).
24. Symposium on History Teaching in the New Europe, Brugge, Belgium, 9–13 December 1991, Symposium on History, Democratic Values and Tolerance in Europé in Sofia, Bulgaria, 19–22 October 1994 and Seminar on History Teaching and Confidence-Building: The Case of Central and Eastern Europe, in Smolensk, Russian Federation, 26 April 1995–28 April 1995 in *Against Bias and Prejudice: The Council of Europe's Work on History Teaching and History Textbooks* (Strasbourg: Council of Europe, 1995).
25. Former East-European countries became members in the Council of Europe in the following years: 1990 Hungary; 1991 Poland; 1992 Bulgaria; 1993 Estonia, Lithuania, Slovenia, Czech Republic, Slovakia, Romania; 1995 Latvia, Albania, Moldova, Former Yugoslav Republic of Macedonia, Ukraine; 1996 Russian Federation, Croatia; 1999 Georgia; 2001 Armenia, Azerbaijan; 2002 Bosnia and Herzegovina; 2003 Serbia.
26. Full and Comprehensive Implementation of the Recommendations Concerning Education for International Understanding, Co-operation and Peace and Education Relating to Human Rights and Fundamental Freedoms, 1974, and the Preparation of the Elaborated Version of the Integrated Action Plan for the Development of International Education: General Conference, 27th session, 1993, 1–7; Records of the General Conference, Twenty-eighth Session, Paris, 25 October to 16 November 1995; Resolutions (Paris: UNESCO, 1996), 63.
27. *Declaration and Integrated Framework of Action on Education for Peace, Human Rights and Democracy*, Report of the 44th Session of the International Conference on Education endorsed by the General Conference of UNESCO at its twenty-eighth session, Paris, November, 1995 (Paris: UNESCO, 1995), 1–9.
28. Nygren, "UNESCO and Council of Europe Guidelines".
29. *Report of the Director-General on the Activities of the Organization in 1947* (Paris: UNESCO, 1947), 13; Council of Europe Consultative Assembly, first session, 10 August–18 September 1949. Reports part IV, sittings 12–15 (Strasbourg: Council of Europe, 1949), 750.
30. Records of the General Conference, 16th session, Paris, 12 October to 14 November 1970, v. 1: Resolutions (Paris: UNESCO, 1971), 51. In the 1990s, after many years of work, UNESCO published *General History of Africa, History of Civilizations of Central Asia, General History of Latin America, The Different Aspects of Islamic Culture* and *General History of the Caribbean.*

31. Otto-Ernst Schüddekopf, *History Teaching and History Textbook Revision* (Strasbourg: Council of Europe, 1967), 11, 126.
32. Resolution (85) 6 on European Cultural Identity 1985.
33. Nineteenth Session of the Standing Conference of European Ministers of Education, Education 2000: Trends, Issues and Priorities for Pan-European Co-operation (Strasbourg: Council of Europe, 1997).
34. Recommendation 1283 (1996) on History and the Learning of History in Europe 1996; Recommendation Rec (2001) 15 on History Teaching in Twenty-first Century Europe 2001; Resolution Res (2003) 7 on the Youth Policy of the Council of Europe (2003).
35. Records of the General Conference, 31st Session, Paris, 15 October–3 November 2001, v 1: Resolutions (Paris: UNESCO, 2002), 62.
36. Ibid., 64.
37. Recommendation No R (85) 7 of the Committee of Ministers to Member States on Teaching and Learning about Human Rights in School (1985), 3.
38. UNESCO in 1978 issued a declaration against racism: Records of the General Conference, 20th Session, Paris, 24 October–28 November 1978, v. 1: Resolutions (Paris: UNESCO, 1979), 60–65. See also Comprehensive Report by the Director-General on the World Situation in Fields Covered by the Declaration on Race and Racial Prejudice and Recommendations Made with a View to Promoting Implementation of that Decision: General Conference, 21st Session (Paris: UNESCO, 1980), 45; Records of the General Conference 22nd Session, Paris, 25 October – 26 November 1983: Resolutions (Paris: UNESCO, 1984), 74–76. UNESCO had highlighted these problems earlier but during the 1980s its stance against discrimination in all forms was developed and clarified. Discrimination against women was condemned in a UN convention in 1979: The Convention on the Elimination of All Forms of Discrimination against Women (1979).
39. Records of the General Conference 23rd Session Sofia, 8 October–9 November 1985, v. 1: Resolutions (Paris: UNESCO, 1985), 104. See also Sunil Amrith and Glenda Sluga, "New Histories of the United Nations", *Journal of World History* 19: 3 (2005): 266–269.
40. Recommendation 1283 (1996) on History and the Learning of History in Europe (1996).
41. "Recommendation Concerning the Safeguarding and Contemporary Role of Historic Areas" in Records of the General Conference, 19th Session, Nairobi, 26 October–30 November 1976, v. 1: Resolutions (Paris: UNESCO, 1977), annex 1, 1–29.
42. Ibid., 28.
43. Records of the General Conference 23rd Session Sofia, 8 October–9 November 1985, v. 1: Resolutions (Paris: UNESCO, 1985), 61–63.
44. Convention for the Protection of the Architectural Heritage of Europe (European Treaty Series no. 121) (Strasbourg: Council of Europe, 1985), article 15.
45. "Recommendation Concerning the Safeguarding", Recommendation No R (98) 5 of the Committee of Ministers to Member States Concerning Heritage Education (Strasbourg: Council of Europe, 1998).
46. Nygren, "International Reformation of Swedish History Education"; Nygren, "UNESCO and Council of Europe Guidelines"; Nygren "The Contemporary Turn".
47. Nygren, "International Reformation of Swedish History Education".

48. *Aktuellt för historieläreren* 1:16 (1970).
49. This was formulated remarkably like what the history teacher, Rudolf Hultkvist from the syllabus group, came away with from one of the Council of Europe's conferences: *Aktuellt för historieläreren 1: 16 (1970); Läroplan för gymnasieskolan Lgy70: Supplement 71: II Historia 10* (1970).
50. Synthesis of Member States Reports on the Application of the Recommendation concerning Education for International Understanding, Cooperation and Peace and Education relating to Human Rights and Fundamental Freedoms (Paris: UNESCO, 1983), 11, 20, 35.
51. Nygren, "UNESCO and Council of Europe Guidelines".
52. UNESCO's recommendation on international understanding, http://web.archive.org/web/20060106092002/http://www.skolverket.se/sb/d/469/a/1843 (accessed 25 March 2015).
53. Nygren, "International Reformation of Swedish History Education".
54. Nygren, "UNESCO and Council of Europe Guidelines".
55. Brita Nilsson, Allan Degerman, Vilhelm Scharp, Waldemar Lendin, Sixten Björkholm, Birgit Rodhe, Erik Brännman, Hans Lennart Lundh, Ivar Seth, Rudolf Hultkvist, Gunnar Ander, Bert Mårald, Göran Behre and Ola Lindqvist were all active internationally at meetings arranged by the Council of Europe and UNESCO, and they contributed to the debates in the teaching journals.
56. *Tidning för Sveriges läroverk* [Swedish Upper Secondary Teachers' Journal] (1954), 780, 785.
57. Nygren, "International Reformation of Swedish History Education".
58. Rapport över verksamheten som gymnasieinspektör 1977–1978, F IIIa, Undervisningsavdelning för skolan (S-avd) 1970–1982, National Archives, Sweden.
59. *Aktuellt för historieläreren* 1 (1970), 13–18.
60. International Symposium on the Occasion of the 40th Anniversary of the Associated Schools Project: The Associated Schools Project: A Review of its Expansion and Development (Paris: UNESCO, 1993).
61. The initial 15 countries were Belgium, Costa Rica, Ecuador, the Federal Republic of Germany, France, Japan, the Netherlands, Norway, Pakistan, Sweden, Switzerland, the UK, the USA, Uruguay and Yugoslavia. ASP included 33 schools in 15 countries in 1953, and in 2003, 7400 schools in 170 countries, according to *UNESCO Associated Schools Project Network (ASPnet): Historical Review 1953–2003* (Paris: UNESCO, 2003), 3, 25.
62. European Cultural Convention, Paris 19.XII. 1954, European Treaty Series, no. 18 (Strasbourg: Council of Europé, 1954). The orientation towards teachers was obvious in all of the Council of Europe's recommendations and meetings concerning history education from 1953 to 2002.
63. International Symposium, 25–27; *Global Review of UNESCO's Associated Schools Project Network*, Centre for International Education and Research, University of Birmingham, UK (Paris: UNESCO, 2003), 3.
64. *Global Review of UNESCO's Associated Schools Project Network*, 32.
65. Nygren, "International Reformation of Swedish History Education"; Nygren, "UNESCO and Council of Europe Guidelines".
66. However, that students treated the essay subjects from a more global point of view is doubtful; in 1952, "Japan During the Last Hundred Years" could very well be a political historical analysis of Japan's build-up before and participation in World War II.

67. Total number of students writing exam essays: 1947, 4,164; 1949, 4,167; 1950, 4,315; 1951, 4,315; 1952, 4,763; 1953, 5,022; 1954, 5,307; 1955, 5,845; 1956, 6,566; 1957, 7,319.
68. *Historielärarnas förenings årsskrift* (1957): 10–11.
69. *Tidning för Sveriges läroverk* (1957): 228.
70. Karen Mundy, "Educational Multilateralism in a Changing World Order: UNESCO and the Limits of the Possible", *International Journal of Educational Development* 19 (1999): 31–48.
71. Nygren, "International Reformation of Swedish History Education"; Nygren, "UNESCO and Council of Europe Guidelines".
72. Nygren, "UNESCO and Council of Europe Guidelines"; Nygren, "The Contemporary Turn".
73. Some 40 of 156 girls (26 percent) had a man's name in the title, while 25 out of 122 boys (20 percent) had a man's name in the title.
74. Nygren, "UNESCO and Council of Europe Guidelines".
75. I. Bergstedt, *Negerfrågan i Förenta Staterna* (Gävle: Vasaskolan, 1969), 5.
76. Citations from R. Ovaska, *Sydafrika förr och nu* (Gävle: Vasaskolan, 1971), 14 and M. Holmberg and P. Lööf, *Sydafrika* (Gävle: Vasaskolan, 1987), 11.
77. Nygren "UNESCO".
78. M. Petterson, *Händelserna på himmelska fridens torg 1989* (Gävle: Vasaskolan, 1992)
79. G. Svensson, *Hitlers ungdomsår* (Gävle: Vasaskolan, 1969), 3.
80. H.-H. Helldahl, *Vietnamkriget* (Gävle: Vasaskolan, 1998)
81. O. Brännström, *Baltutlämningen* (Gävle: Vasaskolan, 1985), 6.
82. M. Ranudd, *Finska vinterkriget 1939–1940* (Gävle: Vasaskolan, 1987).
83. M. Olsson, *Zigenare* (Gävle: Vasaskolan, 1991), 2
84. In Bengt Schüllerqvist's previous study of a teacher's experiences, it was held that "the benefactors of humankind" replaced national heroes: Bengt Schüllerqvist, *En lärares bildningsgång: En biografisk studie av ideal, tradition och praxis i svensk läroverksmiljö* (Uppsala: Pedagogisk forskning i Uppsala 143, 2002), 155.
85. *Appendix to grade report,* Södra Latin, Stockholm.
86. Nygren, "International Reformation of Swedish History Education"; Nygren, "UNESCO and Council of Europe Guidelines"; Nygren, "The Contemporary Turn".
87. *A Handbook for the Improvement of Textbooks and Teaching Materials*, 104–107.
88. Duedahl, "Selling Mankind", 130.
89. Nygren, "International Reformation of Swedish History Education".
90. In 1969 some 2 percent of the students writing individual projects in history wrote about "negro problems" when studying the US Civil Rights Movement, *School Archives, City and Municipal Archives.*
91. See Gunnar Myrdal, *An American Dilemma: The Negro Problem and Modern Democracy* (New York: Harper & Brothers, 1944), reprinted in 1962 and 1969.
92. In higher education, more international contemporary and multicultural university curricula was noted by David John Frank, Suk-Ying Wong, John W. Meyer and Francisco O. Ramirez, "What Counts as History: A Cross-National and Longitudinal Study of University Curricula", *Comparative Education Review* 44:1 (2000): 46f. Arie Wilschut has described how the development of history didactics could follow similar and different directions in European countries. See Wilschut, "History at the Mercy of Politicians and Ideologies".
93. At the time, history teachers debated the problem of fostering both democratic ideas and understanding of foreign countries governed by undemocratic rulers

and ideologies. However, writers agreed that international history was important: *Aktuellt för historieläreren* 2 (1970): 10–14; ibid. 3 (1970): 26–29; ibid. 4 (1970): 26–29; ibid. 2 (1971): 25f. See also Nygren, "UNESCO and Council of Europe Guidelines".

94. Nygren, "UNESCO and Council of Europe Guidelines". For the discussion about how the subject of art should deal with culture, see *Tidning för Sveriges läroverk* (1935): 234 and ibid. (1939): 285f, 325f; Nygren, "International Reformation of Swedish History Education".
95. Katérina Stenou, *UNESCO and the Issue of Cultural Diversity: Review and Strategy, 1946–2004* (Paris: UNESCO, 2004), 3f.
96. *A Handbook for the Improvement of Textbooks and Teaching Materials,* 78f.
97. Records of the General Conference, 31st session, Paris, 15 October–3 November 2001, v. 1: Resolutions (Paris: UNESCO, 2002).
98. Nygren, "UNESCO and Council of Europe Guidelines".

11
UNESCO and the Improvement of History Textbooks in Mexico, 1945–1960

Inés Dussel and Christian Ydesen

Introduction

In the context of World War II, education for peace seemed not only necessary but also urgent. One of the first goals proclaimed by UNESCO was to cultivate "unity in diversity" in order to achieve a better cross-cultural relationship and cooperation between diverse human communities. A central focus area was the improvement of geography and history textbooks in the member states because, as was seen by UNESCO in those years, most entrenched forms of prejudice are established in the formative years of childhood. Education had to move from reflecting nationalistic attitudes to promoting a spirit of mutual understanding.[1]

This focus meant that UNESCO, from its very inception, pursued an agenda of becoming a clearing house for the analysis and revision of textbooks and to offer consulting services to member states, as well as to continually urge them to examine their textbooks.[2] A special program for textbook improvement as a means to promote international understanding was launched in 1946, and it included the organization of a number of seminars and conferences with delegates from member states and expert meetings on the subject of international understanding in general and the improvement of textbooks in particular.[3] The work done on these seminars was subsequently transformed into guidelines for textbook revisions and a number of publications with the aim of inspiring member states to pursue programs of textbook revision.[4]

For Mexico, which had been among the first countries to join UNESCO and the first one to establish permanent representation at the organization, the revision of textbooks had been a policy strategy for quite some time, and had particular resonances in the context of its own political and educational situation.[5] As early as 1924 the need to improve textbooks was considered by the Sixth Scientific Pan American Conference of Lima, and in 1938 Mexico joined the treaty signed by Brazil, Argentina and Uruguay in 1933 providing for the periodic revision of their national history textbooks, "deleting all passages likely to arouse hostility against any American nation".[6] This agreement was the first international accord to deal with the problem of textbook improvement. Mexico also signed the convention of the Inter-American Conference for the Maintenance of Peace meeting in Buenos Aires in December 1936, considering, among other problems, the revision of textbooks as a means to promote international understanding.[7] It also established a National Commission for the Revision of Study Plans and Textbooks in 1944 under the leadership of Jaime Torres Bodet (1902–1974), a seminal figure who will appear repeatedly in this chapter.[8] Torres Bodet was then the secretary of education of Mexico, and organized the commission to align the educational system to the new political times of "national unity" and deradicalization of the legacy of the Mexican Revolution. Hence, by 1946 Mexico had accumulated considerable experience in this endeavour.[9]

Thus a picture can be drawn of merging – and intertwined – historical lines of development centered on the improvement of textbooks for promoting international understanding. The case of Mexico offers the possibility of creating new knowledge at both a theoretical and an empirical level with relevance for the field of education history in general and the history of international organizations in particular. Theoretically, the case creates the opportunity to understand the complex dynamics of how traveling ideas, knowledge and practices intermesh with international, national and local practices. This is a research agenda closely linked with the "spatial turn" in history of education research.[10] The spatial turn transcends the notion of place, being so intimately connected with the idea of nations as subjects and national narratives, and moves toward the notion of networks of knowledge and people that flow through and across spaces that might or might not be nationally bounded.[11] Empirically, the case contributes to a deeper understanding of the historical impact of UNESCO and the development of Mexico's education policies and practices in the early years after World War II, as well as the international connectors and mediators associated with it.

This chapter is based on archival work done both in Paris and Mexico City, and it will focus on the participation of key Mexican educators and historians in the UNESCO initiative for the revision of school textbooks in the 1940s and 1950s in order to determine the contextual spaces in which they operated and had an impact – within both UNESCO and Mexico. In this undertaking we draw on Martin Lawn's work on the international experts of the International Examinations Inquiry in the 1930s, whom he describes as highly visible in their national communities "but invisible in their internationalism".[12] It is our intention to bring light to the trajectory of relevant actors who moved around national and international arenas, and helped define educational problems and policies. Our second analytical tool is to compare the UNESCO discourse on textbook revision in general and the UNESCO guidelines on textbook revision in particular with Mexican policies, initiatives and actual textbooks, being aware that UNESCO was not an abstract entity but a particular field of agents and discourses that, at that time, included important Mexican actors.[13]

Our analytical approach may be summed up in the following research questions: Which routes and spaces did the UNESCO discourse and guidelines on textbook revisions follow in order to reach Mexico? Were these routes and spaces travelled one way, from Paris to Mexico, or were loans and ideas circulated in both directions? How did UNESCO's policy and guidelines merge or intertwine with already existing policies and practices?

UNESCO–Mexico relations

In 2005, when UNESCO was celebrating its 60th anniversary, Pablo Latapí, a renowned Mexican educator then ambassador to UNESCO, said that "there was plenty to celebrate" in this relationship.[14]

He recalled how Mexico had been present since the first organizing conferences, and was among the first countries to sign the Constitutive Act. Its delegate, Manuel Martínez Báez (1894–1987), was elected vice-president of the Executive Council. Mexico was particularly involved in the Project on Basic Education, on which meetings were organized in Paris, Nanking and Mexico City – the latter in November 1945. During those formative years, its participation did not cease to increase: Mexico's protagonism was evident in the organization and development of the Second General Conference in Mexico City, held in November–December 1947. In December 1948, at the Beirut conference, Torres Bodet was elected director-general of UNESCO from 1948 to 1952. He had already been Mexico's delegate, as secretary of education, at the 1945 London conference for the establishment of UNESCO. In 1951, with the creation of CREFAL in Pátzcuaro, Michoacán, still in operation, the presence of UNESCO in Mexico and of Mexico in UNESCO became solidified.

In relation to textbooks, Mexico had already been the host of an Inter-American Conference on the Problems of War and Peace in February 1945, which recommended explicitly the revision of textbooks. Peace, one of the recommendations said, "cannot rest solely on economic measures", hence the need to "recommend to the governments of American republics that everything that, directly or indirectly, supports racist or totalitarian theories or that might compromise friendly relationships between the states of the continent be suppressed from official textbooks used in schools".[15] There had been recent meetings and commissions to revise textbooks, not only to appease political passions in the country but also as part of a will to improve US-Mexican relationships, and it is therefore no surprise that in 1946, on the occasion of the meetings of a UNESCO preparatory commission, Mexico submitted a three-page "Proposal for a textbook on world history" in which it said: "the teaching of world history must (henceforth) be subordinated to the supreme human ends of international justice and brotherhood".[16] Although the proposal was not passed due to the controversies around the idea of promoting world history at the expense of national history, it shows that Mexico was already an active player trying to exert influence on the newly established organization.

However, Mexico's links with UNESCO within the fields of international understanding and textbook revisions are not only visible at the top diplomatic level. Following the routes out of the UNESCO building in Paris, a comprehensive circular letter was sent to all member states in March 1949. It contained a copy of a model plan for textbook revisions, UNESCO's guiding principles and criteria for undertaking such a revision, and a bibliography organized according to language. The letter invited the member states "to study their own textbooks from the point of view of their effect on international understanding" and to report to UNESCO by the end of September.[17] It is unclear whether Mexico actually submitted a report to UNESCO in response to the letter but the model plan for the analysis of school textbooks seems to have attracted some attention in the country. In April 1949, acting director Pablo Campos Ortiz of the UN's Information Office in Mexico wrote to UNESCO to obtain a copy of the model plan.[18]

Information also flowed in the other direction. On the seminal 1947 Sevres seminar, Mexican education policies in practice attracted attention. The Mexican "cultural missions" combined with rural schools were emphasized as an example of providing education for all sections of the community,

> where school and teacher develop and maintain effective and understanding relations with all sections of the surrounding community, the results are likely to provide a demonstration which will greatly affect learning and development of social relations and intergroup understanding outside the specific community concerned – first, in regional relationships and, later, in the field of international understanding[19]

And in August 1951, UNESCO wrote to the member states to obtain a systematic record of education for international understanding initiatives. At the receiving end in Mexico was Sergio Berdeja of the Asociacion Mexicana por la ONU (UN) who was asked to send "information on any experiments known to you which have been designed to help children of school age to understand other countries and to appreciate the social, economic and cultural interdependence of the world and the need for international collaboration".[20]

Plus if we look at the UNESCO seminars mentioned earlier, professor of history at the University of Mexico, Ernesto de la Torre Villar (1917–2009), participated in a working group on world history at the 1950 Brussels seminar.[21] José Muro-Mendez, professor of contemporary history at the Superior Normal School in Mexico City, participated in the 1951 Sevres seminar and in the 1952 Seminar on Education for World Citizenship.[22] As a follow-up to the seminar, a regional meeting entitled National, Continental and World Aspects in the Teaching of History in the Americas was held in Puerto Rico in April 1954, and Mexico was again among the countries invited to send participants.[23]

Thus a picture can be drawn of significant channels of interchange between Mexico and UNESCO, something that increased significantly with the election of Torres Bodet as director-general in 1948. However, all was not bliss in the relations between Mexico and UNESCO. The diverging political goals and ideas about what UNESCO should and could be were most visible during the years in which Mexico occupied the directorship. In his memoirs, Torres Bodet recalled this time as "the bitter years at UNESCO", which he equated to "being in a desert".[24] He said that during the four years, "which I devoted exclusively to UNESCO – I would have to fight, day after day, to attempt to get resources that, day after day, would be denied with the utmost correct and reluctant official deference. UNESCO was," he continued in a disillusioned tone

> after all, a luxury for its wealthier patrons. And they tried to maintain that luxury at the lowest possible price. How many words I had to say, how many trips I had to make, and how many reports I had to write to get some small increments, always diminished by a series of obstacles, wisely planned so that the budget would never be fully disbursed! In that fight, I was ultimately defeated. But it had to be fought, and I did that.[25]

In an analysis of Torres Bodet's participation in UNESCO from the point of view of diplomatic history, a historian once stressed Bodet's ability as a diplomat but also his disappointment with the weight that the Cold War logic and the international imbalance between countries imposed on the establishment of educational and scientific programs.[26] According to the same study, Torres Bodet was appalled to see how political convenience overdetermined what could have been large projects of international cooperation and the promotion of universalistic values, and was also upset by the weakness of the political power of UNESCO, torn between powerful blocs.

But what was it that Torres Bodet brought to UNESCO? What were the experiences he had developed in Mexico, and how did they dialogue with the initiatives on textbooks?

Textbook revision in Mexico and Jamie Torres Bodet

The idea of national textbooks had been present in Mexico at least since the foundation of the Secretary of Education (Secretaría de Educación Pública,

SEP) in 1921. José Vasconcelos (1882–1959), one of the most important intellectuals of his time and the first secretary of education, considered that books were the vehicles of civilization, and that the state had to promote public policies to ensure that they became available, for which he advocated national publishing houses and popular libraries. Literacy was a scam if it was not accompanied by a multiplication of books that would enable literate citizens to put into practice their new knowledge.[27]

Vasconcelos proposed having a publishing house at the SEP that was to print classical texts to be distributed freely among schoolchildren.[28] In the years that followed, SEP was to become a "systematic publisher of textbooks" but Vasconcelos' zeal for erudite and classical works was replaced by "useful readings", such as "Cuore" by Edmondo D'Amicis, valued for its patriotism and abnegation.[29]

In the period 1934–1940,[30] the peak of the radicalization of the legacy of the Mexican Revolution under the leadership of General Lázaro Cárdenas, and when state education became synonymous with "socialist education", SEP established a publications department called Popular Publishing Commission (Comisión Editora Popular), which printed millions of textbooks to be distributed – either freely or at the low cost of 7 cents – to all schoolchildren.[31] The books were not specifically commissioned by SEP, as would later be the case, but were judged to be in line with the goals and philosophy of socialist education. The judgment was made by a newly organized Commission for the Revision of Textbooks and Reference Books that was to regulate the offers of private publishers and recommend which textbooks could be used in classrooms. This was hotly contested by authors and the publishing house.[32]

From 1940, with the government of Manuel Avila Camacho, there was another turn in education, away from radicalization and towards "national unity". Under the umbrella of World War II, Avila Camacho slowly moved out of the ideal of a socialist education and promoted instead an ideology of reconciliation and realignment with the USA. During this period, textbooks became a battlefield between opposing views. On the one hand, there were the defenders of socialist education, which persisted in commissions such as the one for textbooks.[33] On the other hand, there were the *avilacamachistas*, who were in favor of a conciliatory narrative, an "education without hate", and an internationalist, pan-American ideal that included the USA. Since 1940 there was a Commission for the Revision of Textbooks which basically regulated and censored which texts were to be used in schools.[34]

When Torres Bodet became secretary of education in 1943, he took as one of his goals to promote a revision of the textbooks so as to complete the shutting off of the socialist textbooks and also to order and regulate the production and circulation of books among schoolchildren.[35] He established a new Commission for the Revision and Coordination of Educational Plans,

Study Programs and School Textbooks in 1944. This organization, which included the revision of curricular texts and textbooks altogether, set the guidelines for textbooks, following the revision of curricula and study plans, which were to be produced privately but approved and subsidized by the state. The guidelines were to include pedagogical recommendations for textbooks, which would follow up-to-date, active and progressive pedagogies.[36] The textbooks were to turn away from "inappropriate advice" for children, probably a euphemism for indoctrination, and ground teachings on concrete and clear examples, combining imagination with the essential needs of daily life.[37]

Torres Bodet stated as the main goals of education those of peace, democracy and social justice, which were in line with the values soon to be adopted by UNESCO. His thoughts about textbooks are clearly expressed when it comes to the writing of national histories, a sensitive issue about which he was well informed. In his speech to the First Conference of Round Table for the Study of the Problems of the Teaching of Mexico's History in May 1944, which took place three months after the launching of the Commission for the Revision of Plans and Textbooks and shows how closely he wanted professional commissions to collaborate with state policies. He was quite eloquent about the effects of "washing out" all traces of conflicts and passions.

"It is good that hate is cancelled out of the history books of our Fatherland," Torres Bodet wrote.

> It is good, too, that a cleansing campaign is launched to strip the negative pages off the texts. As Secretary of Education, I will applaud all that is made in that direction, but as a public servant and also as a man, I will always make sure that, in our quest for getting rid of our animosities, we do not end absurdly confusing judgments with prejudices, and putting a hypocritical and flickering veil over the grievances of the past – that are history and, as history, teachers – that would give the new generations an improper impression of our life and that, by disfiguring the arduous affairs that they tried to solve, would put Mexico's heroes in the awkward position of protagonists without content and beings who fought against phantoms.[38]

The perils of a "consensus history" that would produce a "harmless, smooth, and harmonized history" that does not satisfy anyone, as would later threaten the History of Mankind project of UNESCO, were evident to Torres Bodet.[39] However, probably more important for him was that history textbooks contributed to fostering a national identity that had to connect with emotions, and that goal could not be reached through pasteurized histories. For him, historians had to "understand the people, feel the people and feel it simultaneously in the spontaneity of the masses and the specific quality of the heroes".[40] He was an admirer of the Italian philosopher Benedetto Croce and was close to historians Arnold J. Toynbee and Oswald Spengler's notion that the history of the world was one of interconnections.[41] For Torres Bodet, history does not provide any immediate design for the future, but neither can it endorse an extreme objectivity. It teaches lessons that are an indispensable aid to understanding the present, and to conceiving of the future without prophetic dogmatisms.[42]

How did professional historians react to these initiatives by the secretary of education? There was widespread concern about the teaching of history, and that is why they joined SEP's efforts for the revision of programs and textbooks. The already mentioned First Conference for the Study of the Problems of the Teaching of Mexico's History was organized by prominent historians such as Silvio Zavala and Rafael García Granados.[43] In their resolutions from 1944, they concluded that textbooks had to be updated, including the latest anthropological and historical knowledge, and that they had to take into account psychopedagogical research on Mexican children and youth already developed by different departments of SEP.[44] They also recommended to SEP that textbooks of universal history that have an inaccurate view of Mexico should be taken out of schools.

The conclusions and recommendations of the conference were channeled through the History Commission of the Institute of Pan American History and Geography, led by Silvio Zavala (1909–2014) from 1947 to 1965, which [the Commission] can be seen as a strong academic partner of the efforts of the secretary of education to improve the teaching of history. Zavala's work, which will be discussed in the next section, was also related to the UN and UNESCO, maybe in more powerful and lasting terms than Torres Bodet's.

However, Torres Bodet's experience in this period pre-UNESCO certainly helped to envision the relevance of textbooks. In 1949, under his directorship, the model plan mentioned earlier was suggested to member states in order to analyze textbooks. It had some of the features of the Mexican Commission – a mixed composition of educators and specialists who carried out analyses of specific disciplines – and proposed some guiding principles, such as accuracy, fairness, worth, comprehensiveness and balance, worldmindedness and international cooperation, which expressed some of the concerns of the Mexican debate about saving the national needs for a unified identity.[45] However, textbooks were considered as a key point of entry for an educational reform that would promote international understanding. In his opening remarks to the UNESCO Conference on History Textbooks in Brussels in 1950, Torres Bodet said that the question of school textbooks was one "of which we must not underestimate either the gravity or the difficulty, for, while the teacher's prestige with his pupils may often be considerable, the textbook always comes to them invested with all the authority of the printed word". And he continued:

> Not merely is it hard to cast doubt on its assertions, but what is more serious, it presents a particular point of view and a particular picture of the world as self-evident, and thus all too often sows the seed for a rank crop of hasty reflexes, snap judgments and emotional reactions. The teacher can arouse and foster the critical sense of his pupils; by its very nature the textbook tends to be dogmatic, conclusive, official – hence both its compilation and the way it is used should receive our most vigilant attention.[46]

The work of Silvio Zavala and the value of world history

The second thread to follow on from how UNESCO contributed to, and was shaped by, Mexican education, particularly in the field of school textbooks, is the work done by Silvio Zavala, who was a central figure among Mexican historians throughout the 20th century.[47]

Zavala worked for the UN in 1945–1946, and then had a close involvement with UNESCO. He helped prepare the Second UNESCO Conference in Mexico City in 1947, producing papers and speeches for the secretary of education. He went to the Paris Congress of Historical Sciences in 1950 and was strongly involved in the History of Mankind project developed by Julian Huxley, Joseph Needham, Lucien Febvre and others, particularly when the Brazilian historian Paulo Carneiro came to direct the project.[48] He participated as a group leader in the Sevres seminar in 1951, and, after some years of trying unsuccessfully, he was named as the permanent delegate of Mexico to UNESCO from 1956 to 1958.[49]

However, his intense participation in UNESCO did not overshadow his involvement with Mexican academic and cultural politics. Instead, it can be said that it gave him a world-class experience that would benefit him greatly locally. There is no doubt that during the late 1940s and early 1950s Zavala was an important leader in the public debates on the teaching of history in Mexico, and with that background he came to participate in UNESCO's meetings and projects. It was not his friendship with Torres Bodet, which does not seem to have been close, as seen in his letters of those years, that drove him to UNESCO during the latter's tenure as director-general.[50]

By 1948 he was already a consultant for the UN. Having worked in the New York office of UNESCO, he was asked to write a handbook about how to include the history and goals of the UN in the teaching of civics and history for the 1951 Sevres seminar, an offer that he felt livened up old concerns about teaching international understanding but that he rejected because he lacked the time to do it properly.[51] In this letter he advanced a thoughtful opinion about the challenges that UNESCO's projects were facing at that time:

> It is very important to keep the problem of generations present. There is a will to teach peace to the generation that grows, but this is done by a generation that has not been able to reach it for itself. This undermines its authority in front of young minds. I don't know at the moment which could be the historical or pedagogical answer to this problem.[52]

Zavala seemed to be genuinely interested, both as a committed citizen and official and as a historian, about this quandary. How can a country overcome a civil war as the one that Mexico had suffered after the revolution? How could peace be achieved in territories where just a few years earlier there was only hate and destruction? Could the teaching of history help in that process? In the posing of these problems, some sort of distance to the optimistic rhetoric of UNESCO can be seen, but at the same time he seemed sympathetic to its goals and purposes.

These concerns are probably what kept him involved in pedagogical projects. In a 1982 interview, when asked about his commitment to the renovation of the teaching of history, he listed four actions: his participation at the commission of SEP for the revision of the teaching of national and international history; his involvement with the Pan American History Commission; his attendance at the Sevres seminar in 1951; and the writing of a universal history in three volumes with the Mexican-Italian professor Ida Appendini, also a member of the History Commission in 1944.[53] It is clear that most of his educational projects took place in the 1950s and in relation to UNESCO and international initiatives.

Probably the field in which he obtained the most experience, and which changed his participation in UNESCO, was the World History project. He was convinced that if history was to serve the purposes of international understanding and peace, it should stress the connections and interdependences. His writing of a textbook on universal history for secondary schools and also his proposal to introducing US history in primary and secondary schools as a separate subject were influenced by his participation in the Pan American Institute and UNESCO.

At the same time, he was not in favor of diluting the appeal and relevance of national histories, which was probably related to his being an official of Mexican historical institutions that were so central to the legitimation of the Mexican state.[54] This is visible in one of his articles, "Historical Museums and International Understanding", published in the quarterly review *Museum* published by UNESCO in 1954. Zavala argued against giving up national history museums in favor of museums dealing with world history in the endeavor of promoting international understanding. Rather, he defended the value of exhibitions and "international rooms showing interesting aspects of the history of other peoples".[55] In that respect he kept a cautious distance from the UNESCO discourse where world history – although controversial among many member states – took a priority position.[56]

As already indicated, a significant experience in Zavala's involvement with UNESCO was his participation at the 1951 Sevres seminar.[57] He was asked to act as one of the four group leaders, and in that capacity he had to organize the discussion, give turns and produce a final report entitled "Teaching History to Pupils above 15 years of Age". The group had to read and discuss a document about world understanding and the teaching of history.[58] Zavala kept lots of papers relating to this experience, and they are available in the special collection he donated to the National Library of Anthropology and History in Mexico City. The folder on the Sevres seminar consists of 367 individual sheets and documents, and among them stand out his own personal notes, either in the program or in small or large papers, that went from Spanish to English and French. Being the group leader, he took care to note down the names of the speakers, put most of his own comments between brackets, and occasionally drew diagrams.

These notes provide valuable hints about how he positioned himself in the discussion of the seminar. In the preliminary program he wrote down, close to the topics to be discussed, whether he considered them "optimistic-unrealistic" or "realistic". He made a note on history going "from the historian to the history teacher (analysis of the process)". On a separate sheet he noted down longer comments: "Excessive or premature optimism that is put in relation to change in the content of history: less dynasties and wars, more social and cultural history... How does the scientific and cultural history of UNESCO place in front of conflicts [sic]. The contact of cultures and techniques can also destroy."[59] This "less optimistic" or "realistic", as he called it, view of history differed from others expressed by UNESCO at that time. Mexican and Latin American history were his main referents, and he made it clear that contact and connection between cultures meant a very destructive result for some people.

On another sheet, undated, he wrote about the discussions in the group he coordinated: "a field trip on the historical consciousness of contemporary world". As a historian, he must have felt he had a privileged insight into how different actors thought about history and saw history's role in the reconstruction of the world. He underlined the differences between those who consider history as a servant of the present, and those who denounce that as propaganda. He wrote between brackets, after a discussion of the relationship between history and current events, which he saw were equated with "troubles": "[difference between historical treatment of current events and problems of its inclusion in history or civics courses. It is the first that is essential.]"[60] Besides the debates about if and how to include current events, he believed that it was the historical point of view that could make a difference for their [current events] role in international understanding.

At other times his notes contained a critical reflection on the methodology of the seminar, and also on UNESCO's strategies, as he wrote that "at times there is an *entretien* on method. Some other times exchange of reports. Or even an effort to reach resolutions acceptable in common." Later he wrote, quite disappointed: "If only we could add to the debates *examples* of practice."[61] He wrote down, and then crossed out, another critical comment: "Possibilities of the group as a thinking instrument and need for first class rapporteurs to bring to light results of good sessions." Could the groups be "a thinking instrument", or were they destined to be an exhibition of divergent positions, whose value would be defined elsewhere, by their political clout?

In the notes he appeared much more cautious and skeptical about what schools and textbooks could achieve on their own. For example, he wrote:

> School can encourage guidelines that promote an understanding among peoples. In this task, the school can find limitations on the part of the social milieu that enclose her. The efforts of the schools in favour of international understanding, to be more efficacious and fruitful, should find a favourable echo on the broader society.[62]

However, it is important to recall, as was done in Torres Bodet's case, that by the time he went to Sevres, Zavala had already been involved in several projects involving the revision of textbooks. A bilateral commission with the USA for the revision of textbooks and the teaching of history in both countries had been set up just after World War II.[63] The commission found its relevance because the depiction of the other nation in both US and Mexican textbooks had been a central controversy for years. On the Mexican side, some of the controversy is captured in an interesting quote from the president of the University of Mexico in 1948:

> You know we are really a magnanimous lot down here. Every time we think of what we lost, of what we should have kept in Arizona, New Mexico, and Texas, then we realize what Mexico really could have been. We have the terrible sinking feeling that somehow our destiny went beyond us, escaped us. But we are giving that up; we are going to be what we can. We have great wealth and we hope to use it for the public welfare.[64]

The work of the commission was not easy, and was a continuation of the efforts already made with other Latin American countries.[65] The US–Mexico Bilateral Commission was also grounded on the broader work done by the Pan American Institute of Geography and History, in which Zavala was prominent. From 1947 its goal was to propose a basic plan for the teaching of history, attending to the requirements of historical knowledge and research, and also of the training of history teachers. Each country was to produce information that would feed the work of the Pan American Commission.[66] So it can be said that Zavala already had some years of experience coordinating a group of historians from different countries and traditions, and trying to come together in a common view of an international – in this case, pan-American – history for mutual understanding.

The bilateral commission consisted of historians from both countries and held a series of meetings in Mexico City, New Orleans, Philadelphia and Austin.[67] UNESCO took a great interest in the commission, and in October 1952, René Ochs from UNESCO's Education Department wrote to Prof. Daniel Cosio Villegas in response to a letter from Villegas in which he offered to report on the bilateral revisions of Mexican and US textbooks. Zavala, who participated as chairman of the History Commission of the Pan American Institute of Geography and History, was also involved in the work. In December 1952, Ochs wrote to Zavala, whom he had met at the Sevres seminar, and asked him to comment on a draft report on active bilateral consultations on the improvement of history textbooks, which was being prepared for a meeting of technical advisors to be held at UNESCO's headquarters that month. Ochs also asked to be updated on Zavala's work. On the very same day he wrote a similar letter to Prof. Villegas, showing his keen interest in the work of these key actors.[68]

However, the programs analyzed by the commission reflected conflicts and biases that would have been difficult to overcome in the bilateral work. For example, in the history program of 1946, still current in 1951–1952, there was an explicit reference to the conflict with the USA – a unit for fourth grade entitled "The North American Invasion and the Dismembering of the Territory", with a lengthy discussion of the foreign policy of the USA since the Monroe Doctrine, the annexation of Texas, California and New Mexico, and the resistance and heroism of the *Niños Héroes* (Heroic Cadets) who fought against the US invasion at the Palacio de Chapultepec in 1847. The teaching should, the program claimed, "mention the essential difference that exists between the current Inter-American concept of the good neighbour and the spirit that oriented U.S. foreign policy in the case of Texas", something that would be repeated in the fifth grade. There was also a paragraph devoted to "the extension of the country at the end of the war, according to the Treaty of Guadalupe. Material and spiritual meaning of this loss. Brief idea of the life of the Mexican population that continued inhabiting in the estranged territories."[69] This phrasing sustained an idea of the Fatherland that was bound to territory and history, not to political entities. It did not seem to endorse a pasteurized version of the relationship between the USA and Mexico.

So was the impact of the History Commission of any relevance to Mexican textbooks in the following years? It is noteworthy that none of the members of the commission, except for Appendini and Ramírez, were teachers or writers of textbooks at the time they joined. However, after their participation in the US–Mexican Bilateral Commission, and Zavala's work for the Sevres seminar in 1951, Appendini and Zavala wrote a three-volume textbook together on universal history in 1953, which was extremely successful in secondary schools – in fact it ran to 38 editions.[70] The book was praised for its consideration of political, economic and cultural history, which stood out as historiographically sound and pedagogically stimulating. It provided moderate judgments on historical events, and the narrative took care to underscore the several dimensions that were at the basis of historical changes.

For example, in relation to the USA, the textbook praised the American Revolution as a landmark for the rights of men, but discussed the Monroe Doctrine of "America for the Americans", against European interventions in, or colonization of, the continent. Later it said that the

> United States initiated a period of intervention in the Latin American nations which could not find their interior peace or which strategically and economically had to be dominated by the Union. After a period of tutelage, the United States returned their autonomy to Cuba and Nicaragua; they evacuated Haiti and Santo Domingo; and reserved their rights in the case of Puerto Rico. They have recently given independence to Philippines.[71]

And as part of the conclusions of the chapter, the text stated that the "Monroe Doctrine has set a barrier to ulterior foreign interventions in American territory, although occasionally it has helped the dominance of an American country over another of the same continent".[72] There was no mention of the conflicts that Mexico had with the USA, most significantly during the war in 1846–1848, but also the invasion of Veracruz in 1914, a more recent episode in a "long history of conflictive co-existence", as Mexican historian Beatriz Ulloa has called it.[73] Hence this textbook seemed in line with what was discussed in the commission of a pan-American history that took into account international events and also provided knowledge and awareness of the complex processes that took place in US territory. It had indeed a pan-American ambition, which could be seen in its discussion of national cases from South to North America. However, references to more present conflicts, such as the ones with the USA, were diluted and delivered in a lukewarm tone that might have intended to appease.

What about other textbooks of that period? According to historian Josefina Vázquez, who reviewed Mexican textbooks to see how they portrayed the USA, most of the official and private books of the period 1940–1960 showed the same two critical points: the independence of the

USA, generally celebrated, and the 1846–1848 war, which resulted in the annexation of Texas to the US territory and the selling of the territories of New Mexico and California. While since the 1950s history textbooks included a critical appraisal of what Mexican actors did in that war, there was nonetheless a characterization of US intervention as a "robbery", "a coward and unjust aggression", which was a natural result of "the US imperialist drive". The USA had prepared itself for this "formidable business" and was presented as a greedy neighbor.

In textbooks from 1950, 1956 and 1958, Vázquez finds that there are references to other conflicts that took place in the 1920s: the invasion of the port of Veracruz in 1914 "without any declaration of war", the persecution of Pancho Villa by General Pershing, and the consequences of the Great Depression in 1929. The official textbooks of the 1970s give a more complex narrative that celebrates the USA's industrialism and entrepreneurship, yet they say that there were two main victims of this expansion: Mexico and the indigenous people.[74] They also pointed to the weakness, mistakes and even disloyalty of the Mexican governments of that time, balancing the view between external aggressions and internal political contradictions. Vázquez concludes that there was ambivalence in the consideration of the USA in official textbooks, which she relates to the will to value some of the US traditions – for example, its liberal constitution – and to the perception of a threat by the enormous power of the USA and its interventionist policies.

In many respects, then, what Zavala said about the quandary of how a generation that had not been able to achieve peace could teach peace was reflected in these visions about the USA. The history of conflicts could not be erased by the needs of an international understanding, and the demand for a world history was taken within some priorities established by the narratives of the national state. The bilateral commissions, so cherished by UNESCO strategies, found, at least in the case of Mexico, a clear limit in national histories. Was UNESCO's project for textbooks on international understanding short-lived? In terms of world history and interconnected narratives, it seems that – to say the least – it achieved little. Yet there were other products of this initiative which would have more lasting consequences. One of them, the free national textbook, will be discussed in the next section.

Where all roads meet – CONALITEG and the free compulsory textbook

If the international narratives were not as effective in turning around the teaching of history, the relevance of textbooks was indeed adopted by Mexican politicians and taken to a higher level. For Mexican textbooks, the groundbreaking year was 1959, when the Mexican National Commission for the Free Textbook (Comisión Nacional de Libros de Texto Gratuitos (CONALITEG)) was created. This time, Torres Bodet was again at the center of the initiative, which was more ambitious than all the previous ones. According to historian Loyo Brambila, it was in this period that Torres Bodet could use all the experience and contacts he accumulated during his UNESCO years. Grounding on his knowledge of textbooks across the world, he said:

> The school textbook is the fruit of a long, complex, profound cultural evolution. It springs out of very deep historical experiences. It represents the synthesis of a teaching, literary, scientific and even political alchemy. In young States, textbooks generally suffer from immaturity, improvisations, shrinkings, or, on the contrary, of sudden arrogance.[75]

The bet was high, but Torres Bodet was confident that, with the strong backing of the Mexican president, Adolfo López Mateos, he could produce a national narrative for all Mexican children. A risky move was to appoint Martín Luis Guzmán, a controversial figure, renowned historian about the Mexican Revolution and a long-term editor at publishing houses, as its first director, where he stayed until his death in 1976.[76] The textbooks should promote the love for the Fatherland through "the thorough knowledge of the great facts that have given ground to the democratic revolution in our country".[77]

If they were only "free" in the beginning, another legal text soon established them as compulsory. Textbooks were considered by the López Mateos administration as part of a broader strategy to achieve more control over public education.[78] It was this last fact that stirred up violent opposition from Catholic sectors, which considered them to be part of a new indoctrination. However, there was a strong consensus that such a policy was needed to expand literacy and schooling. The total number of textbooks printed in 1959 was 625,000, for a school system that had 5,350,478 students enrolled in primary schools. There was an evident shortage of books, and the policy soon won over the opposition. In 1960 the first edition was printed with a print run of 15 million.[79]

The commission was supposed to solve the bad quality and high costs of textbooks for Mexican families, particularly for the poorer ones, and thus help expand the enrolment in basic education.[80] It soon became a symbol of public schooling, of a common curriculum and of an integrated ideology, which is still strong today. In 1960 some 19 books for students and two for teachers were printed. They included paintings by famous artists such as David Alfaro Siqueiros and Raúl Anguiano that were commissioned exclusively for the textbooks, and this decision reflects what had already been discussed in 1944 about the need to consider images as important pedagogical tools.[81]

To what extent do the actions of CONALITEG reflect an impact of the work done by UNESCO's commissions on textbooks? On the one hand, Torres Bodet had been not only a witness but an active promoter of the revisions of textbooks during his tenure at UNESCO, which built on his prior experience in Mexico but certainly added new dimensions, such as the need for international cooperation, the difficulties in building a world history and the need for strong political backup. He considered that textbooks had to express the "soul" and culture of the people, and thus invited historians, writers, painters and teachers to produce them.

However, this impact was also felt in the production of norms and scripts that were set by CONALITEG for textbook authors. The norms and scripts were almost 80 pages long, and included detailed contents for each grade and subject, in what can be termed as an "operationalization" of curriculum that contained methods and suggested activities as much as disciplinary content.[82] For each year and subject there was a list of content knowledge, skills, habits, capacities and attitudes. Along with the textbooks, there were recommendations for the "Exercise Notebooks" for the students, which were to be produced together with the textbooks and as a complement to teaching. As pedagogical principles, the guidelines combined paidocentrism – that is, the idea that teaching has to be organized around the psychology of the child, and patriotic and moral imperatives.

Are there traces of UNESCO's handbook for textbook improvement in these guidelines? In relation to content, the answer is ambivalent. The norms included 28 civic and moral "values", of which only four were valid for the whole basic education: social solidarity, justice, cult of the patriotic heroes and symbols, and obligations towards the family, the school and the Fatherland. There is no mention of peace or international understanding. The guidelines for the first years of schooling list a series of contents for history and civic education that are, above all, patriotic. Respect for the family, the school and the Fatherland are repeated in several paragraphs, but it is noteworthy that there is also an emphasis on understanding, tolerance, justice, mutual respect and help.[83] The only mention of international relationships is the following: "The material of the text has to be in agreement with the Political Constitution and the international accords of the country."[84] At third grade there is also a reference to presenting the international organizations in which Mexico participates. At fourth grade it is stated as one of the goals of the teaching of history that students acquire the knowledge that "the whole world has the imperative need to live in peace".[85] There also seems to be an absence of conflict, particularly in the first grades: as general indications for the authors of the first and second year of schooling, it is said that "compositions that are negative, depressive, or strange to the interests of children, should be avoided".[86] Yet at fourth grade there is a mention of the US invasion of Mexico in 1846 and "its disastrous consequences".[87] The reference to the Niños Héroes and the war with the USA is significant: the symbolic landmarks of Mexican patriotism are very present.

At fifth grade there is a unit about the USA in the 20th century, where the suggested readings are paragraphs from the Organization of American States

(OAS) and UN charts, placed at the same level as the Mexican Constitution. Also, there is a recommendation to focus on the "parallel lives" of American heroes, such as José Martí from Cuba, José Luis Mora from Mexico, Domingo Faustino Sarmiento from Argentina and Justo Sierra from Mexico. There is also a suggestion of a writing exercise on Mexico's contribution to peace and social justice through its politics of non-intervention and its affiliation with the OAS and the UN.[88] The UN and UNESCO are mentioned as relevant actors in the contemporary world, yet the main historical and civic narrative stays within the scope of patriotic history and morals.

Concluding remarks

This chapter has identified the existence of multiple trading spaces where Mexico and UNESCO interchanged ideas, knowledge and practices. Some of these spaces entailed historical entities and some had a bilateral character with UNESCO on the sideline as an interested and keen observer. Crucial to the existence of these spaces were key agents who were able to act and navigate persuasively and decisively in both the national and the international arenas.

Considering the merging lines of historical development, we have shown that Mexico was no newcomer to the field of using textbook revisions as a means to promote international understanding. The picture emerging from this analysis is that Mexico was a country with significant confidence in this field and perhaps even a notion of being a culture-exporting country – a country from which other member states could learn. The fact that UNESCO found it relevant to draw on historians such as Zavala and Villegas as experts seems to testify and substantiate such an interpretation. Although the importance of key agents cannot be exaggerated, institutions such as the US-Mexican commission and CONALITEG, not to mention UNESCO itself, also seem to have played a pivotal role. It seems that the process of creating institutions provided the agents with backing and support. Thus forming or joining an institution might be conducive to propagating and disseminating new educational ideas, knowledge and practices. Forming or joining an institution was a way of creating a space of likemindedness that opened up new opportunities of exchanging ideas, knowledge and practice.

So what does the reading of Mexico's participation in UNESCO and its involvement with textbook policies say about the history of education in that period, and also about the history of international organizations? On the one hand, from this recollection, it becomes obvious that Mexican educators and historians brought with them in 1945–1948 some experience and initiatives regarding textbook revision, and that UNESCO's proposal and guidelines arrived in a place already sown with these concerns. Also, it can be said that this important experience might also have been constitutive of UNESCO's earlier policies.

The routes from UNESCO's headquarters in Paris to Mexico were not, then, traveled only one-way, but there were multiple loans and borrowings that also went through New York, New Orleans, Rio de Janeiro, Buenos Aires, Sevres, Brussels and Beirut, to name just a few places where these actors, true mediators of knowledge and policy strategies, travelled and met. Apart from the Mexican national commission, the network through which UNESCO's impact could flow was developed through some "deeply worn channels" that were the historical and political institutional powers in Mexico, such as the secretary of education, the embassies, and the centers for historical research and graduate teaching.[89] However, there were also new institutional creations, such as CONALITEG, which reflected the new relevance of textbooks for national integration.

This is, paradoxically, what might have been UNESCO's more lasting effect on Mexican education. We say "paradoxically" because the initiative for promoting international understanding through textbooks was supposed to take away school texts from the dominance of national histories. Yet, in the case of Mexico, the lesson learned by the actors who were deeply involved in this initiative was the opposite: if textbooks were a key to fostering identity, then they had to be subsumed within the needs of the nation state. Torres Bodet took to a new level the importance of textbooks when he made them centralized, free and compulsory reading for all Mexican children.

That this narrative had indeed traces from UNESCO's values and rhetoric is also noteworthy. Mexican nationalism in the late 1950s and early 1960s was of a different kind than the 1940s. It included patriotic values such as the love of the Fatherland, but it also made room for international organizations, and for the rule of law and human rights as demonstrated by the unsuccessful Mexican "Proposal for a textbook on world history" in 1946. In that respect it was a modernized patriotism, one in tune with the post-war world.

Also, a distinction between Torres Bodet and Silvio Zavala should be made. Zavala did not participate in CONALITEG – he was by then the ambassador to France – but he certainly could have participated had he wanted to, as he had before despite his intensive traveling and his prolonged stays in Europe and the USA. His 1982 comments about when he stopped intervening in the teaching of history in basic education are clear in that he sees his 1953 textbook about world history as his last contribution.[90] The two actors, then, show different ways of engaging with UNESCO policies, and make visible the multiple threads that were woven between Mexico's educational experiences and UNESCO's life in those years.

Notes

1. Falk Pingel, *UNESCO Guidebook on Textbook Research and Textbook Revision*, 2nd ed. (Paris/Braunschweig: UNESCO, 2010), 7ff.

2. W. Laves and C. Thomson, *UNESCO: Purpose, Progress, Prospects* (London: Dennis Dobson, 1957); P. Luntinen, "School History Textbook Revision by and under the Auspices of UNESCO", *Internationale Schulbuchforschung* 10 (1988): 337–349; UNESCO, *Better History Textbooks* (Paris: UNESCO, 1950).
3. Document C/9, "Looking at the World through Textbooks", quoted in the *Journal of the First Session of the General Conference of UNESCO*, 1946, UNESCO Archives. The key conferences were Seminar on Education for International Understanding, Sevres, France, 1947; Meeting of Experts on the Improvement of Textbooks, Paris, 1950; Seminar on History Textbooks, Brussels, 1950; History Textbooks Meeting, Germany, 1951; a seminar on geography teaching for international understanding, Montreal, Canada, August 1950; Seminar on the Teaching of History as a Means of Developing International Understanding, Sevres, France, 1951; Seminar on Education for Living in a World Community, Woudschoten, Zeist, the Netherlands, 3–30 August 1952; Working Party on Public Information for the Promotion of International Understanding, Copenhagen, 1956; and International Expert Meeting on Improvement of Textbooks for the Objectives of East-West, West Germany, 1962.
4. UNESCO, *A Handbook for the Improvement of Textbooks and Teaching Materials as Aids to International Understanding* (Paris: UNESCO, 1949); UNESCO, *Better History Textbooks*; UNESCO, *Geography Teaching for International Understanding* (Paris: UNESCO, 1951); C.P. Hill, *Suggestions on the Teaching of History – Towards World Understanding* (Paris: UNESCO, 1953); J.A. Lauwerys, *History Textbooks and International Understanding – Towards World Understanding* (Paris: UNESCO, 1953).
5. A. Martínez Palomo, "México y los inicios de la UNESCO" in *México en los orígenes de la UNESCO*, ed. P. Alvarez Laso (México, DF: El Colegio Nacional, 2011), 9.
6. UNESCO, *A Handbook*..., 101–102; see also p. 39. J.A. Lauwerys, *History Textbooks*..., 40. H.J. Krould and H.F. Conover, *Textbooks their Examination and Improvement – A Report on International and National Planning and Studies* (Washington, DC: The Library of Congress, 1948), 93
7. UNESCO, *A Handbook*..., 39; Charles G. Fenwick, "The Inter-American Conference for the Maintenance of Peace", *The American Journal of International Law* 31:2 (April 1937): 222; Samuel G. Inman, *Inter-American Conferences 1826–1954: History and Problems* (Washington, DC: University Press of Washington, 1965), 178.
8. Jaime Torres Bodet (1902–1974) was a renowned poet, politician and diplomat who played a significant role in Mexican politics during the 1940s and 1950s. From the 1920s he occupied several positions in public offices. From 1929 to 1940 he acted as secretary or second secretary to the Mexican embassies of Madrid, Paris, Buenos Aires and Brussels; in 1940, he became sub-secretary of foreign affairs until 1943, when he was appointed secretary of education to appease a conflict with the Teachers' Union. In 1946, under a new presidency, he became secretary of foreign affairs until he was elected director of UNESCO, a position he held from 1948 to 1952. From 1954 to 1956 he was the Mexican ambassador in France, and from 1958 to 1964 he was again in charge of SEP. He continued in diplomatic and educational posts until his death in 1974.
9. C. Greaves Laine, "Política educativa y libros de texto gratuitos. Una polémica en torno al control por la educación", *Revista Mexicana de Investigación Educativa* 6: 12 (2001), http://www.redalyc.org/articulo.oa?id= 14001203 (accessed 20 March 2015).
10. See, for example, Ian Grosvenor, "Geographies of Risk: an Exploration of City Childhoods in Early Twentieth-Century Britain", *Paedagogica Historica* 45:1

(2009): 215–233; M. Lawn, ed., *An Atlantic Crossing? The Work of the International Examination Inquiry, Its Researchers, Methods and Influence* (Oxford, UK: Symposium Books, 2008); E. Fuchs, "History of Education beyond the Nation? Trends in Historical and Educational Scholarship" in *Connecting Histories of Education – Transnational and Cross-Cultural Exchanges on (Post-)Colonial Education*, ed. B. Bagchi, E. Fuchs and K. Rousmaniere (New York: Berghahn Books, 2014), 11–26.

11. Ivan Christensen and Christian Ydesen, "Routes of Knowledge: Towards a Methodological Framework for Tracing the Historical Impact of International Organizations", *European Education* autumn (2015); Barney Warf and Santa Arias, eds., *The Spatial Turn: Interdisciplinary Perspectives* (New York: Routledge, 2009); Vera E. Roldán, "Para 'desnacionalizar' la historia de la educación: reflexiones en torno a la difusión mundial de la escuela lancasteriana en el primer tercio del siglo XIX", *Revista Mexicana de Historia de la Educación* 1:3 (2013): 171–198.
12. Lawn, ed., *An Atlantic Crossing?*, 25.
13. UNESCO, *A Handbook...*
14. P. Latapí, "60 años de la UNESCO: Un aniversario en el que México tiene mucho que celebrar", *Perfiles Educativos* 28:111 (2006): 113.
15. Quoted in M. Martínez Báez, "Los orígenes de la UNESCO" in *México en los orígenes de la UNESCO*, ed. P. Alvarez Laso (México, DF: El Colegio Nacional, 2001), 29.
16. Luntinen, "School History...", 339.
17. 371.671:327.6 A 318 Bibliography on the improvement textbooks as an aid to international understanding, Part II from 1/III/49 up to 31/XII/49, Circular letter signed by Jamie Torres Bodet with three appendixes, UNESCO Archives.
18. Ibid., letter dated 27 April 1949.
19. 327.6:37 A 074 (44) "47" Education for International Understanding – Seminar – Sevres 1947, Part IV, Reports, foreword, Working Papers etc., Working Papers of the Seminar on Education for International Understanding, 1947, Series B, no. 10 "Some Persistent Problems in the Development of Inter-group Understanding", section II: The relation of the school to the community; and some problems of the transition of the adolescent from school to work, UNESCO Archives.
20. 327.6 A 55 "51" Techniques on Education for International Understanding – Questionnaire 1951, Letter dated 1 August 1951, UNESCO Archives.
21. UNESCO, *Better History Textbooks*, 29.
22. 327.6 A 0 74 (492) "52" 15, Seminar on Education for Living in a World Community Netherlands 1952 Participants, List of Participants dated 21 August 1952, UNESCO Archives.
23. 327.6 A 0 74 (492) "52" 19 Seminar on Education for Living in a World Community Netherlands 1952 Follow up, Newsletter to former seminar participants, dated 5 July 1954, UNESCO Archives.
24. Quoted by A. Loyo Brambila, "Caminos entreverados: Cultura y educación en Jaime Torres Bodet" in *Entre paradojas: a 50 años de los libros de texto gratuitos*, ed. R. Barriga Villanueva (México, DF: El Colegio de México – SEP- CONALITEG, 2011), 128f.
25. Quoted in Palomo, México y los inicios de la UNESCO, 18.
26. A. Enríquez Verdura, *Jaime Torres Bodet y la UNESCO: Los límites de la cooperación internacional*, B.A. thesis (Mexico: Centro de Estudios Internacionales: El Colegio de México, 1997).
27. Brambila, "Caminos entreverados", 127.

28. "The peaks of universal thought" in E. Loyo, "El Sembrador y Plan Sexenal. La formación de los nuevos campesinos (1929–1938)", in *Entre paradojas: a 50 años de los libros de texto gratuitos*, ed. R. Barriga Villanueva (México, DF: El Colegio de México – SEP- CONALITEG, 2011), 100; Greaves Laine, Política educative...; L. Martínez Moctezuma, "Los libros de texto en el tiempo" in *Diccionario de Historia de la Educación* (multimedia version, CIESAS-UNAM-CONACyT, 2003) http://biblioweb.tic.unam.mx/diccionario/htm/articulos/sec_29.htm (accessed 20 March 2003).
29. Loyo, "El Sembrador y Plan Sexenal", 100ff.
30. The Mexican political system is presidentialist, and each president is elected for a six-year term. It is quite common that *"sexenios"*, as the six-year periods are called, act as clear markers of changes in policies.
31. Laine, "Política educativa y libros de texto gratuitos"; Rosa Nidia Buenfil Burgos, *Cardenismo. Argumentación y Antagonismo en Educación* (Mexico, DF: DIE-CINVESTAV-CONACYT, 1994); E. Ixba Alejos, *El Estado mexicano: ¿artifice del libro de texto? Origen y hechura de la primera generación de libros de texto (1959–1964)* (Doctoral Thesis, Doctorado en Ciencias con Mención en Investigación Educativa, Departamento de Investigaciones Educativas del CINVESTAV, México DF, 2014), 33.
32. Ixba Alejos, *El Estado mexicano*.
33. That is seen, for example, in the fact that the books approved in 1941 for the teaching of economy and social matters included *Political Economy* by A. Leontiev from the Academy of Sciences of the USSR, *Principles of Political Economy* by A. Segal from the Marx-Engels Institute, and *Theory and Practice of Socialism*, approved and printed by the SEP, according to Ixba Alejos, *El Estado mexicano*.
34. Ixba Alejos, *El Estado mexicano*.
35. Having acted as private secretary to José Vasconcelos in his tenure at SEP in 1921–1923, he cherished the project of publishing high quality, multiplying books that would enrich popular culture and that would support the work done by teachers and schools, according to Brambila, "Caminos entreverados", 127. He created the Biblioteca Enciclopédica Popular – Popular Encyclopedic Library – which printed literary classics from 1944 to 1948, according to Lorenza Villa Lever, *Cincuenta años de la Comisión Nacional de Libros de Texto Gratuitos: cambios y permanencias en la educación Mexicana* (México, DF: CONALITEG, 2009), 46. This experience would be important in his second term as Secretary of Education and the creation of CONALITEG.
36. J. Torres Bodet, *Educación y concordia internacional. Discursos y mensajes* (México, DF: El Colegio de México, 1948), 143–144.
37. These lists were of relative efficacy, or not too efficient. For example, 198 books that had been rejected by the Commission were still being printed in 1959, and were sold clandestinely in schools, according to Villa Lever, *Cincuenta años de la Comisión Nacional de Libros...*, 41. That same year, 65 books were authorized. Most of them had small editions, at its utmost of 15,000 books.
38. Bodet, *Educación y concordia internacional*, 22–23.
39. Poul Duedahl, "Selling Mankind: UNESCO and the Invention of Global History, 1945–1976", *Journal of World History* 22:1 (2011): 130.
40. Bodet, *Educación y concordia internacional*, 54.
41. "The world is one", he said at his opening lecture of the 7th Meeting of the Mexican Congress of History held in Guanajuato, 15 September 1945: "And as Goethe put his ear on the flickering breast of Europe to hear the palpitations of

the 1789 Revolution, hoping to understand what claimed, with its systolic and diastolic gigantic beatings, the heart of that freedom that beat in Paris, so now us, all of us, without comparing ourselves to the poet born in Francfort, have lived multiple hours these years far from the cities and streets in which our persons strolled, with our imagination alert to the loudness of German airplanes over the skies of London, to the cannon shots in Stalingrad, or to the progress of the allied troops in Normandy" (Bodet, *Educación y concordia internacional*, 51). See also Duedahl, "Selling Mankind", 101.

42. Bodet, *Educación y concordia internacional*, 21.
43. J. Lomelí Quirarte de Correa, "La influencia de los Congresos en la enseñanza de la historia" in *La enseñanza de la historia en México*, ed. R. Ramírez et al. (México, DF: Instituto Panamericano de Geografía e Historia – Comisión de Historia, 1948), 298.
44. R. Ramírez et al., ed., *La enseñanza de la historia en México* (México, DF: Instituto Panamericano de Geografía e Historia – Comisión de Historia, 1948), 91.
45. UNESCO, *A Handbook*...
46. UNESCO, *Geography Teaching for International Understanding*, 10
47. As said before, Silvio Zavala was born in 1909 and passed away in 2014. Besides being a prolific and renowned historian, he held many posts in academic and diplomatic institutions, and was probably one of the exemplar cases of the close relationship between the discipline of history and the construction of the modern state in Mexico, according to Enrique Florescano, *El nuevo pasado mexicano* (México, DF: Cal y Arena, 1991). He was one of the intellectual leaders of El Colegio de México, which has formed generations of historians and diplomats in the country. He was the director of the National Museum of History from 1946 to 1954, and president of the History Commission of the Pan American Institute of Geography and History (1947–1965). He was also quite active in the International Council on Museums, and, being the director of the National Museum of History, usually included visits to museums in his trips and a broader concern with the historical education of the public through radio talks and movies. On the other hand, his institutional work at El Colegio de Mexico was central. This center was founded as an offshoot of the Casa de España en México, which started as a center to attract intellectual refugees from the civil war in Spain, according to Stephen Niblo, *Mexico in the 1940s: Modernity, Politics, and Corruption* (Wilmington, DE: Scholarly Resources Books, 1999), 36. It had the support of the Rockefeller Foundation and Harvard University, and besides Zavala there were other prominent intellectuals, such as José Gaos, Daniel Cosío Villegas and Alfonso Reyes. Zavala founded and directed the Center for Historical Studies from 1940 to 1956 and was president of El Colegio from 1963 to 1966.
48. Duedahl, Selling Mankind; P. Bakewell, "An Interview with Silvio Zavala, by Peter Bakewell and Dolores Gutiérrez Mills", *Hipanic American Historical Review* 62:4 (1982): 559.
49. UNESCO, *Bilateral Consultations for the Improvement of History Textbooks* (Paris: UNESCO, 1953), 37; A. Reyes, *Fronteras conquistadas: correspondencia Alfonso Reyes-Silvio Zavala, 1937–1958* (México, DF: El Colegio de México, 1998), 256.
50. See Colección Silvio Zavala, Serie: UNESCO, Cajas 1–4, Biblioteca Nacional de Antropología e Historia (BNAH), Mexico, where there are several letters between the two, always in a formal tone.
51. Letter 15 May 1951 from B. Cohen, Secretary Adjunct for Public Information to Silvio Zavala, Colección Silvio Zavala, ASZ /Serie: UNESCO, Caja 1, Exp. 3, 12-November-1946 al 01-Abril 1952, fjs 288–287, BNAH.

52. Letter 22 May 1951, from Silvio Zavala to B. Cohen, Colección Silvio Zavala, ASZ /Serie: UNESCO, Caja 1, Exp. 3, 12-November-1946 al 01-Abril 1952,fj. 288–289, BNAH.
53. Bakewell, "An Interview with Silvio Zavala ... ", 561.
54. Florescano, *El nuevo pasado mexicano.*
55. Silvio Zavala, "Historical Museums and International Understanding", *Museum* 7 (1954): 96.
56. Duedahl, "Selling Mankind".
57. The strategy of international seminars had many advantages for UNESCO. In a 1951 report on the Brussels Seminar, it says that such form "creates a temporary international community in which the members live in close contact with the problems of different languages, national attitudes and cultural habits ... personal contacts" according to UNESCO, *Geography Teaching for International Understanding,* 5. The small groups should promote free discussion and develop a workshop technique to ensure a lively exchange of ideas.
58. The document, UNESCO/ED/90, "Towards World Understanding – Some suggestions on the Teaching of History", which was prepared by a commission that met in Paris in October 1950, composed of Margaret Miles (UK), Arturo Morales Carrión (Puerto Rico), Merle Curti (USA, Wisconsin), Louis François (France), Hans Fussing (Denmark), Abdul-Foutouh Radwan (Egypt), S. Sen (India) and Louis Vernier (Belgium).
59. UNESCO/ED/92, dated in Paris, 13 February 1951, Colección Silvio Zavala, ASZ /Serie: UNESCO, Caja 1, Exp. 3., 12-November-1946 al 01-Abril 1952,fj.unnumbered, BNAH. The comments were intended as points of discussion or questions. We have transcribed them as they appear in his notes.
60. Colección Silvio Zavala, ASZ /Serie: UNESCO, Caja 1, Exp. 3, 12-November-1946 al 01-Abril 1952,fj 147, BNAH.
61. Ibid.
62. Ibid.
63. 371.671 A 81 Bilateral Arrangements for the Improvement of Textbooks Comite Mexique-Etats Unis Pour L'Amelioration des Manuels d'Histoire and UNESCO 1953, 37, UNESCO Archives.
64. Quoted in George D. Stoddard, "Education in UNESCO", *The Phi Delta Kappan* 30:3 (1948): 74.
65. A similar bilateral agreement between Mexico and Brazil concerning the periodic revision of geography and history textbooks was signed in December 1933, although it was not ratified until December 1947. The agreement called for a revision of history textbooks aimed at weeding out texts prone to create "aversion against any American people" (our translation), according to 371.671 A 81 Bilateral Arrangements for the Improvement of Textbooks, Convention culturelle entre le Brésil et le Méxique, 28 décembre 1933 ratifiée de 3 décembre 1947, UNESCO Archives. Moreover, the agreement stipulated a recurring revision of each country's geography textbooks, taking into account the most recent statistical information and the provision of an "approximate insight into the wealth and production capacities of the American states" (ibid., our translation). It is nonetheless unclear if this commission was ever established. It seems that the moment of the ratification of the agreement was coincidental with the launching of the Pan American Institute of Geography and History, which set out similar goals.

66. The Rockefeller Foundation gave a substantial donation to support the commission. In a letter to Alfonso Reyes, then president of El Colegio de México, Zavala asked him to handle the USD13,500 that was needed to pay for three research contracts and an international meeting in 1955, according to a letter from Silvio Zavala to Alfonso Reyes, 9 February 1955, in Reyes, *Fronteras conquistadas*, 238.
67. UNESCO, *Bilateral Consultations for the Improvement of History Textbooks*, 37 and 371.671 A 81 Bilateral Arrangements for the Improvement of Textbooks Comite Mexique-Etats Unis Pour L'Amelioration des Manuels d'Histoire, UNESCO Archives. The Mexican delegation (composed by Silvio Zavala) consisted of Silvio Arturo Zavala Vallado (1909–), Mexican National Museum of History, Ida Appendini, Secondary Education (1898–1956), Carlos Bosch-Garcia (1919–1994) from the Mexican Historical Society, José Bravo Ugarte from Académie d'Histoire, Daniel Cosio Villegas (1898–1976) from the University of Mexico and Rafael Ramirez, Primary Education (G.I. Sanchez) (Carlos E. Castaneda). The US delegation (composed by Arthur Whitaker) consisted of Maurice Aherns, Corpus Christi University, Ray Billington (1903–1981) from Northwestern University, (Bailey Carrol), Herbert P. Gambrell (1898–1982) from Southern Methodist University, Lewis Hanke (1905–1993) from the University of Texas, Bruce Meador, secretary to the committee, I. James Quillen (1909–1967) from Stanford University, Trevor K. Serviss, Paul Lewis Todd, editor, social studies, Arthur P. Whitaker (1895–1979) from the University of Pennsylvania (Vaughn Bryant) and Paul Lewis Todd, editor, social studies.
68. 371.671 A 81 Bilateral Arrangements for the Improvement of Textbooks, UNESCO Archives.
69. Ramírez et al., *La enseñanza de la historia en México*, 51. See also p. 60.
70. The volumes are 1. Orient, Egypt, Greece. 2. Rome, Middle Ages, Islam. 3. Modern and Contemporary History
71. I. Appendini and S. Zavala, *Historia Universal: Moderna y Contemporánea* (México, DF: Porrúa editors, 1953), 318–320.
72. Appendini et al., *Historia Universal*, 320.
73. B. Ulloa, *De Fuentes, historia, revolución y relaciones diplomáticas* (México, DF: El Colegio de México, 1911), 84.
74. J. Vázquez, "La imagen de Estados Unidos en los libros de texto mexicanos" in *Imágenes recíprocas. La educación en las relaciones México-Estados Unidos de América. Quinta Reunión de Universidades de México y Estados Unidos*, ed. P. Ganster and M. Miranda Pacheco (México, DF: Universidad Autónoma de México Azcapotzalco, 1991), 82–83, 90.
75. Quoted by Brambila, "Caminos entreverados", 132.
76. The historian and writer Martín Luis Guzmán (1887–1976) had been part of Pancho Villa's revolutionary army in 1914, and had to go into exile twice, once after being imprisoned as a villista, and afterwards because of his opposition to the Calles government in 1924. He lived in Spain and the USA, where he became a journalist and writer. In 1929 he published one of the most recognized "revolutionary novels", *La sombra del caudillo*, in which he criticized the caudillismo of Mexican political life since the 1910s. Upon his return to Mexico in 1936, he founded a publishing house, Ediapsa, which provided him with important knowledge for his position at CONALITEG. From 1953 to 1958 he was the Mexican ambassador to the UN. As is seen in this list of high public posts, Guzmán was also a man of state, as much as Torres Bodet, although with a military past.

77. Decree of 12 February 1959 creating CONALITEG, quoted by J. Vázquez de Knauth, *Nacionalismo y educación en México* (México, DF: El Colegio de México, 1970), 221.
78. Laine, "Política educativa y libros de texto gratuitos".
79. Villa Lever, *Cincuenta años de la Comisión Nacional de Libros...;* A. Rangel Guerra, "La impronta de Jaime Torres Bodet en la creación de libros de texto gratuitos" in *Entre paradojas: a 50 años de los libros de texto gratuitos,* ed. R. Barriga Villanueva (México, DF: El Colegio de México – SEP- CONALITEG, 2011), 153–154. CONALITEG has been the subject of several studies, such as Villa Lever, *Cincuenta años de la Comisión Nacional de Libros...;* Barriga Villanueva, *Entre paradojas*; Ixba Alejos, *El Estado mexicano*. Most of the time it has opened a national bid for textbooks to which authors submit their manuscripts. Initially CONALITEG tertiarized the production of the textbooks, but since 1961 it has had its own production system.
80. The members of the commission were Juan Hernández Luna as general secretary; Arturo Arnáiz y Freg, Agustín Arroyo, Alberto Barajas, José Gorostiza, Gregorio López Fuentes and Agustín Yañez as vocals; Ramón Beteta, Rodrigo de Llano, José García Valseca, Dolores Valdés and Mario Santaella as representatives of public opinión; and Soledad Anaya Solórzano, Rita López, Luz Vera, Dionisia Zamora Palleres, René Avilés, Federico Berrueto, Arquímedes Caballero, Ramón García Ruiz, Luis Tijerina Almaguer, Celerino Cano, Isidro Castillo and Jesús M. as pedagogic assistants.
81. Ramírez et al., *La enseñanza de la historia en México*.
82. The level of prescription can be seen in the following sentence for the second grade: "Expressions will be short; sometimes they will include coordinated sentences, but in any case will subordinate clauses be included. Lessons will include gradually in extension and level of difficulty; in no case they should go beyond 3 pages each" according to CONALITEG, *Normas y guiones técnico-pedagógicos a que se sujetará la elaboración de los libros y cuadernos de trabajo para los grados primero a sexto de la educación primaria* (México, DF, undated), 149. See also pp. 139–222.
83. CONALITEG, *Normas y Guiones...*, 148.
84. CONALITEG, *Normas y Guiones...*, 149.
85. CONALITEG, *Normas y Guiones...*, 177.
86. CONALITEG, *Normas y Guiones...*, 149.
87. CONALITEG, *Normas y Guiones...*, 179.
88. CONALITEG, *Normas y Guiones...*, 202.
89. J. Nespor, *Knowledge in Motion. Space, Time and Curriculum in Undergraduate Physics and Management* (New York: Routledge, 1994), 15.
90. Bakewell, "An Interview with Silvio Zavala..."

12
UNESCO's Role in East Asian Reconciliation: Post-war Japan and International Understanding

Aigul Kulnazarova[1]

Introduction

History is an important academic discipline that has been a key concern for states. The task of writing *independently* about history is complicated because interpretations of the past are often "imagined, studied and constructed" under the influence of states and ruling governments to aid them with the prolongations of national identity and political order.[2] As such, history can through its teaching materials and narratives impinge on the lives, memories and relations of the community of peoples. In particular, history education plays an increasingly important role in upholding sustainable peace, security and humanity. While acknowledging this fact, it is yet difficult or impossible for different nations to reach or bridge international understandings about their contested histories and conflicting collective memories. The discretion of history education and textbooks, which serve as a principal basis for any successful post-conflict reconstruction, would today seem even greater in the context of past East Asian historical interpretations. This chapter, while dealing with the issues of painstaking reconciliation between Japan and the Republic of South Korea (hereafter Korea), aims to reflect the role of UNESCO in the region's most complicated disputes relating to history textbooks. In what way has UNESCO contributed to the improvements in regional international understandings and textbook revisions since 1945? Does it today play a sufficient role in those improvements?

Back in the late 1940s, post-war Japan, as a defeated state, had to deal with the problems of war memory and national education both domestically and internationally. As history shows, governments often tend to teach history in a way that inspires national pride, identity and patriotism in the younger generations, and the Japanese post-war approach was no exception.[3] John Stuart Mill, a prominent British political thinker, had expressed such an approach more explicitly: "The strongest of all is identity of political antecedents: the possession of a national history and consequent community of recollections."[4] For Japan, this was a particularly important political agenda: to situate history education in the center of state affairs in order to explain its wartime antecedents' conduct in an objectively justifiable and acceptable way and, most importantly, to teach the rising youth to identify themselves with the national history rather than launching a showdown with the past. For this reason, for many decades in Japan, "the transfer of knowledge from one generation to the next through textbooks is controlled not only by scholarly quality criteria and by pedagogical standards, but also by political interests".[5]

In her critique and review of post-war Japanese educational reconstruction, Yoshiko Nozaki argues that "Postwar Japan is fertile ground for understanding the ways in which issues of war memory and education unfold in a given society and the major part societal forces play in constructing and implementing meanings of the national past."[6] Not only did the state involve itself in this process, often aiming to control it for the sake of its political interests, but also the other forces that played a decisive role in making the national history a regional public discourse. Nozaki writes:

> For decades, against relentless government pressure, a significant number of Japanese scholars, educators, and citizens have sought to reflect on, conduct research on, and teach about the Asia-Pacific War from critical – and often cosmopolitan – peace-and-justice perspectives. They have challenged – or at least put up a good fight to challenge – the dominant, normative right-wing nationalist interpretations of the war.[7]

Despite the prolonged challenge, launched against the state-controlled teachings of war history, education and textbooks remain a big impediment for Japan in its struggle to reconcile with Korea. Regardless of new efforts and approaches that have been undertaken in recent years by the Japanese and Korean sides to resolve the issues of their distorted past in a more open and comparative way, there are still conflicts and disputes about war memories and history textbooks that continue to negatively impact international understanding and cooperation among nations in East Asia.[8] Many issues remain unsettled in the region. However, no one could possibly argue that Japan has not endeavored to reconstruct its relations with the rest of the world in the period that followed World War II, mainly as part of the growing UNESCO activities in the country and increasing public awareness.

After 1945, Japan was occupied for nearly eight years by the Supreme Commander for Allied Powers (SCAP). However, it was mostly UNESCO that raised Japan's new expectations, hopes and horizons in transforming itself to a peaceful nation. In retrospect, it seemed almost inevitable that UNESCO would not be able to offer a desired alternative for the Japanese through its constitutionally embedded principles of fostering peace in the minds of every man. Unquestionably, UNESCO's programs for textbook revisions framed important guidelines for states not only to teach history without nationalistic prejudice and historical injustice but also to eliminate causes of war through the promotion of international understanding among nations. UNESCO undertook these necessary measures. Nonetheless, it was never easy or comfortable to implement them in Japan. The following sections attempt to illustrate the reasons for UNESCO's limited impact on Japanese history education reforms that eventually affected the protracted process of East Asian historical reconciliation.

The contested history of East Asia

The contemporary disputes of history education and textbooks in East Asia that emerged from the time of Japanese intensified colonialism resulted in the annexation of the Korean Peninsula at the beginning of 20th century. The continuing political and historical controversies also were a result of war crimes and border issues originating from the Sino-Japanese War and World War II. Unsettled problems of compensation, official apologies, trials and unrealized expectations of mutual understandings no doubts contributed to the subsequent and still existing tensions in the field of historical reconciliation and education in East Asia.

In 1910, Japan seized the Korean Peninsula, thus beginning the intensive occupation of this territory that continued through to the end of World War II. As part of its military and colonial strategy, Japanese imperialism conducted policies that aimed to eradicate Korean culture by forcing Koreans to adopt Japanese names, use only the Japanese language and practice Shintoism. Many Koreans were displaced from their homes and were forced into labor in Japan.[9] Over 100,000 women, many Korean scholars claimed, were sexually enslaved for the needs of the Japanese Imperial Army during World War II, recently becoming known as the issue of "comfort women".[10]

In the early post-war years, the Japanese history textbooks and other teaching materials actually provided a variety of perspectives on Japanese history, including accounts of World War II in the Asia-Pacific region. That was perhaps due to the US presence in Japan, on the one hand, and the willingness of the Japanese state to re-enter the international community through UNESCO, on the other. While the USA indeed insisted on educational reforms, it was UNESCO that urged the improvements to history textbooks and teaching materials. However, it should be noted that educational reforms, conducted by the post-war Japanese state, did not always comply with the country's political goals. The Japanese Government was not quick to admit the war crimes conducted in the countries that fell under its control. As a result, this impediment in action created one of the most complex textbook issues in the world's history of history education. The fact is that portrayals of the Asia-Pacific War in most of the school history textbooks published after 1950 "employed euphemistic language that presented the path to war and its course as a seemingly natural occurrence devoid of discussions of individual or collective responsibility", which often provided no direct reference to the dead and the historical memories created by the war.[11] Some analysts perceive such a conical and biased approach to teaching the history of past war as an outcome of the impact of the USA, which played "a significant role, whether intended or not, in shaping the process of historical redress and reconciliation in the Northeast Asian region". In particular, Shin, Park and Yang believe that a "soft" US policy towards the Japanese lies behind the geopolitics of the post-war world. "Unlike the Nuremberg trial, the Tokyo trials focused on the actions that most directly affected the Western allies – the attacks on Pearl Harbor and mistreatment of prisoners of war – and largely ignored crimes committed against Asians," the historians claim. Yet when the importance of Japan as a bulwark against communism in Asia increased, "issues of Japan's historical responsibility, unlike in Germany, were largely overlooked or ignored".[12] Other scholars stress that Japan's failure in the early post-war years to apologize to the formerly colonized nations was another side of the problem, as the Japanese continued to terrify their neighbors.[13]

It is striking that even the enthusiasm and desire of the Japanese to transform to a peace-loving and friendly nation, which was notably observed in the post-war UNESCO non-governmental movement, did not grant Japan a greater opportunity to get initiate its reconciliation with Korea and other Asian countries, as in the German case. Both domestic and international forces played their role in the belated process of East Asian reconciliation. On the one hand, the Japanese state, constantly backed up by the geopolitical interests of its main US ally in the region, continued to present history from a deep nationalistic perspective. In Japan, as in most countries, the Ministry of Education was still in control of screening and examination of school textbook content and factual material in accordance with the Textbook Examination Standards, following deliberations within the Textbook Approval and Research Council. Under this textbook system, set by the Ministry of Education, each school – both public and private – could only choose a history textbook from the list recommended by the ministry.[14]

Not until 1965 did a distinguished Japanese historian, Prof. Saburo Ienaga, for the first time challenge the screening system, claiming that the review process conducted by the ministry was unconstitutional and illegal. He criticized the government for its constant interference with textbook content and its ignorance of the "dark" sides of war.[15] In total, three suits were filed by Ienaga in the following two decades, and the court finally decided positively about the legality of the ministry's screening process, but made strong remarks about its excessive measures undertaken toward parts of the Ienaga's textbook, particularly the demand to delete the reference to the so-called Unit 731 case – a secret military medical unit of the Imperial Japanese Army that conducted biological and scientific research through human experimentation during the Sino-Japanese War and World War II.[16] Although the court's final decision in 1982 was something of a compromise, it still overruled the government's continued interference with the history education content, and the legal case thus had an international resonance and impacted further developments in the field. After a new wave of protests from South Koreans, in August 1982 the Japanese Government finally promised that it would pay due attention to its neighbors' criticisms and improve the contents of its history textbooks. Soon after, the Ministry of Education issued a new set of guidelines for screening textbooks, urging educators, historians and publishers to promote international understanding and cooperation through the books' content. This case shows that the interpretation of history was of vital concern to the Japanese state, and in the course of dealing with it "the national narrative [became] directly connected with politics ... serv[ing] as a base for the self-understanding of the state and the legitimization of the political order".[17]

On the other hand, although the creation of UNESCO arguably made it possible to reassess national efforts toward the improvements of international understanding by making "bilateral and regional agreements concerning textbooks", it was not yet time for Japan to reconcile with Korea only on the basis of their disputed history teachings and textbook revisions.[18] First it was a domestic problem for Japan. However, following the Ienaga court case, it eventually turned into a regional and international dispute. Internationally, the Japanese had already impressed the world by their loyal attachments to UNESCO's ideals regarding peace, which in practice brought to action the hundreds of UNESCO associations and clubs in the country. However, the universal ideals of peace and regional reconciliation were easily separable in post-war Japan because UNESCO offered not only the desperately wanted ideological alternative but also the place through which the Japanese could re-enter the world stage. The tensions thus existed between the goal of international community and the organization of nation states.

UNESCO's initial impact on Japan

When UNESCO was established in 1945, Japan as "the ex-enemy country" had no chance of being formally associated with the organization. Among Japanese officials, educators and active citizens, however, there was a shared and dominating belief that UNESCO with its spiritual foundations of peace and mutual understanding could be the only way for Japan to re-enter the world community. The power of peace nurtured "in the minds of men" became a leading factor in Japan promoting the organization's ideals. "To ensure permanent peace of the globe, the humanism which each one of us has in our hearts must be welded into one body. This is the only key to solve this problem," one of the country's leading newspapers explained as many Japanese peoples' motivation toward the standalone activities, independent from those initiated by UNESCO's headquarters. "UNESCO is the very organization intending to realize this humanism through Education, Science and Culture which the peoples on this earth of ours have in common. The activity of this organization therefore, is considered to be the most efficient means by which lasting peace of the world can be ensured," the national newspaper, *Miyako Shimbun,* speculated.[19]

By 1948 there were more than 100 Cooperative Associations and Clubs of UNESCO established at universities and high schools across the country.[20] For the promotion of these and other activities relating to UNESCO, the Ministry of Education increased its funds "from a meager figure of ¥3,000 in 1948 to ¥1,500,000 for 1949."[21] This nationwide movement could not, of course, remain unnoticed by the UNESCO's headquarters in Paris, which saw it as a positive sign indicating a change to the Japanese nation in its path to peace and reconciliation. The question of UNESCO–Japan relations was often exchanged between UNESCO and the SCAP Office in Tokyo. In November 1948 *The New York Times* highlighted a note regarding UNESCO's decision to extend its work in Japan. "With the approval of Gen. Douglas MacArthur, the United Nations Educational, Scientific and Cultural Organization is planning to expand the field of its work to include Japan," it announced, adding that already "a modest beginning has been made toward offering opportunities for re-education". On the Japanese side, *The Nippon Times* stated that an office for UNESCO would be established in Tokyo and led by Dr Shi-mou Lee. The purpose of the new office would be "to serve as a channel for the exchange of scientific, educational and other cultural information" as well as to "promote cultural relations between Japan and the rest of the world."[22]

The significance of UNESCO's possible acceptance was broadly deliberated among the Japanese public and media from the time when the state was in the preparatory stage to its UNESCO membership. In January 1951 *The Asahi Shimbun* released a critical editorial explaining that the "participation in the UNESCO depends upon the world's estimate of how far the Japanese, once the enemy to most of the participant nations, have been enlightened in the peaceful spirit of UNESCO. From this point of view, the people's strong consciousness in the UNESCO movement is desired."[23]

During the continuation of these public discourses, in March 1951 *The Nippon Times* added more deeply to the self-deprecating analysis of the Japanese UNESCO activities during that period. "The wartime evils done by the Japanese military still remain a great barrier to Japan's restoration to a position of trust and confidence as pointed out by the Philippines," the newspaper reminded its readers. "And in the opinion of the British, the Japanese are not as yet ready to undertake UNESCO activities. These views must give rise to serious self-reflection by the Japanese people, and they offer a challenge to prove that the new Japan believes firmly and sincerely in the fundamental freedoms, human rights, justice and peace." However, there was just one problem: that "the UNESCO movement in Japan today is at low ebb – this impression is stronger because it was so enormously active several years ago. At present the Japan UNESCO committee is practically a ghost organization, existing in name only."[24] *The Mainichi* echoed its counterpart by suggesting that it was regrettable to see how various UNESCO activities in Japan had slowed down since the outbreak of the Korean War. Some of the questions that were rising in Japanese minds more intensely in the days prior to UNESCO's membership – Are we sufficiently prepared to enter UNESCO? Have UNESCO activities in Japan been adequate? – at the same time show how anxious and keen not only the state but also the public were about this new window that could offer Japan a desperately wanted route onto the world stage.[25]

On the official side of the public discourse regarding Japan's entry into UNESCO, the National Diet – the legislative organ – issued a report that highlighted the relevant deliberation at the Diet's Holding Affairs Committee meeting of the House of Representatives on 11 May 1951, where a committee member, Mr Kitazawa, while sharing his concern about the obstacles to deepening "the national understanding between nations of the world", for the reason of existing fundamental differences "between Oriental and Occidental nations", came up with a question: How can we overcome this obstacle? The education minister's answer followed with no delay as something of a state directive: "It is the Japanese people's mission to assimilate the Western culture without forfeiting the native Oriental culture". The parliamentary debates also illustrated skepticism of, and even existing discontents among, the Japanese political elites towards the UNESCO's program of education for international understanding. In view of the intensifying Cold War, the contradictions between the two worlds' ideologies – democracy and communism – certainly had impact on any political discourse of the time, in which some Japanese parliamentarians saw a serious obstacle "to the realizability of the ideal of UNESCO, namely the banishment of war from the world". The education minister still articulated: "To promote the intellectual and spiritual understanding among these countries is a field of meaningful works for UNESCO", but it was obviously unclear how would have UNESCO reconciled "Democracy and Communism stand[ing] on the principles of liberty and equality respectively which are contradictory to each other".[26] It was neither easy considering the USSR's abstention from UNESCO until 1954. The minister's concluding remark: "Since love can thrive only on the soil of mutual understanding, UNESCO intrinsically has the nature to reconcile the two principles on the basis of universal solidarity of human beings" did not convince enough the members of the Japanese Diet who were concerned with UNESCO's ability to build a peaceful world.[27]

In the Diet, UNESCO was rather seen as an opportunity to regain trust, status and reputation among the member states of the UN system. It was in that context that, at the sixth session of UNESCO's General Conference in summer 1951, which formally admitted Japan to its membership, the head of the Japanese delegation, Tamon Maeda, stated that the "Spirit of UNESCO is the guiding principle for Japan, which is on the path of rebuilding itself as a peace-loving and democratic state".[28]

Following the year of Japan's acceptance, the government in August 1952 established the Japanese National Commission for UNESCO, which actively coordinated all UNESCO headquarters projects and published works relating to the promotion of education for international understanding and cooperation. The UNESCO clubs and associations also moved under the national commission's auspices, where they continued their activities by disseminating knowledge about human rights, democracy and mutual understanding. Books and essays, translated into Japanese and published by the national commission during its initial years of formation, included *Humanism and Education in East and West* from 1953, the *Race Question* series from 1955, *School Textbooks in Japan: A Report of Survey from the Standpoint of Education for International Understanding and Cooperation* from 1957 and *The Treatment of the West in Textbooks of Japan: A Historical Survey* from 1958.

Implementation of Education for International Understanding in Japan

Since its establishment in 1945, UNESCO aimed to develop the universal standards for educational policies and practices by promoting international understanding, peace, friendship, human rights, democracy and fundamental freedoms. Embedded in the principles of the UN Charter and the Universal Declaration of Human Rights, UNESCO's program of education for international understanding was foremost concerned with helping "school authorities and teaching staffs make mutual understanding among peoples a major emphasis in all kinds of education".[29] To advance the program in the field of teachers training, UNESCO sponsored international seminars for teachers and specialists to facilitate open discussions and urge for new methodological developments to emphasize international understanding and quality in education. These seminars were gatherings of international experts who, for a period of four or five weeks, discussed a single educational problem on the basis of papers prepared beforehand. The seminars always had a selected thematic focus, such as "education for international understanding", "education of children between 3 and 13 years of age", "teaching about the UN and the specialized agencies", "adult education in the rural communities of Asia" and "teaching of geography as a means of developing international understanding".[30] During 1947–1950 many seminars and workshops were organized in different countries, to name few – in Austria, Belgium, Brazil, Canada, France, India, the UK and the USA. But two of them, held in Sevres (1947) and Brussels (1950), particularly stood out because they launched UNESCO's most important programs at that time, one focusing on *Education for international understanding* and the other on the *Improvement of textbooks, particularly of history books and of teaching material*. The Japanese official delegates were not among the participants since Japan was not a member of UNESCO.

Instead the Japanese Ministry of Education organized its own annual workshops for secondary schools' teachers at different locations throughout the country and preparations for which started from the late 1940s onward. Each event had education for international understanding on the agenda, and many of the participants in UNESCO's international seminars had been asked to attend the workshops to utilize their experience as a kind of follow-up activity. In January 1950 the Ministry of Education provided a rough description of some of its observations and activities concerning UNESCO's Resolution on Japanese Activities for 1950. It emphasized the fundamental importance of the resolution's realization in Japan through the training of educational groups and seminars for teachers. In addition, the ministry ensured that the Elementary and Secondary Education Bureau of Japan would sponsor special UNESCO seminars for teachers in the country.[31]

The first series of seminars took place from June to November 1950. These were organized in eight major regions of Japan with each seminar lasting for seven days and dealing with various subjects. Among them was a special group project devoted to education for international understanding. At about the same time a study group of teachers of social studies in secondary and elementary schools was formed and affiliated to Kyoiku Daigaku (a teacher training university), the Liberal Arts University of Tokyo and Ochanomizu Women's University. Dr Lee, the head of UNESCO's Office in Tokyo, in his reports to the director-general, said that he had been consulted by the members of this group on regular basis. The study group recommended a range of activities as part of the program for international understanding, such as the creation of "pen pals clubs", the promotion of international understanding in school magazines, and a special course in the 10th grade on the international efforts for world peace.[32]

The practical implementation of these recommendations began swiftly. In 1951, Lee reported to Paris that in several Japanese schools during a campaign entitled "How can we deepen pupils' international understanding in lower and upper elementary school?" the members of UNESCO clubs came up with the idea of giving donations for the relief of poor Korean children. The UNESCO Student Club of Utsunomiya University had raised money for relief in Korea, and Takao Matsuyama, the representative for student clubs, sent a check with a letter that said:

> We Japanese who had bitter experience in the past war have great sympathy for the miserable conditions in Korea. Although the donation we are sending you is only a small amount, we sincerely hope that it will be added to a fund for the relief of Korean children as a token of good will and fervent desire for peace among young generation.[33]

In 1951–1952, several other schools and universities in Japan also raised money for the relief of Korean children. These campaigns showed some evidence of bracing reconciliation efforts among private groups of Japanese youth and educators. As a follow-up, in his report of 3 December 1952 to UNESCO's headquarters, Tadakatsu Suzuki, the secretary-general of the Japanese National Commission for UNESCO, wrote that these campaigns attached "great importance to education for international understanding in Japan", and expressed his state's enthusiasm for cooperating with UNESCO's Korean Emergency Educational Assistance Program by providing fellowships and facilities for study abroad to Korean teachers, students and technicians. "We are anxious to do what we can for our immediate neighbours along the lines of the recommendations the [UNESCO] mission will make,"[34] – Suzuki wrote in the report. Japan was one of the very few countries in Asia at that time which played an active role in UNESCO's educational campaigns. The nature of those campaigns abroad although endeavoring to promote friendship and assistance to the Korean people eventually did not advance into a greater reconciliation process. Education is, no doubt, the best way to bridge and build mutual understanding among peoples, but only if its aims and teaching materials promote peace, cooperation and human rights. In the case of post-war Japan, the apprehensions evidently grew out of its national interests to regain the trust of the world community and the UNESCO's guidelines to nurture peace in the minds of men, regardless of the cultural, historical and social differences that have always existed between nations. It cannot be also said that in the initial post-war years, Japan's one-sided approach to UNESCO's campaigns was to some extent affected by the *status quo* of the post-war international relations. For the Japanese, international understanding and friendly cooperation merely meant bridging Eastern and Western civilizations, and did not seemingly and explicitly emphasize the need for reconciliation with Asian neighbors. Such a partial approach to this important task was largely overlooked by UNESCO as well.

UNESCO and "better history textbooks" in Japan

"History is at once potentially the most divisive and the most unifying of school subjects; none lends itself more readily to the fomenting of prejudice and hostility, or to the fostering of fellow-feeling with all humanity".[35]

This was the core argument of UNESCO's program of education for international understanding for beginning the campaigns of textbook revisions and improvements.

> the bias toward nationalism is obvious: apart from its possible exploitation for chauvinistic purposes, the common practice is to concentrate attention through most of the school life of the pupil on the history of his own country, and treat other peoples chiefly in so far as they impinged on national interests.[36]

With this understanding of deeply rooted nationalistic traditions in societies, in 1946, UNESCO proclaimed, as its key policy priority, the improvement of school history textbooks. At the first General Conference it passed an important resolution for the creation of a Programme for the Improvement of Textbooks and Teaching Materials as Aids in Developing International Understanding. In accordance with the program's principles, UNESCO called upon its member states to review history and civics textbooks to ensure that they indeed promote international understanding, tolerance and peace. Within the next couple of years the organization published a detailed plan of action. It recommended that member states should carry out studies of their own textbooks and initiate mutual or bilateral textbook studies. The core of this plan was to create a clearing house of textbook improvement information at the UNESCO headquarters for its member states.[37] As prescribed, the organization's main role was "to assist member states in developing policies, norms, and standards for the provision of textbooks and other learning materials which facilitate quality education".[38] The 1950 Seminar in Brussels reached at the delegates' agreement to revise and improve their national history textbooks in line with UNESCO's recommendations.

The Japanese official delegates did not take part in this important seminar but similar steps toward the improvement of textbooks had in fact been given serious consideration by the SCAP authorities in Japan and by UNESCO's headquarters. As early as May 1949 the UNESCO Expert Committee on Japanese Questions, formed of Australian, Chinese, Filipino and US representatives (Mr Bunce of SCAP led the committee), met for four days in Tokyo to discuss Japanese measures in accordance with UNESCO's textbook recommendations.[39] In its report to the executive board in August 1949, the committee strongly advised that the "textbooks in use in Japan should be examined to ascertain how closely their contents coincide with UNESCO objectives".[40] Unfortunately the examination of Japanese textbooks was not carried out by UNESCO at this early stage because it met severe "criticism from those who oppose the execution of activities either in Germany or Japan".[41] Although UNESCO's textbook activities formed an important part of the improvements to international understanding among nations, they could not be extended to Japan because of the existing political tensions among UNESCO's member states.

The Committee of Experts on Japanese Questions nevertheless decided to "encourage the experiments being undertaken in international education in Japan and to collect information on the results already obtained by international education with a view to promoting international understanding".[42] Impressively, the Japanese state's involvement in the book activities also accelerated within just one year in 1950, such that the Japanese Ministry of Education put out 40 publications on UNESCO with the print run of each varying from 1,500 to 100,000 copies. The Universal Declaration of Human Rights was one of these publications, and was the most widespread. It was published in English on one page and in Japanese on the other side.[43]

Since 1953 the national commission had carried out two independent surveys on textbooks and teaching materials within the primary and secondary schools curricula. In 1954 the results of the first survey were released as "School Textbook in Japan, 1953 – A Report of Survey", and in 1956 this was followed by "Report of a Survey of School Textbooks in Japan 1954".[44] Based on both reports, and from the viewpoint of education for international understanding, in 1958 the commission summarized the investigations in a separate publication derived from the treatment of Asian and European cultures in 39 series of social textbooks in Japanese elementary and secondary schools. The results were divided into nine sections according to school level, selected themes and academic disciplines, such as geography, history and social studies. Four separate analyses were undertaken for the history textbooks only. First, the surveys conducted in lower and upper secondary schools showed that the teachings of national and foreign (world) histories were not always cross-taught in Japanese schools, so for the purpose of education for international understanding only the textbooks teaching world history were analyzed. Second, European history was the main focus in the Japanese textbooks, whereas Asian history featured less and was taught from a pre-Modern perspective. For example, a large number of pages were devoted to ancient Chinese history in the textbooks. As the authors of the survey reported, such unbalanced treatments of world and regional histories in the Japanese textbooks were due to the fact that "Japan after the Modern Times is built up under the influences of West European culture and civilization". Therefore there was no methodology developed to study "the original histories of India and China after their Modern Times" because "Japanese [saw] these countries with western eyes". While considering such treatment as "a good deal", the authors, for the purpose of building better international understanding, admitted "the necessity of comprehending the fundamentals of a fair attitude and of renewing their treatment of South-Eastern Asian countries which have become recently self-governing nations".[45]

Another major report, which the Japanese National Commission for UNESCO published in 1958, was *The Treatment of the West in Textbooks of*

Japan: A Historical Survey, which emphasized the modification of textbooks in Japanese high schools since the 19th century Meiji Era. The same approach was evident in this report as well – little or no teaching about Asian modern history, but much about the West. In general these studies, conducted in post-war Japan, ignored or overlooked the modern history of Japan's relations with Korea and other neighbors at all levels of schooling. Although UNESCO's history textbook campaigns formed part of Japan's educational reforms, the complex field of historical reconciliation was for the most part undermined. As a result of state policy, history textbooks in Japan to this day, according to many critics, especially outside Japan, have a major problem in reflecting objectively the war memories and the shared past with Korean, Chinese and other Asian nations, which limits the younger generation's awareness and opportunity to learn about their history in a more critical and unbiased way. This shows that even if UNESCO had an impact on the Japanese post-war transformation, this impact had its limitations in practice, especially when it came to the revision of history textbooks.

In general, Japanese school textbooks today are often limited to the presentation of historical facts in strict chronological order. They include very little analysis or interpretation of past events, particularly about the period between the two world wars. "History textbooks currently constitute a delicate political issue involving an intertwined mix of international and domestic concerns in countries around the world," Prof. Hiroshi Mitani of the University of Tokyo, the author of a history textbook used in Japanese junior high school, writes bitterly.

> The intensity of the disputes over them waxes and wanes in response to the will of governments and civic movements, but they seem likely to remain a potential source of discord among nations for the foreseeable future. This is because of the tendency to view each country's history textbooks as being written basically as "national history".[46]

Altogether workshops, seminars and survey projects initiated by UNESCO within the field of textbook revisions was never sufficiently effective in Japan, as evidenced by the fact that there were still government-sanctioned textbooks, whereas the first UNESCO-commissioned seminar in Brussels recommended that governments should entrust the writing and publishing of history textbooks to private rather than state-sponsored parties. Even now some history textbooks continue to distort the accounts of war crimes and imperialism. Pronged experience of the Japanese with history textbook revisions is a strong reminder of how much work remains before UNESCO can achieve its mission of constructing peace in the minds of men.

Education, particularly, history education, is always influenced by the political currents of ruling government. Therefore the political climate plays an important role in any potential changes in textbooks concerning international understanding. "In particular, history and social studies schoolbooks cannot be isolated from the political and social background that shape historical awareness and the understanding of citizenship within a certain society," Pingel writes as a historian who was practically involved in the process of norms creation for textbook improvements.[47] In fact, the Japanese case of dealing with the textbook revisions and improvements shows that it is almost impossible to teach historical insights and disseminate their knowledge through textbooks, which do not aid the political interests of the given state.

Conclusion

The discussion in this chapter has demonstrated mixed connections between historical reconciliation, education and UNESCO's activities in Japan. In sum, if UNESCO had an impact on East Asia's most complicated disputes over historical reconciliation, it was limited and excluded from its guidelines of education for international understanding and textbook revisions.

The issue of textbook revisions in post-war Japan was a difficult one. Not only external pressure from the neighboring countries but also opposing internal forces affected the content of history education for decades. In fact, no serious "textbook revision process [was] initiated in (East) Asia, nor did research here develop into an acknowledged academic sub-discipline".[48] Despite the Japanese efforts to conform with UNESCO's guidelines and recommendations for international understanding, Japanese history education and textbooks domestically remained ambiguous and partial. Japan always enthusiastically and quickly responded to and joined new programs initiated by UNESCO, but the organization's insufficient role or inability to get involved in East Asian reconciliation unlike, for example, in West Germany became apparent too. UNESCO did indeed have significance for Japan's reconstruction in the post-World War II period, as the only bridge between the country and the international community, and the provider of a useful platform for national educational reforms. On the other side, Japan was the birthplace of a non-governmental UNESCO movement that won credit in the eyes of the rest of the world. However, the documents studied reveal that it did not help Japan to win its closest neighbors, and this is the question that historians and reconciliation activists today cannot afford to ignore.

The problem of history education and textbooks in East Asia cannot be solved without teaching about colonialism and war crimes and, most importantly, without their recognition within the state's educational system as an integral part of Japanese international history. Because the issues of the past and their interpretations continue to affect the lives of people in East Asia, including the Japanese, whose state policies tend to resist mutual historical understanding, Japan is still unable to promote alone deeper cooperation measures with its neighbors. It is true that the improvement of history education and textbooks does not depend on the complexity of historical issues in a given country, but rather on the movement of history dialogue to a more dynamic stage among all the involved parties.

In the decades of meetings, seminars, conferences and programs that have addressed the issues of better history textbooks and international understanding, UNESCO's measures were clearly not sufficient in East Asia. Hence the organization's mission is thus far from being completed. For Japan, the launch of UNESCO associations and clubs was not just a truistic venture, and the time during which the Japanese altered themselves often influenced the prevailing nature and direction of events, both domestic and international.

Notes

1. I should like to thank the Academic Council for the UN System (ACUNS) for an opportunity to present these ideas first at the ACUNS-ASIL Workshop held at The Hague Institute for Global Justice in summer 2014, and the Danish Research Council for generous funds to support the larger project on UNESCO, of which the current chapter forms a part. I am also grateful to Alistair Edgar, the executive director of ACUNS, and Poul Duedahl, the director of the Global History of UNESCO Project, for their helpful comments on earlier drafts of this chapter.
2. A. Kulnazarova, "The Formation of the Post-colonial Indian State in Soviet Historiography: The Interplay Between the Ruling Ideology and the Writing of History", *Journal of Management and Social Sciences* 5:1 (2009): 1.
3. C.P. Hill, *Suggestions on the Teaching of History* (Paris: UNESCO, 1953).
4. John Stuart Mill, *Considerations on Representative Government*, ed. Currin V. Shields (New York: The Liberal Arts Press, 1958), 229.
5. F. Pingel, "Can Truth be Negotiated? History Textbook Revision as a Means to Reconciliation", *Annals of the American Academy of Political and Social Science* 617 (2008): 182.
6. Y. Nozaki, *War Memory, Nationalism and Education in Postwar Japan, 1945–2007: The Japanese History Textbook Controversy and Ienaga Saburo's Court Challenges* (New York: Routledge, 2008), xii.
7. Y. Nozaki, *War Memory...*, xiii. Generally, on the textbook controversy, see H. Bando, "History Teaching and Historiography: The Textbook Controversy", *Historical Studies in Japan VII: 1983–1987* (Tokyo: Yamakawa Shuppansha, 1990); C. Barnard, *Language, Ideology and Japanese History Textbooks* (London: Routledge, 2003); K.B. Pyle, "Japan Besieged: The Textbook Controversy: Introduction", *Journal of Japanese Studies* 9:2 (1983): 297–300; Y. Zhao and J.D. Hoge, "Countering Textbook Distortion: War Atrocities in Asia, 1937–1945", *Social Education* 70:7 (2006): 424–430; T. Yayama, "The Newspapers Conduct a Mad Rhapsody over the Textbook Issue", *Journal of Japanese Studies* 9:2 (1983): 301–316; Chunghee Sarah Soh, "Politics of the Victim/Victor Complex: Interpreting South Korea's National Furor over Japanese History Textbooks", *American Asian Review* 21:4 (2003): 145–178; C. Rose, "The Textbook Issue: Domestic Sources of Japan's Foreign Policy", *Japan's Forum* 11:2 (1999): 205–216; T. Sedden, "Politics and Curriculum: A Case Study of the Japanese History Textbook Dispute", *British Journal of Sociology of Education* 8:2 (1982): 213–225.

8. Recent examples of new reconciling efforts include the application of a transnational approach by the joint writing of Japanese, Korean and Chinese historians of several transnational history textbooks. In this regard, Pingel notes that "These books cannot replace the obligatory history books; however, it would be a big step forward if students and teachers could use books that offer a wider view and do not only reflect a well-known and canonized national narrative". For more on transnational history textbook projects, see F. Pingel, "Can Truth be Negotiated...", 181–198.
9. Soon-Won Park, "The Politics of Remembrance: The Case of Korean Forced Laborers in the Second World War" in *Rethinking Historical Injustice and Reconciliation in Northeast Asia: The Korean Experience,* ed. Gi-Wook Shin, Soon-Won Park and Danqing Yang (Abingdon, UK: Routledge, 2007), 55–74.
10. Chunghee Sarah Soh, "The Korean 'Comfort Women' Tragedy as Structural Violence" in *Rethinking Historical Injustice and Reconciliation in Northeast Asia: The Korean Experience,* ed. Gi-Wook Shin, Soon-Won Park and Danqing Yang (Abingdon, UK: Routledge, 2007), 17–35; B-S.L. Yoon, "Imperial Japan's Comfort Women from Korea: History & Politics of Silence-Breaking", *The Journal of Northeast Asian History* 7:1 (2010): 5–39.
11. Julian Dierkes, *Postwar History Education in Japan and the Germanys: Guilty Lessons* (New York: Routledge, 2010), 5.
12. *Rethinking Historical Injustice and Reconciliation in Northeast Asia: The Korean Experience,* ed. Gi-Wook Shin, Soon-Won Park and Danqing Yang (Abingdon, UK: Routledge, 2007), 3.
13. T. Berger, "The Constitution of Antagonism: The History Problem in Japan's Foreign Relations" in *Reinventing the Alliance: US-Japan Security Partnership in an Era of Change,* ed. John Ikenberry and Takashi Inoguchi (New York: Palgrave, 2003), 63–90; T. Christensen, "China, the U.S.-Japan Alliance, and the Security Dilemma in East Asia", *International Security* 23:4 (1999): 49–80; J. Lind, "Apologies in International Politics", *Security Studies* 18:3 (2009): 517–556.
14. For more details about the Japanese textbook approval and screening, see H. Mitani, "Japan's History Textbook System: Creation, Screening and Selection", http://www.nippon.com/en/in-depth/a00701/. Japan's Ministry of Foreign Affairs additionally provides an in-depth explanation of the school textbook system in English, Chinese and Korean languages. See "Japan's School Textbook Examination Procedure".
15. Nozaki, *War Memory*...See also S. Ienaga, "The Historical Significance of the Japanese Textbook Suit", *Bulletin of Concerned Asian Scholars* 2 (1970): 3–12 and S. Ienaga, "The Glorification of War in Japanese Education", *International Security* 18:3 (1993): 113–133.
16. The unit was based near the city of Harbin in northeastern China and was disguised as a water-purification unit. In August 2002 the Tokyo District Court for the first time recognized that Imperial Japanese Army units, including Unit 731, used biological weapons in violation of international conventions, and for that the Japanese state is responsible. However, the court denied Chinese plaintiffs compensation, filed between 1997 and 1999 by some 188 survivors of the war crimes. According to some historians, as many as 250,000 people may have been killed by the infamous Unit 731 as part of its biological experiments in the 1930s and 1940s, when much of China was occupied by Japanese troops.
17. Sven Saaler, *Politics, Memory and Public Opinion: The History Textbook Controversy and Japanese Society* (Munich: Iudicium, 2005), 120.

18. The program on history textbook revisions was initially adopted by the First General Conference of UNESCO in 1946.
19. *Miyako Shimbun,* 17 September 1947.
20. No document was found by the author indicating the exact numbers of both cooperative associations and clubs in earlier period of the UNESCO Movement in Japan. Nagahiro Yamashita, the first analyst of the movement, for example estimated the number of UNESCO cooperative associations and clubs reaching 105 at the end of 1948. However in his critical analysis, published in *Kokusai Rengo,* Yamashita pointed out that in spite of the rapidly increasing number of the cooperative associations, "the UNESCO Movement was still in its infancy and not properly understood even by its followers in Japan", see "General Headquarters Supreme Commander for the Allied Powers, Military Intelligence Section, General Staff Allied Translator and Interpreter Section: Publications Analysis, 9 November 1948", *UNESCO Programme – Japan, Part I up to 31 March 1950,* X07(520)/ X07.7(52), Number 220 (UNESCO, Paris, 1948): 6, UNESCO Archives. Adding to that, in 1952 the Secretary-General of the Japanese National Commission for UNESCO, Tadakatsu Suzuki, reported about the operation of some 103 cooperating associations and 13 supporting organizations, united under the Federation of UNESCO Associations in Japan, as well as 170 UNESCO student clubs set up in various colleges and universities. According to him, "this spontaneous development of the movement on a national scale serves to indicate the profound interest of the Japanese people in Unesco", see "Report on the Visit to UNESCO Headquarters to the Director-General by Mr. Tadakatsu Suzuki, 3 December 1952", *Relations with Japan – National Commission, Part I up to 31/XII/1953,* X07.21(520)NC/UNESCO/XR/NC/68 (UNESCO, Paris, 1952): 1, UNESCO Archives.
21. "Dr. Lee Shi Mou to Dr. Jaime Torres Bodet: Report on UNESCO Programme in Japan, 4 August 1949", *UNESCO Programme – Japan, Part I up to 31 March 1950,* X07(520)/UOJ/49/Rep.2 (UNESCO, Paris, 1949): 7, UNESCO Archives.
22. "Newspaper clippings etc.", *Supreme Commander for the Allied Powers Civil Information and Education Section, Public Opinion & Sociological Rsch Div.,* General Subject File (1945–1952), Record Group 331, Box No. 5938/ MI-215, Enclosure "A" (College Park, Maryland, 1948–1949), US National Archives.
23. *The Asahi Shimbun,* 31 January 1951.
24. *The Nippon Times,* 17 March 1951.
25. *The Mainichi,* 19 March 1951.
26. "Diet Report No. 80: Deliberation at the 10th Diet. Holding Affairs Committee meeting of the House of Representatives, May 11, 1951", *Supreme Commander for the Allied Powers Civil Information and Education Section, Public Opinion & Sociological Rsch Div.,* General Subject File (1945–1952). Record Group 331, Box No. 5938 (College Park, Maryland, 1951): 1–3, US National Archives.
27. Ibid. It should be also noted that in those years, communism was still seen as a powerful political force and attractive ideology not only by the political elites but also among intellectual circles in post-war Japan. In 1951 the SCAP administrators critically observed in their report to Washington that "Among Japanese intelligentsia there is much... reluctance to take an irrevocable stand against communism... Thus many of the professors and writers with great influence on younger minds are neither exposing the fallacies and evils of communism nor helping to propagate democracy". As they continued there were also many other Japanese organizations, mass media and individuals who were fighting against

communism, and that "SCAP has been encouraging and advising" them, see "General Headquarters Supreme Commander for the Allied Powers, Plans for fiscal year 1952, 22 April 1951", *Supreme Commander for the Allied Powers Civil Information and Education Section, Public Opinion & Sociological Rsch Div.*, General Subject File (1945–1952). Record Group 331, Box No. 5938 (College Park, Maryland, 1951): 106, US National Archives.

28. "UNESCO. General Conference, 6th, 18 June – 11 July 1951", *Records of the General Conference, Sixth Session, Proceedings* (UNESCO, Paris, 1951): 114–115, UNESCO Archives.
29. "UNESCO: Five Years of Work (Prepared by the Mass Communications Department of UNESCO), January 1951", *Supreme Commander for the Allied Powers Civil Information and Education Section, Public Opinion & Sociological Rsch Div.*, General Subject File (1945–1952). Record Group 331, Box No. 5938 (College Park, Maryland, 1951): 4, US National Archives.
30. Resolutions adopted on the Report of the Programme and Budget Commission. The Programme of UNESCO: Chapter 2. – Education. Schools and Youth", *Records of the General Conference. Resolutions*, resolution 3C/IX.2.514 (UNESCO, Paris, 1948): 19, UNESCO Archives.
31. "Rough description of some of Education of Ministry's observations and activities concerning the Resolution of the General Conference of UNESCO on its activities in Japan in 1950, 26 January 1950", *Supreme Commander for the Allied Powers Civil Information and Education Section, Public Opinion & Sociological Rsch Div.*, General Subject File (1945–1952). Record Group 331, Box No. 5938 (College Park, Maryland, 1950), US National Archives.
32. "Dr. S. M. Lee, Unesco Representative in Japan, to Dr. Jaime Torres Bodet, Director-General: Report on UNESCO Programme in Japan, February 1952", *UNESCO Programme in Japan, 1949–1950,* X07.7 (52)/UOJ/52/Rep. II (UNESCO, Paris, 1952): 1–7, UNESCO Archives.
33. "Extract from Report on UNESCO Programme in Japan by Dr. S. M. Lee, UNESCO Representative in Japan: Donations Towards Fund for Distressed Children," *UNESCO Programme in Japan, 1949–1950,* X07.7 (52) Ref. UOJ/52/Rep. II (February) (UNESCO, Paris, 1952), UNESCO Archives.
34. "Report on the Visit to UNESCO Headquarters to the Director-General by Mr. Tadakatsu Suzuki, Secretary-General, Japanese National Commission for UNESCO, 3 December 1052", *Relations with Japan – National Commission, Part I up to 31/XII/1953,* X07.21(520)NC/UNESCO/XR/NC/68 (UNESCO, Paris, 1952): 2–3, UNESCO Archives.
35. UNESCO. *Organizing Programmes of Education for International Understanding* (Paris: UNESCO, 1965), 32. When UNESCO began to develop its program on textbook revisions, it took into account past experiences, particularly those undertaken by the League of Nations' International Committee on Intellectual Cooperation, which was responsible for international textbook revisions between 1925 and 1945.
36. "UNESCO. General Conference First Session, Held from 20 November to 10 December 1946", *Records of the General Conference, First Session, Proceedings* (UNESCO, Paris, 1947): 151, UNESCO Archives; *A Handbook for the Improvement of Textbooks and Teaching Materials* (Paris: UNESCO, 1949).
37. *Better History Textbooks* (Paris: UNESCO, 1950), 21–22.
38. Falk Pingel, "Old and New Models of Textbook Revision and Their Impact on the East Asian History Debate", *The Journal of Northeast Asian History* 7:2 (2010): 15.

39. This first Expert Committee on Japanese Questions was set up by the director-general of UNESCO in accordance with the resolutions of the Third Session of the General Conference in 1949. Although no Japanese observers were invited to the committee it was anticipated that in future the committee membership would be increased to seven, but restricting it to the delegates of the member states of UNESCO.
40. "Report of the Expert Committee on Japanese Questions, submitted to the Executive Board for Decision, 2 August 1949", *UNESCO Programme for 1949–1950,* X07.7(52) 008.41, 17 EX/10 (UNESCO, Paris, 1949): 1, UNESCO Archives.
41. "J.W.R. Thomson, Adviser to the Director-General, to Dr. Shi-Mou Lee, 11 August 1949", *UNESCO Programme for 1949–1950,* X07.7(52)008.41/X.R./110190 (UNESCO, Paris, 1949), UNESCO Archives.
42. "Committee of Experts on Japanese Questions, Third Session: Report to the Director-General, 20 February 1950", *UNESCO Programme for 1949–1950,* X07.7(52)008.41/19 EX/9 Addendum (UNESCO, Paris, 1950), UNESCO Archives.
43. "Report of Conference: UNESCO Committee of Experts, Tokyo, 14 December 1950", *UNESCO Programme for 1949–1950,* X07.7(52)088.41 (UNESCO, Paris, 1950): 5, UNESCO Archives.
44. "Miyazaki to Evans, dated 30 March 1957", *Relations with Japan – Participation Programme: Part I up to 31/XII/1959,* X07.21(529)AMS/Annex II (UNESCO, Paris, 1957), UNESCO Archives.
45. *School Textbooks in Japan: A Report of a Survey from the Standpoint of Education for International Understanding and Cooperation* (Tokyo: Japanese National Commission for UNESCO, 1958), 97–98.
46. H. Mitani, "Japan's History Textbook System: Creation, Screening and Selection", http://www.nippon.com/en/in-depth/a00701/. See also H. Mitani, "Writing History Textbooks in Japan" in *History Textbooks and the Wars in Asia: Divided Memories,* ed. Gi-Wook Shin and Daniel C. Sneider (New York: Routledge, 2011).
47. Falk Pingel, *UNESCO Guidebook on Textbook Research and Textbook Revision* (Paris: UNESCO, 2010), 62.
48. Pingel, "Old and New Models of Textbook Revision ...", 15.

Figure PV.1 Afghan boys play football near where one of the Buddhas of Bamiyan once stood. Subsequent to the 2001 destruction of the two giant standing Buddha statues by the Taliban, UNESCO in 2003 simultaneously inscribed the cultural landscape of the Bamiyan Valley onto the World Heritage List and the List of World Heritage in Danger Conducting UNESCO's Nubia Campaign, Abu Simbel, Egypt, 1966. (© United Nations).

Part V
Practising World Heritage

UNESCO launched the History of Mankind project in 1950 to promote the fact that people all live in the same world, belong to the same humanity and have a common history. The subsequent publication of a series of volumes on the humanity's shared past were written by scholars from all over the world, highlighting the interdependence of various cultures and accentuating their contributions to the common cultural heritage. Simultaneously, UNESCO worked to safeguard the common heritage of war-devastated countries, during armed conflicts and as part of its expert mission program. Later it expanded the concept of common heritage to include nature and wildlife.

The organization's work in the field of world heritage became world famous with the Nubia campaign, launched in 1960. The purpose was to move the temples of Abu Simbel to keep them from being swamped by the Nile after the construction of the Aswan High Dam. During the campaign, 22 monuments and architectural complexes were relocated. This was the first and largest in a series of campaigns that included campaigns in Venice in Italy, Borobudur in Indonesia and Angkor in Cambodia.

This work led to the adoption, in 1972, of the Convention Concerning the Protection of the World Cultural and Natural Heritage. A World Heritage Committee was established and the first sites were inscribed on the World Heritage List in 1978. Since then other important legal instruments regarding cultural heritage and diversity have been adopted by UNESCO member states.

Today more than a thousand sites are listed, most of them cultural heritage sites, with Italy being the country with the greatest number, closely followed by China. On the one hand, the World Heritage Programme is now UNESCO's most widely known and prestigious undertaking. On the other, it is also – from time to time – criticized for having a negative, unintended impact and for being incapable of truly protecting heritage sites in danger.

13
UNESCO and the Shaping of Global Heritage

Aurélie Élisa Gfeller and Jaci Eisenberg

Introduction

By the turn of the millennium, heritage had taken center stage at UNESCO.[1] Today the organization's flagship heritage instrument, the 1972 Convention Concerning the Protection of the World Cultural and Natural Heritage (generally known as the World Heritage Convention), and this convention's list of sites of "outstanding universal value", the World Heritage List, constitute its most publicly identifiable mission. UNESCO's heritage instruments and projects have greatly influenced national heritage policies worldwide, and it is for this reason that this chapter details the rise of heritage as part of the organization's mandate, the mutations in its understanding and promulgation of heritage, and the effects of its heritage policies on the ground.

While UNESCO's involvement in heritage built on the efforts of its precursor, the League of Nations' International Committee for Intellectual Cooperation (ICIC), this point is only briefly raised here as a basis for comparison. ICIC's work in the arts led to the establishment, in 1926, of an International Museums Office, which organized congresses and guidelines for preservation.[2]

Our story begins in earnest in the post-war period when, different from the League of Nations, UNESCO's initial involvement in heritage entailed internal institutional strengthening and active support for the creation of complementary governmental and non-governmental international organizations. Once this capacity was established, from the 1960s onwards, UNESCO tested its ability to carry out more ambitious heritage projects, against a complicated context full of political turmoil, economic modernization and environmental emergencies.

The international campaign to save the temples of Nubia was UNESCO's first large-scale rescue effort and as such merits extended exploration. The organization coordinated the massive international funding and expertise required to move the Pharaonic temples of Egyptian and Sudanese Nubia out of the path of the imminent flooding from the new Aswan Dam.

We then move to survey the many international heritage-related recommendations and treaties which took shape during this period, marked by an early high with the World Heritage Convention, and a later pique enshrining non-European forms of heritage. By way of a conclusion, we explore the positive, negative and often ambiguous efforts of UNESCO's heritage efforts on the ground.

Institutional landscape

The first international organization created after World War II to deal exclusively with heritage – the International Council of Museums (ICOM), established in Paris in 1946 – was a direct outgrowth of interwar experiments. Though not officially part of UNESCO, ICOM had close links with it. An institutionalized network of experts, it set out to assist UNESCO in its museum-related activities.[3]

In its founding years, UNESCO strengthened its institutional mandate to cover not only museums but also monument protection, a reaction to the massive wartime destruction and ensuing post-war reconstruction. In 1948 UNESCO's General Conference called for a study to explore establishing an expert committee to aid states in preserving historic monuments and sites.[4] The first meeting of experts on monument conservation, convened by UNESCO in 1949, saw the participants reflecting on their national situations in reports.[5] Stanisław Lorentz – the founding director of the Chief Directorate for Museums and Monuments Protection of the post-war Polish Government,[6] drew upon Poland's tragic history in advocating the reconstruction of destroyed historic buildings.[7] The reports comprised a precious knowledge base regarding national practices, while an exhibition of drawings and photographs offered compelling evidence of conservation approaches worldwide. The experts sought the creation of a permanent expert committee entrusted with promoting information-sharing and expert collaboration; informing UNESCO of the need to establish expert missions; preparing international legislation; and "protect[ing] movable and immovable heritage of universal value" during war.[8] The General Conference established this International Committee on Monuments, Artistic and Historical Sites and Archaeological Excavations (henceforth UNESCO Monuments Committee) in 1950.

The UNESCO Monuments Committee addressed a number of existing and innovative topics, ranging from expert missions and emergencies (the destruction caused by an earthquake in Greece) to medium- or longer-term plans (a conservation manual, microfilm storage facilities and a center to document Egyptian art).[9] However, much attention was absorbed by a question lingering since the interwar period: regulation of archeological excavations.[10] In 1937 the International Museums Office held an International Conference on Excavations. This conference concluded the Cairo Charter, marking the beginning of international regulation against the illicit trafficking of cultural property.[11] In 1956 the General Conference broadened the scope of the Cairo Charter to stolen and illegally exported objects through a Recommendation on International Principles Applicable to Archaeological Excavations.[12]

UNESCO also played a central role in creating new heritage-related international organizations. One such institution was the International Centre for the Study of the Preservation and Restoration of Cultural Property (ICCROM), established in Rome in 1959. Already discussed at the General Conference of 1952, the proposed center sought to gather and disseminate information, coordinate and promote research, procure expertise and train specialists. It was to be closely associated with an existing European institution, which ended up being Italy's Central Conservation Institute (Istituto Centrale del Restauro), so ensuring ICCROM's establishment in Rome. Though created as an independent organization with its own governing bodies, ICCROM retained close ties to UNESCO. For example, from 1960 onwards, it took an active advisory and even coordination role in UNESCO's heritage-safeguarding campaigns.[13]

The other major heritage institution UNESCO helped to established was the International Council on Monuments and Sites (ICOMOS). This institutionalized network of experts was founded in 1965 to spread best practice for architectural conservation.[14] It was intended to be the first heritage-related NGO, and its mandate differed from those of ICOM and ICCROM in that it was restricted to immovable heritage (initially conceived of as "monuments and sites").[15] ICOMOS was designated as an advisory body in the text of the World Heritage Convention. Though it shared this status with ICCROM and the International Union for the Conservation of Nature (IUCN) – a nature-related forum for governments, NGOs, scientists and business – ICOMOS alone subsequently assumed the task of providing recommendations regarding the cultural sites proposed for inclusion on the World Heritage List. Through the national committees founded in an increasing number of countries throughout the world and the international specialized committees established in fields such as the economics of conservation and risk preparedness, ICOMOS also acted as a crucial platform for transnational knowledge-sharing.

Nubia dispersed

The International Campaign to Save the Monuments of Nubia (1960–1980) was the first UNESCO heritage enterprise launched with the intent of drawing global attention to a specific cultural site. It was further notable because it involved an enormous amount of financing, as well as a large publicity campaign aimed at repackaging this seemingly local issue into one of global importance. The implementation of the campaign made heritage action highly visible on a global scale.

The first president of the new Egyptian republic established in 1954, Gamel Abdel Nasser, recognized the pressing need for his nascent republic to modernize: a growing population – which had more than doubled from the turn of the 20th century to after World War II – and industrialization made an adequate supply of power a burning issue. In 1955, Nasser authorized the construction of a new high dam at Aswan, supported by Soviet engineers. The project would replace the inadequate low dam constructed under British rule (1898–1902), one which had already been raised twice (1907–1912; 1929–1934). The new dam was projected to create an artificial lake – Lake Nasser – of 5,000 km^2. It would effectively flood the Nile River Valley, threatening the temples and monuments located along the river in Egypt and Sudan, requiring 50,000 people to be relocated, yet increasing food production by 50 percent.[16]

The construction of the Aswan High Dam also made Nubia a locus of the Cold War. Though the Egyptians and Soviets were unwilling to scale back their plans for modernization to avoid or attenuate the attendant destruction, the Egyptians were cognizant of the impact the dam would have on the Nile River Valley monuments. They were certain that these structures were worthy of some degree of preservation for the common benefit of humanity. This manifested itself in the Egyptian Government's request, in January 1955, for UNESCO to coordinate foreign archeological teams which could remove some structures in order to preserve part of the valley's heritage before the imminent flooding. In 1955, to regulate these removals, UNESCO helped to create a Documentation and Study Centre on the History of Art and Civilization in Ancient Egypt. This early call sparked archeological assistance from only two countries: Germany and Italy.[17]

The documentation center's chief expert, Louvre Museum Egyptologist Christiane Desroches Noblecourt, sounded an alarm for a broader campaign in April 1956. However, the Suez Crisis blocked this initiative from immediately gaining traction: Frenchwoman Desroches Noblecourt and other UN staff members were temporarily evacuated by the US Sixth Fleet. The question was again raised in April 1959, when the recently appointed culture minister for Egypt (then the United Arab Republic (UAR)), Saroïte Okacha, asked UNESCO to coordinate scientific and technical aid for Nubia. Torgny Säve-Söderbergh, a member of the Scandinavian mission to the Sudanese Nubia campaign and author of the 1987 official UNESCO history of the Nubia campaign, ascribed Okacha's request to the influence of James Rorimer, director of the Metropolitan Museum of Art, who had brought the Nubian conundrum to Okacha's attention, as well as that of Desroches Noblecourt, who later curated a meeting between Okacha and UNESCO's assistant director-general, René Maheu, during a stopover. Things moved quickly from that point – one day after the fateful Maheu–Okacha meeting, Okacha learned that UNESCO's director-general, Vittorio Veronese, had agreed to put the proposal to save the temples in front of the executive board. Okacha then went on to secure Nasser's approval, which he did in January 1959.[18]

To entice more parties to join the safeguarding campaign, the UAR Government offered a "counterpiece" – compensation amounting to half of the archeological finds recovered – to those archeological teams proceeding in the threatened zones.[19] Through the 1920s, all archeological digs in Egypt had operated on this principle: every dig was split into two lots, with half going to the Egyptian Antiquities Department in recognition of the rights granted to dig, and half going to the foreign archeological teams in recognition of their expertise and resources. This tradition died out as the colonial influence in Egypt waned. Sudan launched a separate appeal, one which similarly offered a counterpiece, but which differed in underscoring the relative poverty of Sudan and appealing to the unexplored potential of Sudanese archeology. UNESCO's executive board examined the UAR and Sudanese requests in tandem at its 55th session in November–December 1959, agreeing to spearhead the appeals on the grounds that the campaign would have a positive effect on the organization's global reputation.[20]

In preparation for the UNESCO appeal, an Honorary Committee of Patrons was appointed to show important backing for the initiative, and UNESCO's director-general, Veronese, was given the power to form an International Action Committee (IAC) of eminent persons to aid the campaign. Veronese launched the UNESCO campaign for Nubia on 8 March 1960. The event has stuck in many minds because the French culture minister, André Malraux, made a heartfelt plea for humanity to collectively save the Nubian monuments from destruction, giving the world community an opening to act independently of the Cold War quagmire then facing the region.[21]

The UNESCO-run International Campaign to Salvage the Temples of Nubia is often considered in two phases: the securing and transfer of Abu Simbel, which then instilled confidence for a similar campaign for the Philae temples. Several plans were solicited for Abu Simbel, including a proposal which would have left the temples in place and created an underwater park. Eventually, however, Abu Simbel was deconstructed by being cut into blocks and reconstructed 60 meters higher than its original position. Careful attention was paid to respect the original sun exposure of the temples.[22] The success of the Abu Simbel campaign led to a dedicated campaign for Philae, begun in 1972 and finished in 1978.[23] The temples of Philae Island were deconstructed and moved to a different island – Agilkia – near the site of the Aswan Low Dam.

However, these two flagship campaigns did not comprise the entirety of the salvage campaign: most of the remaining temples were artificially rearranged into four other groups and dispatched to sites in the UAR and in

Sudan. Four temples were sent abroad to Italy, Spain, the Netherlands and the USA as acknowledgement for extensive help in the rescue operations. Three temples were deemed unmovable and left onsite. In total, 50 foreign states sent 150 experts and provided USD 40 million in funding to help relocate the temples that were movable. The UAR/Egypt contributed half of the USD 40 million deployed to save Abu Simbel and half of the USD 30 million used to rescue Philae.[24]

The technical expertise and bureaucracy necessary for such a campaign, not to mention the intricacies of financing, made it run for 20 years in total. Relaying the experience of one country may help underscore its complexities: the USA joined the campaign to save the Nubian temples in June 1963, when President John F. Kennedy approved participation as a means to influence the political and economic situation of the Arab world. US participation had strings attached: the Americans stipulated that their financial contribution would be provided directly in Egyptian currency, to be obtained through the sale of US wheat in Egypt. The USA would thus reap economic benefits by selling crops abroad, thus using the proceeds to finance their foreign aid. The US Government sought to earmark their aid for Philae, but in the end the Egyptians used it to dismantle all the movable temples.[25] The USA could attempt to be picky about the use of its funding because it was the primary foreign donor to the first campaign at Abu Simbel: it pledged USD 12 million; all other countries, save Japan, pledged less than USD 1 million.[26] This explains why the USA received one of the complete temples – the extremely desirable Temple of Dendur – in recognition of its aid, which was "won" by the Metropolitan Museum of Art in New York following a nationwide competition.[27]

UNESCO's role in the Nubia campaign changed over the course of the 20-year operation. In 1960 the IAC suggested that UNESCO should run the entire Nubian campaign, but Veronese opposed, retaining only the role of intermediary between the Egyptian Government and foreign archeological teams. Soon it transitioned to organizer and, finally, it ended up in charge of the rescue operation, though informed by at least seven expert committees.[28] At its conclusion, UNESCO's newspaper, *The UNESCO Courier*, was not shy about its impact: the cover of the February—March 1980 issue proclaimed, "Victory in Nubia – The greatest archaeological rescue operation of all time", and the text prominently gave credit to "Unesco and the world community".[29]

UNESCO's rhetoric masks ambiguous motivations and outcomes. For Lucia Allais, the Nubia campaign was not merely an opportunity to preserve pillars of world heritage for the good of humanity. Rather, it was a calculated attempt by Nasser to benefit from East–West tensions in favor of "nation-building" projects: "with Khruschev he built the dam; with UNESCO he salvaged the temples".[30] In this sense, Egypt (and Sudan) moved forward, but in another sense, Egypt moved backwards. The reinstatement of counterpieces opened up the campaign to allegations that promoted and sanctioned neocolonialism. Locally the UNESCO campaign had an ambiguous impact: the outpouring of international support for heritage preservation in the Nile River Valley led Egypt and Sudan to build museums for the monuments retained, but this was only possible with international financial aid.[31]

However, there were also several positive effects from the Nubia campaign. First, as Desroches Noblecourt pointed out before the campaign, "Egyptological studies had been at a standstill in the region above the First Cataract of the Nile... Only a sparse handful of archeological missions were at work...". The UNESCO appeal brought a new wave of scholars to the region.[32] Second, in undertaking the highly mediatized safeguarding campaign, UNESCO increased the number of visitors to these sites, thus raising their prominence. Third, the organization used the campaign as an educational tool, creating the Tutankhamun Treasures exhibit to tour major donor countries from 1961 to 1966.[33]

The Nubia campaign directly influenced UNESCO's emergency responses to threatened heritage, serving as a model for the next "urgent practical request":[34] a plea to save the medieval and Renaissance heritage located in Venice and Florence from the devastating floods of 1966.[35] The campaign further inspired the move to clean and restore the Borobudur Temple in Indonesia (1970–1983),[36] but its most direct legacy is in unlocking aid for pressing heritage preservation projects, such as at the archeological sites at Carthage (Tunisia, 1972), under extreme pressure from rapid urbanization and increasing tourism;[37] the Acropolis (Greece, 1977), because of its advanced age;[38] and the ancient urban center of Moenjodaro (Pakistan, 1974), because of flooding and harsh climatic conditions.[39] Scholars tend to assert that the Nubia campaign was the first instance of cultural diplomacy – UNESCO's area of expertise – being successfully integrated into international technical assistance.[40]

International norm-making

Norm-making permitted UNESCO to consolidate its influence in the heritage field. Its norm-setting agenda initially began as an outgrowth of interwar efforts. The 1954 Convention for the Protection of Cultural Property in the Event of Armed Conflict with Regulations for the Execution of the Convention, concluded at The Hague, was an outgrowth of the Pan-American Union's 1935 Roerich Pact, which had advanced similar goals. The 1954 convention was a watershed because it marked the first international treaty devoted exclusively to heritage.[41]

UNESCO's norm-making agenda continued while it pursued major rescue operations, as a way to deal with the effects of decolonization and expanded membership. In response to these shifts, in 1970 the organization's General

Conference adopted the Convention on the Means of Prohibiting and Preventing the Illicit Import, Export and Transfer of Ownership of Cultural Property.[42] However, the newly independent countries found the convention to be insufficient in practice, leading, in 1978, to the foundation of a new entity: the Intergovernmental Committee for Promoting the Return of Cultural Property to its Countries of Origin or its Restitution in Case of Illicit Appropriation. This committee supported the bilateral negotiations which served as an effective application of the aims of the 1970 convention.[43] A further attempt to strengthen the aims of the 1970 convention came in 1995, with the passage of the Convention on Stolen or Illegally Exported Cultural Property by the International Institute for the Unification of Private Law (Institut International pour l'Unification du Droit Privé (UNIDROIT)). Yet the restriction on appending reservations,[44] combined with the low ratification of the UNIDROIT Convention – 37 as of November 2015, compared with 129 for the 1970 convention – means that the 1970 convention remains authoritative.[45] The regime formed by the 1970 convention and subsequent measures has had a long reach in practice, leading to the return of ill-gotten objects;[46] the adoption of policies in line with the 1970 convention by governments and archeological institutions;[47] and the effective closing of the auction market to antiquities purchased after the passage of the 1970 convention without adequate documentation.[48]

In contrast with the 1970 convention, which embodied a particularistic vision of heritage ownership (heritage belonging to the group that produced it or its descendants), the 1972 World Heritage Convention expressed a cultural internationalist view – one advocating global ownership of heritage.[49] Though adopted by UNESCO, this convention was the outcome of three interrelated initiatives advanced by the US Government, IUCN and UNESCO. These initiatives had sought, respectively, to advance the idea of "A Trust for the World Heritage"[50] and to propose criteria for the creation of "an effective legal system for the protection of monuments and sites".[51] These multiple parties explain why the World Heritage Convention came to encompass not only cultural but also natural heritage. Although three draft documents were elaborated, the UNESCO draft ultimately served as the basis for the World Heritage Convention.[52]

The World Heritage Convention is the only international heritage instrument to have near universal ratification, and the World Heritage List is its main component. The list grew to 1,031 in 2015 from 12 sites in 1978 – 802 cultural, 197 natural and 32 mixed cultural and natural.[53] Numerous actors – governments, economic actors as well as site managers and conservation practitioners – are keen to have heritage sites inscribed on the World Heritage List. Indeed, the World Heritage title offers the additional values of increased protection, enhanced political prestige and public awareness and economic development through international aid and tourism expenditure.[54]

Although Europe and North America have always been overrepresented – 166 out of 320 sites in 1994 versus 492 out of 1,031 in 2015 – there have been consistent efforts since the late 1980s to remedy the Eurocentric bias of the World Heritage Convention and List. In 1992, specific criteria for "cultural landscapes" were adopted to bridge the Eurocentric divide between culture and nature.[55] The 1994 Global Strategy for a Representative, Balanced and Credible World Heritage enumerated several imbalances, including the overrepresentation of European historic towns and Christian monuments, and the underrepresentation of vernacular architecture and the heritage of "living cultures". The substitution of an anthropological, as opposed to an art historical, approach to heritage was advanced as the remedy.[56] In the same year, two expert meetings arranged by governments recommended the inscription of atypical heritage such as "heritage canals" and "cultural routes",[57] and in a separate meeting at Nara (Japan), 45 experts signed a declaration introducing cultural relativism into the "test of authenticity" enshrined in the World Heritage Operational Guidelines.[58] As a result of these conceptual shifts, the World Heritage List rapidly became more diverse,[59] even if wealthy states in the Global North were quicker to make use of these new categories than those in the Global South, due in part to the huge costs involved in preparing the World Heritage List nomination files.

The 1990s were a crucial decade for not only the conceptual evolution of the World Heritage Convention but also the institutionalization of a non-European heritage concept: intangible heritage. UNESCO successively adopted a non-binding Recommendation on the Safeguarding of Traditional Culture and Folklore (1989),[60] a Living Human Treasures System based on a Korean proposal (1993),[61] and a Proclamation of Masterpieces of the Oral Heritage of Humanity Program (1998).[62] These steps culminated in the 2003 Intangible Cultural Heritage Convention.

The various origin stories for the 2003 convention all point to the renewed assertiveness of the non-European world – Bolivia, a Spanish plea in favor of Morocco, Japan – in global heritage norm-making.[63] These developments were all significant at different stages: the plea for Morocco was an important factor behind the Proclamation of Masterpieces program, which involved UNESCO's designation of "masterpieces of the oral and intangible heritage of humanity" to foster identification, preservation and promotion efforts by governments, NGOs and local communities.[64] The decisive shift to codification can be attributed to the 1999 election of a new director-general, Koiichuro Matsuura of Japan. Matsuura upgraded the program to a convention – the 90 items designated as masterpieces in 2001, 2003 and 2005 were subsequently incorporated into the convention's Representative List of the Intangible Cultural Heritage of Humanity.[65] This was in line with his early-mandate prioritization of intangible heritage as one of UNESCO's eight priority programs, while Japan donated USD 3.2 million for both the Masterpieces program and the 2003 convention.[66]

In line with their role in the genesis of the 2003 convention, Global South countries rapidly ratified the Intangible Cultural Heritage Convention and eagerly went on to propose practices for inscription on its lists, with China claiming the lion's share (38 out of 364 items in 2015) – versus Italy for the World Heritage Convention (51 out of 1,031 sites in 2015). Several Global North countries, in contrast, expressed reservations, and some, including Canada, Russia, the UK and the USA, are still not party to this convention. This shows that, after the 1970 Convention on the Transfer of Cultural Property, the 2003 Intangible Cultural Heritage Convention represented a new assault against the prevailing Eurocentrism of UNESCO's heritage norms.

Effects on the ground

UNESCO's World Heritage initiatives have proved to be a double-edged sword in practice, according to the anthropologists, geographers and sociologists who have examined this question in depth. The results are generally positive in economic terms but hard to qualify in human terms. For every site or tradition that sees an increase in tourism and attendant revenues, there is likely to be a local constituency protesting the alteration of the fundamental nature of the site or tradition inscribed on the flagship 1972 Convention's World Heritage List or on the newer 2003 Convention's lists: the List of Intangible Cultural Heritage in Need of Urgent Safeguarding, the Register of Best Safeguarding Practices and the Representative List of the Intangible Cultural Heritage of Humanity.

Ever since the advent of world heritage lists, the validity of a common world heritage has been challenged. This is evident in the case of the giant Buddha sculptures located, until 2001, in the Bamiyan Valley of Afghanistan. In early 2001 the leader of the Taliban called for "the elimination of all non-Islamic statues and sanctuaries from Afghanistan", including the Bamiyan Buddhas. Protests against the imminent destruction of these statues, invoking the universality of "cultural heritage", took place worldwide but were not enough to modify the Taliban's stated intentions. A month after Mullah Mohammad Omar's decree, the Buddhas were destroyed. Pierre Centlivres takes note of the many competing interpretations of the Buddhas' importance. To the Afghan population, the Buddhas were secondary to the attributes of the Bamiyan Valley, remaining only as relics representative of the pre-Islamic pagan tradition in the region. To the Taliban, differently, the Buddhas were sacred, and thus idols.[67] Beyond Afghanistan, Centlivres advances Paul Bernard's assertion that Art (with a capital A) has replaced religion as "the guardian of the value of the sacred" in "the West".[68] By destroying the Bamiyan Buddhas, the Taliban both rejected the "Western" reverence of Art and proved that "the category of 'cultural heritage' hardly existed for the Taliban or was, at best, suspicious".[69] Only after the destruction of the Buddhas, in 2003, were the cultural landscape and archeological remains of the Bamiyan Valley inscribed simultaneously on the World Heritage List and the List of World Heritage in Danger. The nomination file attested to the ultimate triumph of UNESCO's valorization of Art over iconoclasm in that it specifically referenced the destroyed Buddhist icons.[70]

David Berliner sees similarities in the recent events in Mali. The armed conflict ignited in early 2012 between factions seeking to promote autonomy in the north of the country brought pleas from UNESCO to safeguard religious sites in Timbuktu and the Tomb of Askia in Gao.[71] Timbuktu was inscribed on the World Heritage List in 1988 for its role as a center of Islam in the 15th and 16th centuries, and its three great extant mosques – Djingareyber, Sankore and Sidi Yahia.[72] The Tomb of Askia was inscribed in 2004 to commemorate the flourishing imperial trade in the area, as well as the regional mud-building tradition.[73] With the fighting intensifying, on 28 June 2012, UNESCO and the Malian Government proposed that these two World Heritage sites should be placed on the List of World Heritage in Danger.[74] A spokesperson for the Salafi movement Ansar Dine, interviewed by a French news outlet in Mali shortly after UNESCO's decision, responded: "God, he is unique. All of this, it's *haram* ('forbidden' in Islam). We, we are Muslims. UNESCO, it's what?" He then reaffirmed Ansar Dine's avowed destruction of all mausoleums in Timbuktu.[75] In practice, placing the Malian sites on the List of World Heritage in Danger had the opposite of the intended effect: world heritage as a concept was foreign to and unimportant for these radical Islamists, except as a point for protest.[76]

Charlotte Joy elaborates on the extent to which local populations are familiar with UNESCO. Evaluating the situation in Djenné (Mali) almost 20 years after the inscription of the Old Town (1988), she reported that if local populations were aware of UNESCO, it was predominantly "through restrictions: those concerning access to the archeological sites and restricting architectural practices".[77] More generally, the negative impacts of restrictions that follow World Heritage List status are a commonality in existing studies. Shu-Yi Wang relates that the inscription of the ancient city of Pingyao (China) in 1997 prompted the local administration to restrict the city center to tourist-oriented enterprises, effectively ridding the zone of "Stores oriented to the local community".[78] Lisa Breglia describes how a government-sponsored upgrade at Chichén Itza (Mexico) in the early 1980s, made in advance of its nomination and, in 1988, its successful inscription on the World Heritage List, forced the guardians of the site, the *antiguos*, to relocate outside its perimeter.[79] David Harrison points to the ambivalence of local populations who must coexist with the (admittedly few) World Heritage sites commemorating "Shameful episodes". A case in point is Auschwitz-Birkenau, inscribed as a Polish site in 1979. The inscription of the former concentration camp ensures the identity of the site is as "a memorial to infamy".[80]

The process of inscribing a site on the World Heritage List, an inherently international enterprise, often butts up against localized or regional conflicts. Berliner also highlights situations where World Heritage designation aggravates existing border disputes.[81] The Temple of Preah Vihear, inscribed in 2008 on the World Heritage List as a Cambodian site, became a key point of contention in 2011 in the ongoing Cambodian-Thai border dispute. The inscription of the site in Cambodia was a major point of contention for Thailand because ownership had been an open question as far back as 1962, when an International Court of Justice decision placed the temple within Cambodia's borders.[82] The 2008 inscription happened after an earlier agreement between the two countries to work towards a joint nomination. Helaine Silverman argues that the inscription of Preah Vihear led to military violence in 2011 that could have been anticipated, given the difficult relationship between these two countries.[83] Around the same time, in the aftermath of Palestine's 2011 admission to UNESCO as full member, UNESCO's Executive Board decision to consider two sites in the West Bank – the Tomb of the Patriarchs in Hebron and Rachel's Tomb in Bethlehem – as potential Palestinian World Heritage sites provoked the ire of Israel.[84] In this case the World Heritage inscription process became enmeshed with the contentious question of sovereignty in the region. At the time of writing, the matter was still under discussion at UNESCO but no definitive decision had been taken.[85]

The successful inscription of a site can provoke criticism that the past has been rewritten to fit a specific agenda that is not necessarily reflective of reality. With regard to built heritage, Berliner probes the after-effects inscription on the World Heritage List in 1995 of the town of Luang Prabang (Laos). His study shows how UNESCO's efforts to preserve this Laotian royal town have effectively rewritten national history by emphasizing a royal past that is no longer reflective of the Laotian present.[86] Ken Taylor and Jane Lennon pick up the *leitmotif* of rewriting the past through World Heritage List inscription in discussing the consequences of the listing of the Sacred City of Anuradhapura (Sri Lanka) in 1982. They cite a critique furnished by Sudharshan Seneviratne, director of the Anuradhapura World Heritage site. Seneviratne notes that Anuradhapura's World Heritage inscription and later identification as part of the Sri Lankan "cultural triangle" have linked it to the Sinhala Buddhist tradition, when in fact the site also holds importance for China, India and the Mediterranean, based on artifacts discovered onsite.[87] Graeme Evans highlights the sometimes comical incoherence that can result from UNESCO-related tourism at traditional sites. At Chichén Itza, one tour operator offered a "Mayan Millennium" package, which included a New Year's Eve ceremony led by a Mayan high priest. However, Mayan civilization would not celebrate such an event inasmuch as it has its own calendar, separate from that of the Judeo-Christian tradition.[88] On a more hermeneutical level, Bruce McCoy Owens recalls Shelley

Errington's observation that the inscription of the Borobudur Temple Compounds (Indonesia) in 1991 locked the site within one specific frame of meaning, effectively "eliminating other 'frames', other 'lives', other stories about it".[89]

The protection of intangible heritage has led to a unique set of problems. The impetus to protect and preserve through inscription on the List of Intangible Cultural Heritage has publicized some previously traditional spaces and rituals, fundamentally altering their nature. Isabelle Brianso points to shifts in Jemaâ el-Fna Square (Morocco) after its inclusion on the aforementioned list in 2008: a space whose "first vocation" was to "produc[e] conviviality and social bonding" is now one whose existence is married to tourism.[90] J.P. Singh provides a counterpoint in noting that the inscription of Jemaâ el-Fna Square protected it from "encroachment and destruction by developers".[91] Similar to Brianso's critique, Berliner points out that the 2008 inscription of the Kankurang, a Manding secret initiatory rite in the Gambia and Senegal, has completely altered the ritual. It was inscribed to protect it from threats such as modernization and the rise of Islam, but protection through inscription has made this secret ritual public.[92] Susan Keitumetse warns about the potential devaluing of intangible heritage in Africa through listing as Intangible Cultural Heritage.[93]

While these examples highlight the challenges of conceptualizing and implementing a common world heritage, such a status also has positive effects, particularly related to tourism and the attendant economic development, but also to conservation. The World Heritage label has become a *de facto* tourist guide, encouraging many people to visit sites they would not have otherwise,[94] as well as providing spectacular itineraries for travel agencies.[95] The financial windfall from tourism spurred by World Heritage designation, which benefits not only the immediate area of inscription but also the communities surrounding the inscribed sites, is significant.[96] In taking stock of the situation in Lalibela (Ethiopia) almost 30 years after the World Heritage listing of the Rock-Hewn Churches (in 1978), Marie Bridonneau sees significant changes at all levels. In high politics, she cites an increase in government and privately financed tourist infrastructure, as well as how UNESCO World Heritage status has helped the Ethiopian Government to obtain development loans from international financial institutions. At a local level, Bridonneau demonstrates how tourism has become a more reliable and accessible source of financing for entrepreneurs than their local bank.[97] Overall, economists Bruno Frey and Lasse Steiner conclude that protection is one of the strongest positives about World Heritage inscription because it permits countries with inadequate "financial resources, political control, and technical knowledge about conservation" to preserve their history.[98]

However, even the positives engender negatives: a tourism-related perimeter established around Lalibela in 2010 – well after World Heritage inscription – had the effect of displacing residents and turning some of the homes into uninhabited displays.[99] Joy echoes these twin positive-negative effects with regard to Djenné,[100] and Edward Bruner relates similar positive, but different negative, effects with regard to Elmina Castle (Ghana), inscribed in 1979. He shows that the local population is pleased with World Heritage tourism, and that tourism has consequently become the primary decision-making factor in the economy. Yet the tourists – including many African-Americans seeking to trace their family roots – are dismayed by the commercialization of what they anticipate to be a spiritual site.[101]

The most significant positive consequence of a common world heritage is the evolution of UNESCO's understanding of this concept. The shift in emphasis from monumental to intangible heritage, from a European understanding of heritage to one that begins to take into account all cultures of the world, has leveled the heritage playing field by fostering a greater understanding of global interconnectedness. The former ICOMOS World Heritage coordinator, Léon Pressouyre, remarked in 2000 that under such a shifting regime, a Maori mountain could have as much meaning as the mythical Greek Mount Athos.[102]

In so globalizing its outlook and approaches, UNESCO has changed course in its direction of World Heritage sites, moving from the marginalization of local populations to efforts to include them. The inscription of the Borobudur Temple Compounds in 1991, which, among other effects, instituted tickets for visitors, created a feeling of exclusion among the local population.[103] In the aftermath of the eruption of Mount Mérapi in 2010, for instance, UNESCO developed a program for local participation in the preservation efforts at Borobudur, thus turning around and involving the affected populations in ownership of their site.[104]

Conclusion

The ongoing debate about the ultimate value of UNESCO's heritage efforts underscores the centrality of heritage to UNESCO's current mandate and, conversely, that of UNESCO to heritage conservation. Admittedly, dealing with heritage at an international level was and is not a uniquely UNESCO phenomenon – in addition to the fact it grew from precursors, it existed and expanded in a community of related NGOs. Nonetheless, through its various programs and instruments, UNESCO has established itself as the "supreme global arbiter" of heritage.[105] Our discussion shows that the development of UNESCO's actions in this field was a reflection of the strengths and weaknesses of its member states, only natural for an intergovernmental international organization. By providing a groundbreaking inventory of the impact of World Heritage listing, it also suggests that there is no reductionist answer to the question of the net outcome, particularly because positive advances can have negative after-effects, and the categorization on the spectrum of positive–negative depends on the eye of the beholder. Armed with such information, policy-makers, national heritage officials and scholars will be more aware of the complex challenges derived from World Heritage nomination.

Notes

1. In this work, "heritage" is employed as a short form of "cultural heritage". Due to space constraints, the authors made a choice to explore cultural heritage alone.
2. For pre-UNESCO international heritage efforts, see Annamaria Ducci, "Europe and the Artistic Patrimony of the Interwar Period: The International Institute for Intellectual Cooperation at the League of Nations" in *Europe in Crisis: Intellectuals and the European Idea, 1917–1957*, ed. Mark Hewitson and Matthew D'Auria (Oxford: Berghahn Books, 2012).
3. Jukka Jokilehto, *ICCROM and the Conservation of Cultural Heritage. A History of the Organization's First 50 Years, 1959–2009*, vol. 11, ICCROM Conservation Studies (Rome: ICCROM, 2011), 6–7.
4. Noted in Jaime Torres Bodet (UNESCO director-general), letter to Indian National Commission for Cooperation with UNESCO, 15 June 1949, UNESCO Archives, Central registry 1st series, 069 72 A 064(44) "49" part I (hereafter UNESCO 069 72 A 064(44) "49" part I).
5. The invitation letters are in UNESCO 069 72 A 064(44) "49" part I. The participants were from Europe (9), the USA (1), Latin America (2), Asia (2) and the Middle East (1).
6. Stanisław Lorentz, "Zapiski do autobiografii [Notes for an Autobiography]", Kwartalnik Historii Nauki i Techniki [History of Science and Technology Quartery] 24:4 (1979): 746–747.
7. Stanislas Lorentz, Conservation et restauration de monuments en Pologne au cours des années 1945–1949, UNESCO/MUS/Conf.1/10, 15 October 1949, KU Leuven Archives, Raymond Lemaire Papers, box UNESCO 29, folder UNESCO 1949–1980.
8. Réunion d'experts sur les sites et monuments d'art et d'histoire, UNESCO, 17–21 October 1949, [final report], Documents UNESCO/MUS/CONF. 1/1, UNESCO 069 72 A 064(44) "49" part I.
9. UNESCO, Comité international pour les monuments, compte-rendu de la quatrième session tenue à Paris du 21 au 25 septembre 1953 (hereafter UNESCO International Committee, 4th session), UNESCO Archives, Central registry 1st series, 069 72 A 02/06 part IV.
10. UNESCO, Comité international pour les monuments, compte-rendu de la troisième session tenue à Istanbul du 20 au 24 octobre 1952, UNESCO Archives, Central registry 1st series, 069 72 A 02/06 part III.
11. Ana Filipa Vrdoljak, "International Exchange and Trade in Cultural Objects" in *Culture and International Economic Law*, ed. Valentina Sara Vadi and Bruno de Witte (London: Routledge, 2015), http://works.bepress.com/cgi/viewcontent.cgi?article=1035&context=ana_filipa_vrdoljak. 7–8.
12. UNESCO International Committee, 4th session, 7.
13. Jokilehto, *ICCROM and the Conservation of Cultural Heritage. A History of the Organization's First 50 Years, 1959–2009*, 11: 14–15, 32–34, 37–38, 53–61.

14. For the early history of ICOMOS, see Aurélie Elisa Gfeller, "Preserving Cultural Heritage across the Iron Curtain: The International Council on Monuments and Sites from Venice to Warsaw, 1964–1978" in *Geteilt – Vereint! Denkmalpflege in Mitteleuropa Zur Zeit Des Eisernen Vorhangs Und Heute*, ed. Ursula Schädler-Saub and Angela Weyer, *Schriften des Hornemann Instituts/ICOMOS – Hefte des Deutschen Nationalkommitees* (Petersberg: Michael Imhof Verlag, 2015), 115–121.
15. Jokilehto, *ICCROM and the Conservation of Cultural Heritage. A History of the Organization's First 50 Years, 1959–2009*, 11: 11.
16. "Abu Simbel: Now or Never", *UNESCO Courier*, October 1961, 4, http://unesdoc.unesco.org/images/0006/000642/064240eo.pdf (accessed 3 September 2014); Lucia Allais, "The Design of the Nubian Desert" in *Governing by Design: Architecture, Economy, and Politics in the Twentieth Century*, ed. Aggregate (Architectural History Collaborative) (Pittsburgh, PA: University of Pittsburgh Press, 2012), 184; Kwame Anthony Appiah, "Whose Culture Is It?", *New York Review of Books* 9: February (2006): 62; Chloé Maurel, *Histoire de l'UNESCO. Les trente premières années. 1945–1974* (Paris: L'Harmattan, 2010), 283; Torgny Säve-Söderbergh, *Temples and Tombs of Ancient Nubia: The International Rescue Campaign at Abu Simbel, Philae and Other Sites* (London and Paris: Thames and Hudson and UNESCO, 1987), 11, 50, 52.
17. Maurel, *Histoire de l'UNESCO. Les trente premières années. 1945–1974*: 284; Säve-Söderbergh, *Temples and Tombs of Ancient Nubia: The International Rescue Campaign at Abu Simbel, Philae and Other Sites*, 64.
18. Christiane Desroches Noblecourt, *La grande nubiade: Le parcous d'une egyptologue* (Paris: Stock/Pernoud, 1992), 158–165; Maurel, *Histoire de l'UNESCO. Les trente premières années. 1945–1974*, 284; Säve-Söderbergh, *Temples and Tombs of Ancient Nubia: The International Rescue Campaign at Abu Simbel, Philae and Other Sites*, 67.
19. Maurel, *Histoire de l'UNESCO. Les trente premières années. 1945–1974*, 284.
20. Allais, "The Design of the Nubian Desert", 192; Maurel, *Histoire de l'UNESCO. Les trente premières années. 1945–1974*, 284; Säve-Söderbergh, *Temples and Tombs of Ancient Nubia: The International Rescue Campaign at Abu Simbel, Philae and Other Sites*, 9, 71–73.
21. Allais, "The Design of the Nubian Desert", 186–187; Maurel, *Histoire de l'UNESCO. Les trente premières années. 1945–1974*, 284; Säve-Söderbergh, *Temples and Tombs of Ancient Nubia: The International Rescue Campaign at Abu Simbel, Philae and Other Sites*, 74.
22. *Avant-projet des ouvrages de protection des temples d'Abou Simbel*, 3 – documents techniques, 3.1 – conditions naturelles – 3.11: Ensoleillement des temples, 1–4, http://unesdoc.unesco.org/images/0013/001342/134225fo.pdf (accessed 3 September 2014).
23. Lucia Allais, "Integrities: The Salvage of Abu Simbel", *Grey Room* 50 (2013): 19.
24. Shehata Adam Mohamed, "Victory in Nubia: Egypt", *UNESCO Courier*, February–March (1980): 8–9, http://unesdoc.unesco.org/images/0007/000747/074755eo.pdf#74755 (accessed 3 September 2014); Allais, "Integrities: The Salvage of Abu Simbel", 10; Säve-Söderbergh, *Temples and Tombs of Ancient Nubia: The International Rescue Campaign at Abu Simbel, Philae and Other Sites*, 10.
25. Allais, "Integrities: The Salvage of Abu Simbel", 23; Maurel, *Histoire de l'UNESCO. Les trente premières années. 1945–1974*, 285.

26. UNESCO International Campaign to Save the Monuments of Nubia, Executive Committee, 12th Session (Madrid, 26–28 September 1966), Financing of the Abu Simbel Project, Semi-Annual Plan of Financial Operations, Annex 1: Trust Fund for the International Campaign, Contributions pledged and contributions paid as at 31 July 1966 for the Abu Simbel Project (in USD), 20 September 1966, Paris, UNESCO/NUBIA/CE/XII/2, http://unesdoc.unesco.org/images/0015/001548/154825eb.pdf (accessed 3 September 2014).
27. Alexander A. Bauer, "New Ways of Thinking About Cultural Property: A Critical Appraisal of the Antiquities Trade Debate", *Fordham International Law Journal* 31 (2008): 65–80; Appiah, "Whose Culture Is It?", 62–65.
28. Maurel, *Histoire de l'UNESCO. Les trente premières années. 1945–1974*, 285.
29. Shehata Adam Mohamed, "Victory in Nubia: Egypt", 5.
30. Allais, "The Design of the Nubian Desert", 184.
31. Säve-Söderbergh, *Temples and Tombs of Ancient Nubia: The International Rescue Campaign at Abu Simbel, Philae and Other Sites*, 12.
32. Christiane Desroches Noblecourt, "Floating Laboratories on the Nile", *UNESCO Courier*, October 1961, 26, http://unesdoc.unesco.org/images/0006/000642/064240eo.pdf#64245 (accessed 3 September 2014).
33. Allais, "The Design of the Nubian Desert", 180, 82; Maurel, *Histoire de l'UNESCO. Les trente premières années. 1945–1974*, 286.
34. Säve-Söderbergh, *Temples and Tombs of Ancient Nubia: The International Rescue Campaign at Abu Simbel, Philae and Other Sites*, 11–12.
35. John Henry Merryman, "Two Ways of Thinking About Cultural Property", *The American Journal of International Law* 80:4 (1986): 228.
36. Säve-Söderbergh, *Temples and Tombs of Ancient Nubia: The International Rescue Campaign at Abu Simbel, Philae and Other Sites*, 11–12.
37. Singh, *United Nations Educational, Scientific, and Cultural Organization (UNESCO): Creating Norms for a Complex World*, 86–87; Carlo Perelli and Giovanni Sistu, "Jasmines for Tourists: Heritage Policies in Tunisia" in *Contemporary Issues in Cultural Heritage Tourism* ed. Jamie Kaminski, Angela M. Benson, and David Arnold, *Contemporary Geographies of Leisure, Tourism and Mobility* (Milton Park, Abingdon, Oxon, UK: Routledge, 2014), 78.
38. Bhownagar Jehangir, "Save the Acropolis", (1977) http://www.medmem.eu/en/notice/INA00699 (accessed 3 September 2014).
39. Säve-Söderbergh, *Temples and Tombs of Ancient Nubia: The International Rescue Campaign at Abu Simbel, Philae and Other Sites*, 11–12. UNESCO Islamabad, "Safeguarding Moenjodaro", http://unesco.org.pk/culture/moenjodaro.html (accessed 3 September 2014).
40. Allais, "The Design of the Nubian Desert", 180–84; Maurel, *Histoire de l'UNESCO. Les trente premières années*, 1945–1974, 283–97. Also see the recent article by a historian on Nubia: Paul Betts, "The Warden of World Heritage: UNESCO and the Rescue of the Nubian Monuments", *Past & Present* suppl. 10 (2015): 100–125.
41. David A. Meyer, "The 1954 Hague Cultural Property Convention and Its Emergence into Customary International Law", *Boston University International Law Journal* 11:349 (1993): 355–356; Roger O'Keefe, *The Protection of Cultural Property in Armed Conflict* (Cambridge, UK and New York: Cambridge University Press, 2006), 94, 96–98, 135–138.
42. Lyndel V. Prott, "Strengths and Weaknesses of the 1970 Convention: An Evaluation 40 Years After its Adoption" in *The Fight Against the Illicit Trafficking*

of Cultural Objects. The 1970 Convention: Past and Future (Paris, UNESCO Headquarters: UNESCO, 2011), 2.

43. Mounir Bouchenaki, "Return and Restitution of Cultural Property in the Wake of the 1970 Convention", *Museum International* 61:1–2 (2009): 141.
44. Zsuzsanna Veres, "The Fight Against Illicit Trafficking of Cultural Property: The 1970 UNESCO Convention and the 1995 UNIDROIT Convention", *Santa Clara Journal of International Law* 12:2 (2014): 100.
45. Prott, "Strengths and Weaknesses of the 1970 Convention: An Evaluation 40 Years After its Adoption", 9.
46. For an updated list maintained by UNESCO, see http://www.unesco.org/new/en/culture/themes/restitution-of-cultural-property/successful-restitutions-in-the-world/ (accessed 1 October 2014).
47. Government example: the USA has established bilateral cultural heritage agreements with 17 countries, http://eca.state.gov/cultural-heritage-center/cultural-property-protection/bilateral-agreements (accessed 1 October 2014). Archeological institution example: University College London, http://www.ucl.ac.uk/archaeology/research/ethics/policy_antiquities (accessed 10 October 2014).
48. Souren Melikian, "How UNESCO's 1970 Convention Is Weeding Looted Artifacts Out of the Antiquities Market", *Blouin Artinfo*, 31 August 2012, http://www.blouinartinfo.com/news/story/822209/how-unescos-1970-convention-is-weeding-looted-artifacts-out-of (accessed 1 October 2014).
49. For information about the cultural internationalist and particularistic views, see Merryman, "Two Ways of Thinking About Cultural Property", 831–853.
50. Peter H. Stott, "The World Heritage Convention and the National Park Service, 1962–1972", *The George Wright Forum* 28:3 (2011): 283.
51. UNESCO, Meeting of experts to coordinate, with a view to their international adoption, principles and scientific, technical and legal criteria applicable to the protection of cultural property, monuments and sites, Paris, 26 February–2 March 1968, Conclusions of the meeting of experts, UNESCO Archives, SHC/CS/27/7; UNESCO, Meeting of experts to establish an international system for the protection of monuments and sites of universal interest, Paris, 21–25 July 1969, Conclusion of the meeting, UNESCO Archives, SHC/CONF.43/7.
52. Stott, "The World Heritage Convention and the National Park Service, 1962–1972", 285–286.
53. UNESCO, World Heritage Center, World Heritage List, http://whc.unesco.org/en/list/ (accessed 7 November 2015).
54. Marc Askew, "The Magic List of Global Status: UNESCO, World Heritage and the Agenda of States" in *Heritage and Globalisation*, ed. Sophia Labadi and Colin Long (London and New York: Routledge, 2010), 23; Lynn Meskell, "UNESCO's World Heritage Convention at 40: Challenging the Economic and Political Order of International Heritage Conservation", *Current Anthropology* 54: 4 (2013): 483.
55. Aurélie Elisa Gfeller, "Negotiating the Meaning of Global Heritage: 'Cultural Landscapes' in the UNESCO World Heritage Convention, 1972–1992", *Journal of Global History* 8:3 (2013): 483–503.
56. Aurélie Elisa Gfeller, "Anthropologizing and Indigenizing Heritage: The Origins of the UNESCO Global Strategy for a Representative, Balanced and Credible World Heritage List", *Journal of Social Archaeology* 15:3 (2015): 366–386.
57. Aurélie Elisa Gfeller and Jaci Eisenberg, "Scaling the Local: Canada's Rideau Canal and Shifting World Heritage Norms", *Journal of World History* (forthcoming).

58. Operational Guidelines for the Implementation of the World Heritage Convention, WHC/2/Revised, February 1994, http://whc.unesco.org/archive/opguide94.pdf (accessed 18 July 2014); The Nara Document on Authenticity, 1994, http://www.icomos.org/charters/nara-e.pdf (accessed 30 June 2014).
59. Christoph Brumann, "Heritage Agnosticism: A Third Path for the Study of Cultural Heritage", *Social Anthropology/Anthropologie Sociale* 22:2 (2014): 178.
60. Recommendation on the Safeguarding of Traditional Culture and Folklore, 15 November 1989, http://portal.unesco.org/en/ev.php-URL_ID= 13141&URL_DO= DO_TOPIC&URL_SECTION= 201.html (accessed 6 July 2014).
61. Valdimar T. Hafstein, "Intangible Heritage as a List: From Masterpieces to Representation" in *Intangible Heritage*, ed. Laurajane Smith and Natsuko Akagawa (London and New York: Routledge, 2009), 94.
62. Noriko Aikawa-Faure, "From the Proclamation of Masterpieces to the Convention for the Safeguarding of Intangible Cultural Heritage" in *Intangible Heritage*, 14–20.
63. Valdimar T. Hafstein, "Claiming Culture: Intangible Heritage Inc., Folklore©, Traditional Knowledge™" in *Prädikat Heritage: Wertschöpfungen aus kulturellen Ressourcen*, ed. Dorothee Hemme, Markus Tauschek, and Regina Bendix (Berlin: LIT Verlag, 2007), 77–80; Thomas M. Schmitt, "The UNESCO Concept of Safeguarding Intangible Cultural Heritage: Its Background and *Marrakchi* Roots", *International Journal of Heritage Studies* 14:2 (2008): 95–111; Chiara Bortolotto, "Globalising Intangible Cultural Heritage? Between International Arenas and Local Appropriations" in *Heritage and Globalisation*, ed. Colin Long and Sophia Labadi (London and New York: Routledge, 2010), 105.
64. UNESCO, Decisions adopted by the Executive Board at its 155th session, 19 October–6 November 1998, 155 EX/Decisions, 9, http://unesdoc.unesco.org/images/0011/001142/114238e.pdf (accessed 7 August 2014).
65. UNESCO, Proclamation of the Masterpieces of the Oral and Intangible Heritage of Humanity (2001–2005), http://www.unesco.org/culture/ich/?pg= 00103 (accessed 12 December 2015).
66. Aikawa-Faure, "From the Proclamation of Masterpieces to the Convention for the Safeguarding of Intangible Cultural Heritage", 22.
67. Pierre Centlivres, "The Controversy Over the Buddhas of Bamiyan", *SAMAJ: South Asia Multidisciplinary Academic Journal* 2 (2008): 3, 8–9.
68. Ibid., 9.
69. Ibid., 9.
70. World Heritage Scanned Nomination, 208rev, Cultural Landscape and Archaeological Remains of the Bamiyan Valley, 5 July 2003, http://whc.unesco.org/uploads/nominations/208rev.pdf (accessed 3 September 2014).
71. UNESCO, UNESCO Director-General (Irina Bokova) expresses concern about the situation in Mali, 2 April 2012, http://whc.unesco.org/en/news/865 (accessed 3 September 2014).
72. Timbuktu, World Heritage List, http://whc.unesco.org/en/list/119 (accessed 3 September 2014).
73. Tomb of Askia, World Heritage List, http://whc.unesco.org/en/list/1139 (accessed 3 September 2014).
74. UNESCO, "Heritage sites in northern Mali placed on List of World Heritage in Danger", 28 June 2012, http://whc.unesco.org/en/news/893 http://whc.unesco.org/en/news/893 (accessed 3 September 2014).

75. Authors' translation from the French. "Mali: des islamistes détruisent des mausolées à Tombouctou", *Slateafrique.com*, 1 July 2012, http://www.slateafrique.com/90235/nord-du-mali-des-islamistes-detruisent-des-mausolees-de-saints-tombouctou (accessed 3 September 2014).
76. David Berliner, "Opinion: Patrimoine mondial et crimes de guerre", *La Libre Belgique*, 9 July 2012, http://www.lalibre.be/debats/opinions/patrimoine-mondial-et-crimes-de-guerre-51b8ed6ae4b0de6db9c7249f (accessed 3 September 2014).
77. Charlotte Joy, "'Enchanting Town of Mud': Djenné, a World Heritage Site in Mali" in *Reclaiming Heritage: Alternative Imaginaries of Memory in West Africa*, ed. Ferdinand De Jong and Michael Rowlands (Walnut Creek, CA: Left Coast Press, 2007), 156.
78. Shu-Yi Wang, "From a Living City to a World Heritage City: Authorised Heritage Conservation and Development and Its Impact on the Local Community", *International Development Planning Review* 34:1 (2012): 8.
79. Lisa C. Breglia, "Keeping World Heritage in the Family: A Genealogy of Maya Labour at Chichén Itzá", *International Journal of Heritage Studies* 11:5 (2005): 392.
80. David Harrison, "Introduction: Contested Narratives in the Domain of World Heritage" in *The Politics of World Heritage: Negotiating Tourism and Conservation*, ed. David Harrison and Michael Hitchcock (Clevedon, UK: Channel View Publications, 2005), 6.
81. David Berliner, "Opinion: Patrimoine mondial et crimes de guerre", *La Libre Belgique*, 9 July 2012, http://www.lalibre.be/debats/opinions/patrimoine-mondial-et-crimes-de-guerre-51b8ed6ae4b0de6db9c7249f (accessed 3 September 2014).
82. Tom Fawthrop, "Who Does the Preah Vihear Temple Belong To?", *Al Jazeera*, 4 June 2011, http://www.aljazeera.com/indepth/features/2011/05/2011531124449228802.html (accessed 3 September 2014).
83. Helaine Silverman, "Border Wars: The Ongoing Temple Dispute between Thailand and Cambodia and UNESCO's World Heritage List", *International Journal of Heritage Studies* 17:1 (2011): 7, 15.
84. David Berliner, "Opinion: Patrimoine mondial et crimes de guerre", *La Libre Belgique*, 9 July 2012, http://www.lalibre.be/debats/opinions/patrimoine-mondial-et-crimes-de-guerre-51b8ed6ae4b0de6db9c7249f (accessed 3 September 2014). For more detail about UNESCO and the Israel–Palestine issue, see Aurélie Elisa Gfeller, "Culture at the Crossroad of International Politics: UNESCO, World Heritage, and the Holy Land", *Papiers d'actualité/Current Affairs in Perspective: Fondation Pierre du Bois* 3 (2013), http://www.fondation-pierredubois.ch/Papiers-d-actualite/jerusalem.htm (accessed 5 September 2014).
85. UNESCO Executive Board, 195 EX/10, Paris, 11 August 2014, Item 10 of the provisional agenda, Implementation of 194/EX Decision 12 on "The Two Palestinian Sites…", http://unesdoc.unesco.org/images/0022/002291/229194e.pdf (accessed 4 September 2014).
86. David Berliner, "Multiple Nostalgias: The Fabric of Heritage in Luang Prabang (Lao PDR)", *Journal of the Royal Anthropological Institute* 18 (2012): 769, 771.
87. Ken Taylor and Jane L. Lennon, "Cultural Landscapes: A Bridge between Culture and Nature?", *International Journal of Heritage Studies* 17:6 (2011): 550; Sudharshan Seneviratne, "Situating World Heritage Sites in a Multicultural Society: The Ideology of Presentation at the Sacred City of Anuradhapura, Sri

Lanka" in *Archaeology and the Postcolonial Critique*, ed. Matthew Liebmann and Uzma Z. Rizvi (Lanham, MD: AltaMira Press, 2008), 177–195.

88. Graeme Evans, "Mundo Maya: From Cancún to City of Culture. World Heritage in Post-colonial Mesoamerica" in *The Politics of World Heritage: Negotiating Tourism and Conservation*, ed. David Harrison and Michael Hitchcock (Clevedon, UK: Channel View Publications, 2005), 43.
89. Bruce McCoy Owens, "Monumentality, Identity, and the State: Local Practice, World Heritage, and Heterotopia at Swayambhu, Nepal", *Anthropological Quarterly* 75:2 (2002): 282, citing Shelly Errington, "Making Progress on Borobudur: An Old Monument in New Order", *Visual Anthropology Review* 9:2 (1993): 56.
90. Isabelle Brianso, "World Heritage in the Time of Globalization" in *International Cultural Policies and Power*, ed. J.P. Singh (Houndmills, Basingstoke, Hampshire, UK: Palgrave Macmillan, 2010), 174.
91. Singh, *United Nations Educational, Scientific, and Cultural Organization (UNESCO): Creating Norms for a Complex World*, 94.
92. Guy Belzane, interview with David Berliner, "Un véritable boom du patrimoine mondial", TDC: La revue des enseignants 1051, 1 March 2013, 28–29, http://lamc.ulb.ac.be/IMG/pdf/Un_veritable_boom_du_patrimoine_mondial.pdf (accessed 3 September 2014).
93. Susan Keitumetse, "UNESCO 2003 Convention on Intangible Heritage: Practical Implications for Heritage Management Approaches in Africa", *South African Archaeological Bulletin* 61:184 (2006): 166–171.
94. Singh, *United Nations Educational, Scientific, and Cultural Organization (UNESCO): Creating Norms for a Complex World*, 93; Joy, "'Enchanting Town of Mud': Djenné, a World Heritage Site in Mali", 150.
95. Marie Bridonneau, "Le patrimoine, un outil pour le développement?" in *Patrimoine et développement: études pluridisciplinaires*, ed. Michel Vernières (Paris, FR: Karthala, 2009), 95.
96. Berliner, "Multiple Nostalgias: The Fabric of Heritage in Luang Prabang (Lao PDR)", 779.
97. Bridonneau, "Le patrimoine, un outil pour le développement?", 101–103, 105.
98. Bruno S. Frey and Lasse Steiner, "World Heritage List: Does It Make Sense?", *International Journal of Cultural Policy* (2011): 1.
99. Bridonneau, "Le patrimoine, un outil pour le développement?", 107.
100. Joy, "'Enchanting Town of Mud': Djenné, A World Heritage Site in Mali", 150.
101. Edward M. Bruner, "Tourism in Ghana: The Representation of Slavery and the Return of the Black Diaspora", *American Anthropologist* 98:2 (1996): 290–293.
102. Léon Pressouyre, "The Past Is Not Just Made of Stone", *UNESCO Courier*, December 2000, 19, http://unesdoc.unesco.org/images/0012/001213/121326e.pdf#121332 (accessed 4 September 2014).
103. Geoffrey Wall and Heather Black, "Global Heritage and Local Problems: Some Examples from Indonesia" in *The Politics of World Heritage: Negotiating Tourism and Conservation*, ed. David Harrison and Michael Hitchcock (Clevedon, UK: Channel View Publications, 2005), 158.
104. Masanori Nagaoka, "Revitalization of Borobudur: Heritage Tourism Promotion and Local Community Empowerment in Cultural Industries" in *ICOMOS 17th General Assembly* (Paris, France: ICOMOS, 2011), 663.
105. Brumann, "Heritage Agnosticism: A Third Path for the Study of Cultural Heritage", 173.

14
Safeguarding Iran and Afghanistan: On UNESCO's Efforts in the Field of Archeology

Agnès Borde Meyer

People usually associate UNESCO with archeology for only a few events and names, such as the removal of the Abu Simbel temples in Egypt, or the notion of world heritage. Other initiatives and actions, and their impact on archeological activities, are generally lesser known. That is the case for the Asian continent, where the perception of UNESCO's actions by its various partners, and the definitive impact on their own actions, is something we know little about. When studying the institutional and private papers about archeological missions in Central Asia – in my case in Sistan, a geographical area between Iran and Afghanistan – one can therefore be surprised to see what has taken place and was carried out by UNESCO from the 1940s. I, of course, suspected that the organization's initiatives and approaches would have had some sort of impact on national institutions dedicated to archeology, and on scholars in the field, from the first inclusion of UNESCO in the Congress of Orientalists in 1948 and up until the events which disturbed the relations between Iran, Afghanistan and the organization in the 1980s – the evolution of the archeological work and institutions was, after all, significant during this period. But the question is to what extent UNESCO took part in its development.

The first interventions

One of the first involvements of UNESCO in the field of archeology took place at the 21st International Congress of Orientalists in 1948, when Jacques Bacot from the French Oriental Society, an expert on Tibet, suggested the foundation of a Union of Orientalist Societies in partnership with UNESCO, which should form the cornerstone of future international archeological studies in Asia. Similar societies, most of them created in the 19th century and of a national nature, had been of great importance to archeologists so far, introducing the members to colleagues and forming national networks that made, for example, the search for financial help easier. The congress for various reasons declined the proposal, but now the participants had at least considered the idea of cooperation and they discussed its advantages and its drawbacks.[1]

It was therefore not until discussions about an international rule for archeology, and not only historic monuments, took place in the early 1950s that the organization again attracted the attention of archeologists. That would soon have an impact, also in Iran and Afghanistan, even though the contact was made on the initiative of Western scholars.

At the time there was an Iranian Archaeological Service that was part of the University of Tehran and directed by Jean Godard, a French architect and art historian, whereas in Afghanistan there was only a museum in Kabul. An Italian scholar from the Italian Institute for the Middle and Far East and the French Archaeological Delegation in Afghanistan, directed by Daniel Schlumberger helped the afghan curator. However, the service did not work very well so Schlumberger thought that the possible intervention of UNESCO – this "great international organization", as he put it in 1953, which had authority, weight and worked independently of the national institutions – could be the way to rescue Afghan world heritage.[2] His enthusiasm can also be explained by a hope that the sudden interference of UNESCO would put an end to the recurring competition between France and Italy about who should take the lead in excavating and preserving Afghan archeological remains and who had influence at Kabul Museum. In this context, UNESCO was the lesser of two evils.

The building that contained Afghanistan's historical and archeological collections at the time was in a very bad state in the wake of World War II. Its holdings of artifacts came mainly from the French excavations, while another part was a donation from King Ghazi Amir Amanullah Khan, but according to the descriptions by Schlumberger of the French archeological delegation, the building was more a warehouse than a museum. It was full of artifacts in shoeboxes, if they had not been stolen, and nobody knew because there was no inventory and no overview of what might or might not be missing from the collection.[3]

Who was responsible for this situation? From the point of view of the Western scholars and diplomats, particularly in the British community, it was the French: they had a monopole on Afghanistan, which they were not able to apply. According to the French, however, the Afghan Government was responsible because it had not spent any money on the museum's restoration.[4] The government most likely had a different opinion, but the French and German papers studied for this chapter do not reveal how it perceived the matter of responsibility. What the literature does mention, though, is that the king himself suggested asking UNESCO for help, and an official was sent in 1953.

Through the 1950s, UNESCO and ICOM were busy making a set of recommendations with regard to archeological artifacts, and in 1956 a proposal of recommendations on International Principles Applicable to Archaeological Excavations was accepted by the General Conference of UNESCO in New Delhi. It would be the first in a long list of recommendations issued by the organization in the years to come – recommendations that defined and highlighted archeology as something in its own right in order to understand the history of humankind, and not just a discipline assisting philological, linguistic and historical studies.

The recommendations favored a national frame for archeological work because that was how the field was organized in most countries, and they called for the opening of national museums and archeological institutes as well as the training of local archeologists where they were not to be found. That would soon prove to be an obvious occasion to finally give Afghan archeologists a voice and eventually leave all future decision-making with regard to archeological artifacts to representatives of Afghanistan, the country where they were found. It resulted in a mission of UNESCO to Kabul directed by Jean Gabus, then curator of the Archaeological Museum of Neuchâtel.

From 1957 to 1960, Gabus thus stayed in Afghanistan to entirely reorganize Kabul Museum. The first thing he did was to use his influence on the Afghan Government to have the museum's curator, Ahmed Ali Khozad, removed from his post and replaced by Raim Ziai, a historian who had studied in France. Gabus also organized trips for Raim Ziai in Germany, the UK and France. The aim was to quickly give the new director some training in art history and museology. At the same time, Gabus chose a Frenchman, Thibaut Courtois, to catalog the museum's artifacts, and a Syrian scholar to restore them.

In the meantime, with the events in Egypt with the rescue of the Abu Simbel temples, the organization not only demonstrated that its members had the will to realize great projects but also made headlines and archeology the center of public attention and a highly popular discipline. All of a sudden UNESCO was considered – scientifically, financially and politically – a highly credible partner among western archaeologists. It can be seen, for example, in the papers on the foundation of the Association for Iranian Art and Archaeology in 1960–1962, whose founders, Roman Ghirshman and Arthur Pope, were the initiators of the associations' collaboration with UNESCO.[5] Of the two, one was a Russian-born French citizen, who had studied at the École du Louvre and had dedicated his life to archeology in Iran since the 1930s. The second was an American, well known for his contribution to publications and exhibitions on Persian art for many years. With UNESCO they hoped to gain credibility, weight and, "who knows, money?", as one of them said.

It was also primarily Western archeologists who were sent on expert missions. When the first mission to Afghanistan stopped and was followed by another, it was again led by a French scholar, this time Alexandre Lezine, a specialist in Islamic architecture. He stayed in Kabul in 1962 and 1963 in order to estimate the general state of the archeological and historical monuments in the country and to offer recommendations regarding their restoration.[6]

Training students, founding institutions

This capacity of having an impact and making a difference encouraged UNESCO to ask wealthier Western countries to help Iran, Afghanistan and other Asian countries by sending scholars there and to accept students from these countries.[7]

The help was much needed, especially in Afghanistan where there was barely anyone with a knowledge of archeology. German and French papers speak about how the director of UNESCO's first expert mission, Gabus, personally approached the two embassies, requesting grants for Afghan students so that they could study archeology and art history abroad, and he wanted specifically to give them access to the Louvre School, to the University of Strasbourg and to the West German universities in Bonn, Heidelberg and Berlin, as well as to the Deutsches Archäologisches Institut, also in Berlin. He also wanted students to receive grants lasting five years instead those normally offered through UNESCO which usually lasted for between eight and 12 months.[8]

To achieve these goals, Gabus spoke to the national pride of and competition between the Western states. For example, he demonstrated to Klaus Fischer, an archeologist who had ties with the director of the Deutsches Archäologisches Institut, how France, Switzerland and Italy had already done a lot to train young Afghan students, and that Germany, the USA and Denmark, all of which had been accepted by Afghanistan to work on its territory, now finally had a chance to pay it back by training a new generation of Afghan archeologists.[9] With the French ambassador, on the other hand, Gabus highlighted the long tradition of archeological competition between France and Italy in the Asian region by disapproving of the unofficial monopole of archeological training in the area that had more or less belonged to the Italians for some years. According to Gabus, the Italian Institute for the Middle and Far East was trying "with success to entice in Rome all future Afghan archaeologists".[10]

It is difficult to know exactly how many Afghan students attended Western universities in the following two decades due to Gabus' efforts because the archives conflict. The reports made by UNESCO about students in foreign countries gives no information about the scientific disciplines chosen by the students. In 1957, for example, 32 Afghan students went to France alone. Zemar Tarzi, an Afghan archeologist and lecturer at the University of Strasbourg, was one of them. He speaks of seven students who went to Strasbourg alone.

For 1960 the report says there were no Afghan students in Germany, but the papers of the Deutsches Archäologisches Institut speak of three students in 1960 and 1961. The reason for this discrepancy might be that the students were sponsored by West Germany directly upon the request of UNESCO. The final report from Gabus speaks of eight students dispatched to France (3), Germany (2) and Italy (3).[11] Probably fewer than 30 students had the opportunity to study archeology at Western universities with these first grants.

The impact of UNESCO on archeological training is one thing, while the foundation of national institutions for archeology is another. At the General Conference in New Delhi in 1956, UNESCO had decided to carry out a "major project for the mutual appreciation of the cultural values of the Orient and the Occident" strongly supported by the USSR, India, Pakistan, Iran and Afghanistan. One of the first steps was to launch projects that would study the countries involved, first and foremost in Asia, and "to give the national authorities and the scientific institutions the occasion to pursue them".[12] Iran already had archeological institutions, such as the Archeology Department at the University of Teheran and the Asia Institute in Shiraz, but with the help of UNESCO it now established an institute for the study of the literature of Central Asia and an Iranian Committee for Coordination of the Studies on the Civilizations of the Peoples of Central Asia. In Afghanistan nothing happened, and in 1967 the director-general, René Maheu, asserted the necessity of at least a body which would give to the Afghan Government the opportunity to make plans and coordinate archeological excavations in the country with the help of UNESCO. The government decided to attract an international scientific community in Kabul by founding, in 1968, the Regional Center for Kushan History. Two years later yet another UNESCO expert mission, directed by an English archeologist, was sent to the country and arranged for a museum for Islamic art history to be opened in the Qala Ikhtyaruddin citadel in Herat. At the same time the team launched a series of publications of old Afghan writers that could function as useful sources and initiated the restoration of archeological remains by first listing some of the historical monuments in the country.[13]

An international collaborative archeology?

UNESCO continuously, and according to its recommendation of 1956, worked for "international collaboration" and "mixed missions", mainly binational missions, as well as the involvement of specialists from many different countries. It was not easy. The organization had limited powers, and its credibility was not sufficient to put bonds on the priorities of each member state. West Germany, for example, wanted to revive its lost archeological territories, and in 1960 it opened a German Archaeological Institute Office in Tehran, like 30 years before.[14] Also the Italian Institute for the Middle and Far East made its own national strategy by working directly with Iran without the interference of UNESCO, and so did the French, even though they promised space for local as well as foreign scholars and students on the excavation fields of the French archeological delegation in Afghanistan.[15]

The representatives of diplomatic institutions and of the ministries, whose function was first and foremost to emphasize the progresses and successes of the national archeological teams, were particularly skeptical about such collaborative work. The reports kept by the German and especially the French foreign affairs show how they saw foreign intervention and any kind of success accomplished by other nationalities as almost a kind of hostile aggression, as in 1958 when the French ambassador in Kabul claimed that a US survey in Sistan had been a complete failure, and almost indicated that the Americans were disguised spies rather than archeologists, despite the fact that they had produced several articles in high-ranking, international scientific journals.[16] Similar reactions were met by the Italian archeologist Umberto Scerrato, when he and not a French person was chosen as a consultant for Kabul Museum in 1957, or by the director of the German Archaeological Institute, Erich Boehringer, when he traveled to Iran and Afghanistan in 1960 and 1961 to propose a French and German archeological collaboration in Afghanistan and Syria, a collaboration that was eventually rejected.[17]

However, there have been a few successful experiences in the past of binational collaboration, not least that between the British Museum and the University of Pennsylvania, where they worked together on the archeological site of Ur under the direction of Leonard Woolley. For UNESCO, that was the example to follow, and it was not surprising that Woolley was asked to write the first volume of the organization's new prestigious History of Mankind series. But to what extent did UNESCO actually have an influence on the collaboration during the subsequent decades?

In a way, collaboration was seen favorably by scholars due to the constant need for financial support and the lack of a sufficient number of archeologists and other scholars on the excavation fields.[18] It proved totally impossible to sketch a map of the archeological sites in Sistan, for example, when some of them had the ambition.[19] However, collaboration was still to a large degree uncoordinated and a matter of one country's willingness and ability to settle a deal with Iran and Afghanistan. That was one of the reasons why the consultative committee behind UNESCO's East–West Major Project finally in 1965 decided to initiate a new international project – a study of the civilizations of Central Asia – in which all local and foreign scholars with an interest had an equal and real opportunity to meet and take part in a truly international collaborative undertaking, at least during the various conferences on the subject, which took place in the 1960s and 1970s.

The reports found in the German and French archives tell how the participants first joined forces by undertaking "a general review of the work done in their respective countries" on Central Asia.[20] Soon they would also together list the questions they wanted to develop further with scholars from other countries. Richard Hoggart, assistant director-general for social sciences, humanities and culture at UNESCO, later described "the Central Asian project as a good example of the kind of scholarly and cultural co-operation which could be organized by UNESCO".[21]

The Central Asian project consisted initially of a four-year experimental period with meetings of specialists which were in many ways important – not least for the countries of Central Asia. They helped the scholars and their institutions to get used to the reality of regional and international cooperation, and to Western eyes the conferences were also international in the sense that they took place in the various Central Asian countries, such as in the USSR in 1967 and in Afghanistan, India and Tajikistan in 1968. UNESCO would help to finance the organization of the meetings and the publication of the results. The project also resulted in the foundation by UNESCO and the Afghan minister of information and culture of an international center of study on the Kushan era and its impact on Central Asia. The list of participants at the conferences shows that a large number of the scholars, specialized in the various Iranian, Indian or Turfan disciplines, afterwards traveled to Dushanbe or Kabul to take part in the study and the discussions there.[22]

Did the conferences give the scholars a real habit of cooperating? They certainly increased the relations between European and American scholars, and sometimes also with Iranian and Afghan scholars. The professional correspondence between the German archeologists Klaus Fischer from Bonn University and Herbert Härtel, director of the Indian Art Department at the Berlin State Museum, is an example of new networks being created along the way, and shows that real efforts were made to introduce foreign scholars into their national scientific circles by inviting them to give lectures. The lists include many different nationalities, even though most were still Western scholars. The involved scholars received support from their respective governments and institutions, simply because participating in events launched by UNESCO was seen as something positive and something to be proud of. This was also the case in the host country, where the letters of Klaus Fischer show that his work was seen as both a scientific and a national duty.[23]

In the 1970s, several other non-UNESCO groups with similar interests were created – usually non-governmental scholarly organizations. The International Conference of South Asian Archaeologists, founded by Bridget and F.R. Allchin, was one of them, and it favored the participation and introduction of scholars, especially from South Asia. The International Association for the Study of Cultures of Central Asia, founded on the initiative of the Tajik scholar G. Gafurov, was another. The conferences and associations partly inherited their format from the first international conferences, but the archives show that they were first and foremost the outcome of individual more than institutional initiatives in order to make different platforms for people from East and West to meet, even though many of the participants were consultants for UNESCO and had participated in the organization's Central Asia conferences.[24]

Whose history?

New UNESCO-initiated international conferences took place in the 1970s – for instance, in Samarkand on Timurid art, and in Kabul on the Kushan era, and a new project was elaborated at the General Conference in 1976, namely the preparation of a *History of Civilizations of Central Asia*.[25]

The aim of a study of the civilizations of Central Asia was the "presentation of a significant example of the meeting of various cultures".[26] Colin Renfrew in his textbook on archeology reminds us that in the archeological process "remains the major issue of what we hope to learn".[27] The traditional approach would be the reconstruction of the factual history of a vanished civilization, or to draw a more precise picture of a culture, such as the Persian civilization. The efforts of Aurel Stein, Alfred Foucher and Ernst Herzfeld before World War II to find Greco-Buddhist tracks is a good example of this traditional approach to archeology. However, UNESCO wanted to follow another track: "The new project concerning Central Asia has no obligation to follow the classical ways already made by Iranology or Turcology, for example," a report claimed, "but its best interest is the study of the interactions between cultures, the way they met, the relationships they had built together."[28]

In other words, UNESCO's interest followed the lines and purposes of one of its predecessor, its History of Mankind project, which had studied the interactions between cultures, the way they met and the relationships they had built together, eventually leading to a globalized world with international organizations as the end goal of history. The final work was thus a sort of prehistory of UNESCO itself.[29] In that sense it was easy to recognize the explanation of the disinterest of the organization at the time when scholars wished to create an encyclopedia of Islam, because even if such a publication was an international undertaking made by scholars of various nationalities, it had no interest for the organization if the aim was mainly to draw a picture of a conquering civilization, because UNESCO only highlighted "interactions between cultures", mainly of the peaceful kind, and "the relationships they had built together".[30]

It was also how the history was presented when volume after volume of *The History of Mankind* eventually appeared from 1963 to 1976 and promoted a positivist conception of history, where pacifism played a major part, favoring a theory of peaceful diffusion of ideas and artifacts and mutual interdependence. Indeed, the volumes showed that during the empire of Kushan, for example, all the great religions of the period met – Zoroastrianism, Hinduism, Buddhism, Iranian religion – and that its roads had led Iranian Manicheism and Nestorian Christianism as far away as to China, so there was indeed also something true about the organization's take on history and something to study further.[31]

The group behind the study of the civilization of Central Asia in 1971 in fact made such a list of subjects about the various cultures along the Silk Route that were worthy of a closer look. One of them was the Kushan civilization, and another list of ten topics – such as the commercial and political relations with other countries – were made. For each topic, several subissues were suggested.[32] Kushan artifacts and archeological studies suddenly became popular in international scholarly communities. Studying Kushan civilization also permitted the inclusion of Soviet scholars and traveling to the Soviet territories of Tajikistan and Uzbekistan. Such a mission like the excavation, by Daniel Schlumberger and Soviet scholars, of Aï Khanoum, a Greco-Bactrian site in the north of Afghanistan, was often impossible.[33]

At the same time, the Iranian and Afghan states saw in archeology a way to develop a history that could confirm and promote their national aspirations, and in 1969 the Afghan press even asked the director-general, René Maheu, to initiate actions in the country similar to the restoration and protection of the Abu Simbel temples in Egypt, the targets being the minarets of Jam and Herat, and the tomb of Ghazni, "which belong to the living culture of Afghanistan".[34] Such a hunger for any support for the national pride was part of the climate at the time. The same was true in Iran, when Maheu's friend, André Malraux, during an exhibition of about 7,000 years of art in Iran, described the country as the "Greece of Orient" and as the area where the "genius" of humanity had been developed.[35]

Did UNESCO's efforts in the region have other forms of unintended impact – for example, on other archeological fields and studies? Sistan, a wide desert cut across the middle by a political boundary separating Iran and Afghanistan, is an example of exactly that. It had been almost a *terra incognita* before World War II. Because of the political context, the Iranian archeological part was more visited during the 20th century: the Kuh-i-Khwaja sassanid site, first described by the English Aurel Stein in 1916, and by the German Ernst Herzfeld in the 1920s, was studied by Gullini in 1960. The Italian Tosi excavated the prehistoric Shar-i-Shokta site in the1960s, and ten years later Scerrato focused his research on the achaemedid Dahan-i-Gulahman site, whose publication was still more confidential. In the Afghan part of the area, Roman Ghirshman in 1936, then the American George Dales in the 1960s, excavated the prehistoric site of Nad-i Ali, and the American Trousdale the prehistoric Sar-o-Tar site. Schlumberger worked on the Islamic Lashkari Bazar site just after World War II. There were also several surveys, by the Frenchman Joseph Hakin in 1936, the Englishwoman Beatrice de Cardi in 1949, the American Fairservis in 1950, the Italian in the 1960s and above all the German Klaus Fischer in the 1970s, whose publications today are a key reference. First the archeologists hoped to find the "missing link" between the Indus and Iranian civilizations. They also focused on the commercial history of an area which had been part of the Lapis Lazuli and Silk roads. Finally they wanted to understand the history of urban development and of the irrigation systems of this desert area. These works gave to the well-known English orientalist Clifford Edmund Bosworth the material for a large part of his research. However, the Sistan stayed partly unknown after the 1980s, and Ute Franke, in an article about the Herat Museum, stresses the fact that "this relative dearth of scientific investigation stands in contrast to the cultural and political importance of the Eastern provinces".[36] From the point of view of Maurizio Tosi, who excavated the site of Shar-i-Shokta for the Italian Institute for the Middle and Far East, the explanation was exactly its peripheral location far from the occidental and oriental centers.[37] Yet German, Italian and US missions, which began to dig in the desert in the 1960s, were inspired by a "new archeology" which favored interdisciplinarity. Even if some of the missions were based on the idea of tracing any evidence of cultural spreading, similar to UNESCO's approach, some of the excavations, particularly the Achaemenid Dahan-i-Gulahman site (even though the findings were considered as exceptional) were hardly known "outside the circle of some specialists". UNESCO is probably not the only reason for this situation. Nevertheless, by focusing on topics such as the Kushan civilization, UNESCO moved the financial, intellectual, human and political attention away from areas such as Sistan or subjects such as the Achaemenids to the study of Central Asia.

As this chapter shows, UNESCO had a real impact on the archeological studies in Afghanistan and Iran. First and foremost, it gave archeologists the confidence that they would have important – sometimes financial, sometimes only moral – support from an international organization at a time when their own institutions and the Iranian and Afghan governments had hardly anything to contribute. It also gave the two Asian states a way to express their wishes in a collegial way, to refer to common standards for archeological excavation and preservation, and to build or consolidate their national archeological institutions. In spite of traditional competition between Western states, UNESCO was also a real school of collaboration for the community of archeologists and the associated disciplines. In this context it had the opportunity to draw a new framework for archeology in Asia independent from its national institutions. In the 1990s and the 2000s, this experience resulted in the creation of several new, binational and multinational archeological projects and research centers.

On the other hand, UNESCO also founded a history of the area from a point of view which can be questioned – a version of history that could be used as a tool for the promotion of national aspirations, even though it was not UNESCO's intention, and one whose point of view, with its emphasis on cultural exchange, favored one version of history and ignored the complexity of the past.

However, UNESCO had a greater impact in Afghanistan than in Iran. The reason is probably that Iran already had a relatively well-functioning national archeological institution when the organization entered, and consequently the country could act more independently. In the 1980s that was about to change, when the Government of the Islamic Republic of Iran saw archeology as a tool of the shah, because of the celebrations in Persepolis in 1971. But already in 1985 it suddenly founded the Iranian Cultural Heritage Organization with a department in each province and with offices in each major city. The new organization was probably inspired more by the German archeological institutions than by UNESCO, but it still worked within the framework and with the recommendations of UNESCO. Since then the archeological activities in Iran have only increased and are now considered a top priority with regard to national interest. In Afghanistan, on the other hand, UNESCO remained as a potential partner, and even created the International Committee of Coordination Concerning Safeguarding of Afghan Cultural Heritage in 2002. These two stories show that UNESCO knows how to adapt its strategy to the political context of different countries. The only remaining question is, at what price?

Notes

1. *ACTES XXIè Congrès International des Orientalistes, Paris 23–31 juillet 1948* (Paris: Imprimerie nationale, Société asiatique, 1949).
2. Letter from Daniel Schlumberger to Claude Schaeffer, 6 September 1953, Archives des Poste, Kabul, Série B85, Centre des Archives Diplomatiques, Nantes.
3. Ibid.
4. Letter from ambassador Christian Belle to the Ministry of Foreign Affairs, 18 December 1957, Archives des Postes, Kabul, Série B85, Centre des Archives Diplomatiques, Nantes.
5. B2 and B5, Roman Ghirshman Papers, Guimet Museum, Paris.
6. Noël Duval, "Alexandre Lézine (1906–1972)", *Antiquités africaines*, 8 (1974): 7–12.
7. Letter from the German Embassy in Iran to Erich Boehringer, 18 September 1958, B94 22, Teheran, German Foreign Affairs Archives, Berlin.
8. Jeanine Auboyer Papers, Guimet Museum.
9. Letter from Klaus Fischer to Erich Boehringer, 29 July 1960, DAI, 16–01, F, 1956–1960, Kabul, DAI Archives, Berlin.
10. Letter from Ambassador Arnaud d'Andurain de Mäytie to Maurice Couve de Murville, Ministry of Foreign Affairs, August 1960, La Courneuve, Asie-Oceanie, Afgha. 1944–, Dossier 40, Afghanistan 1956–1967, Centre des Archives Diplomatiques, Nantes.

11. 1 of 1 69FR0417, Final report,13/7-8/10/60, UNESCO Archives.
12. "Réunion d'experts pour les études de l'Asie central, Maison de l'UNESCO, Paris, 24–28 avril 1967", Box 8 and 9, Jeanine Auboyer Papers, Guimet Museum.
13. Mission reports from Afghanistan, UNESCO Archives.
14. B 94 22, Correspondence regarding the activities of Erich Boehringer, Director of the German Archaeological Institute, DAI, in Asia. DAI Archives, Berlin.
15. Letter from Klaus Fischer to Erich Boehringer, 29 July 1960, DAI, 16–01, F, 1956–1960, Kabul, DAI Archives, Berlin.
16. Letter from Ambassador Christian Belle to the Ministry of Foreign Affairs, 30 April 1958, Asie Océanie, 1944–1955, Afghanistan, Archives Diplomatiques de la Courneuve, Paris.
17. Agnès Borde Meyer, "Un diptyque franco-allemand en archéologie: la proposition singulière d'Erich Boehringer en 1960, ses prémisses et son impact", *Le Franco-Allemand oder die Frage nach den Herausforderungen transnationaler Vernetzung/Le Franco-Allemand ou les enjeux des réseaux transnationaux* (Berlin: Logos Verlag, 2014), 73–85.
18. Usually, in Iran as in Afghanistan, one scientist, sometimes with one assistant, two when he was lucky, worked in the field, which was rarely easy. The Sistan area, for example, where Joseph Hackin, Roman Ghirshman, Beatrice de Cardi, Georges Dales, Klaus Fischer and their students carried out surveys and excavated, was a great desert with only one road, and with continuous windy conditions, called "the 120 days wind". Each article and report starts with a description of the bad conditions.
19. It was a recurrent wish of the archeologists and the directors of the French Archaeological Delegation in Afghanistan; in Iran, for the German Ernst Herzfeld and the American Henry Breasted before World War II, and later for the American archeologists Richard Frye and Daniel Schlumberger, and for the successive directors of the new Iranian and Afghan archeological services.
20. "Outline of the Report of the Collective Consultation of the Study of Civilizations of Central Asia Held at UNESCO Headquarters, 26 April–5 May 1971", Box 8 and 9, Jeanine Auboyer Papers, Guimet Museum.
21. "Collective Consultation on the Study of Civilization of Central Asia", 1971, SMB-ZA.11/VA 14827, Zentralarchiv, Staatliche Museen zu Berlin.
22. "Réunion d'experts pour les études de l'Asie central, Maison de l'UNESCO, Paris, 24–28 avril 1967", Box 8 and 9, Jeanine Auboyer Papers, Guimet Museum.
23. For example, a letter from Klaus Fischer to the rector of the Rheinische Friedrich-Wilhelms University in Bonn, 27 April 1970, PA 1998 ½, Rheinische Friedrich-Wilhelms University of Bonn.
24. Correspondence with Lohuizen de Leew, Raymond Allchin, and Herbert Härtel; Zentralarchiv Berlin, Museum für indische Kunst nach 1945, 14828 Correspondence Härtel 1974–1981, Jeanine Auboyer Papers, Guimet Museum.
25. Madhavan Palat, *History of Civilizations of Central Asia: Towards the Contemporary Period: From the Mid-nineteenth Century to the End of the Twentieth Century* 4 (Paris: UNESCO Publishing, 2005), 14–15.
26. "Réunion d'experts pour les études de l'Asie central, Maison de l'UNESCO, Paris, 24–28 avril 1967", Box 8 and 9, Jeanine Auboyer Papers, Guimet Museum.
27. Colin Renfrew and Paul Bahn, *Archaeology: Theories, Methods and Practice* (London: Thames and Hudson, 2012), 17.

28. "UNESCO, réunion d'experts pour l'étude des civilisations de l'Asie centrale. Paris, 24–28 avril 1967, Rapport final", Box 8 and 9, Jeanine Auboyer Papers, Guimet Museum.
29. Poul Duedahl, "Selling Mankind: UNESCO and the Invention of Global History, 1945–76", *Journal of World History* 22:1 (2011): 101–133.
30. "UNESCO, réunion d'experts pour l'étude des civilisations de l'Asie central". Paris, 24–28 avril 1967, Rapport final, and "Le nouveau projet d'Asie centrale n'a pas à suivre les voies classiques fort bien tracées déjà par l'Iranologie ou la Turcologie par exemple. Mais il aura tout intérêt à s'attacher plutôt aux interactions entre les cultures, au mouvement qui les a confrontées, aux rapports qui se sont établis entre elles", Box 8 and 9, Jeanine Auboyer Papers, Guimet Museum.
31. Svetlana Gorshenina and Claude Rapin, *De Kaboul à Samarcande: Les archéologues en Asie Centrale* (Paris: Découverte Gallimard, 2001).
32. "Centre régional d'histoire Kouchan, institut afghan d'archéologie", SMB-ZA.11/VA 14827, Zentralarchiv, Staatliche Museen zu Berlin.
33. Gorshenina and Rapin, *De Kaboul à Samarcande: Les archéologues en Asie Centrale.*
34. Kaboul, Archives des Postes, B60, Centre des Archives Diplomatiques, Nantes.
35. *Sept mille ans d'art en Iran, octobre 1961-janvier 1962* (Les Presses artistiques, 1961); *Trésors de l'Ancien Iran* (Musée RATH: Genève, 1966).
36. Ute Franke, *National Museum Herat, Areia Antiqua Trough Time* (Berlin: DAI, Eurasien Abteilung, 2008).
37. Maurizio Tosi, "Excavations at Shahr-i-Sohkta, a Chalcolithic Settlement in the Iranian Sistan, Preliminary Report on the First campaign, oct-dec 1967", *EAST and WEST* 18 (1968).

15
UNESCO and Chinese Heritage: An Ongoing Campaign to Achieve World-Class Standards

Celine Lai

In 1972, when UNESCO announced the Convention for Protection of World Heritage Sites, it was excellent news to archeologists and most historical resources workers in China. Soon after, the Government of the People's Republic of China ratified the convention, and four years later it submitted 15 cultural sites to be inscribed on the then newly implemented World Heritage List.

The inscription was completed by 1987.[1] The timing coincided with the Open Door Policy adopted by Deng Xiaoping, and the World Heritage Sites thus gave a refreshed and renewed understanding of history and heritage. That was something most educated Chinese needed, because it was right after they had experienced one of the most tragic and devastating periods of cultural destruction, notoriously known as the Cultural Revolution (1966–1976). It began as a political movement but was soon turned into an event to deny, and later destroy, the material culture connected to the past of China that the politicians in power considered shameful and responsible for the international humiliation that the country had suffered during the previous century. Countless archeological sites and historical monuments were subjected to contamination and even destruction approved by governmental authorities. This changed when the government recovered from the political calamities toward the end of the 1970s, and slowly resumed its role on the international scene, when declaring ownership of the cultural heritage of world-class standards seemed to be both an act of great symbolic value and a brilliant way to define international branding.[2]

Today most Chinese living on the territory of present-day China share the common view that culture and heritage convey a sense of pride and appreciation, so in that sense UNESCO had a significant impact. But what does world heritage actually mean in China? In this chapter I will address this question from the perspectives of administration, society and political leadership. I will provide a concise indication as to the extent to which China incorporates the concept of heritage into the country's policies, and what its people think about "heritage" or "world heritage".

Heritage, China and UNESCO

Heritage is a modern concept, born in Europe during the 17th century with the growing interest in collecting and studying antiquities. As part of their exploration of the world, archeologists brought back treasures with them from the Egyptian, Near Eastern and Hellenistic sites to Europe. Discussions about the concepts of culture and civilization began to appear in the scholarly studies of these artifacts and led to questions about the origin of Western civilization.[3] Heritage was in other words developed out of a curiosity about the past, and to appreciate and understand historical objects and monuments. It evolved into a package of explanations that connected the past and the present through the testimony of material objects, and in Europe a range of heritage studies based on objects collected from elsewhere appeared in scholarly books and journals, and in the shape of museums and university departments. This definition of heritage, as a package of human knowledge and human achievements that transcend modern divisions of nationalities, or territorial boundaries, was also the understanding of heritage that was eventually adopted the UNESCO.[4]

In China, a passion for traditions and antiquities long existed among the privileged and literati. However, the concept of heritage, which denotes equal ownership of cultural property among citizens, arrived almost at the same time as archeology was introduced. The remains of the old parts of the historical city of Anyang were among the first sites excavated in the 1920s and 1930s by Harvard-trained Chinese archeologists. The site was a Bronze Age settlement located in Henan Province in the northern part of Central China in the middle reaches of the Yellow River, the area traditionally known as the cradle of Chinese civilization. The excavations revealed the earliest evidence of Chinese writing, now known as oracle bone inscriptions, and enabled archeologists to identify the site as the capital of the Yin and Shang dynasties well known from Chinese historical texts. The discovery refuted some of the arguments of the emerging Doubting Antiquity School, which suggested that the early dynasties were probably forgeries composed centuries after the fall of the Shang Dynasty for political purposes, and proved to be essential in affirming the authenticity of many old records and accounts which were criticized and seen as fables, and to a large extent the archeological finds from Anyang restored the credibility of Chinese history.[5]

However, the concept of heritage was developed rather differently in Europe and China because of the fact that heritage was introduced in China at a time when the country was experiencing a crisis about establishing its national identity, so heritage was first and foremost seen as Chinese heritage – a way to create a sense of shared history and identity through archeology.[6]

Until fairly recently, however, there was no clear definition of heritage in the Chinese language. "Heritage" is translated as *yizhan*, which literally means "the inherited property". The same word also signifies legitimate ownership, but it does not necessarily indicate the revival or recurrent use of something inherited from the past, as in the English meaning of the word, as in the English definition, where it is the "history, traditions and qualities that a country or society has had for many years and that are considered an important part of its character".[7]

In fact the term "heritage" or *yizhan* was not commonly used in official Chinese publications or in individual writings until recently. When referring to historical and archeological resources, the Chinese Government usually used the term *wenwu*, which means "cultural relics". Hence the name *Guojia wenwu ju*, which is the central authority that supervises archeological excavations and historical monuments in the country, in the English translation was originally called the State Bureau of Cultural Relics. In many ways, in fact, the Chinese term *chuancheng*, literally "to receive and transmit", would be much closer to the meanings of the present Western definition, but it is still mostly used in the promotion of traditional practices or activities, such as craftsmanship and classical opera. The reason is that the choice of a Chinese version was a direct outcome of the general change in the conception of preserving heritage two decades ago with the implementation of UNESCO's World Heritage Programme.

China was, and still is, a very active member state. In 1987 it had the first sites registered on the World Heritage List, and another dozen cultural sites were subsequently added during the next two decades.[8] In the meantime the term "heritage" as *yizhan* began to appear in legal documents and official guidelines on protecting historical monuments and archeological resources. One of the earliest official documents that adopted the term "cultural heritage" in the place of "cultural relics" was the Rules on the Implementation of the Law of the People's Republic of China on the Protection of Cultural Relics of 1992, which mentioned the site of Anyang – also known as Yinxu or, literally, the Ruins of Yin – as a "heritage" site.[9]

That soon had consequences also for the English translation of *Guojia wenwu ju*. In 2003 the term "cultural relics" was deleted and the bureau was renamed the State Administration for Cultural Heritage, and its subordinate provincial institutes followed suit. As Zhang Bai, former chairman of the Chinese branch of ICOMOS, noted, the adoption of the term "heritage" was intended to indicate a shared context – a symbolic way to show that the country was adapting to international society.[10] This testified to the closer bonds developed with UNESCO after 2000 that drew attention from the policy-making level of the country.

Administration and world heritage

As of 2014, China has 47 inscribed World Heritage sites: 33 cultural, 10 natural and four mixed. The country has no central authority to look after the enlisted sites so their administration relies on the existing governmental agencies, in the sense that the work is headed by the State Administration for Cultural Heritage and supported by its provincial branches and by offices set up in various cities and counties. For example, the administration of Yinxu, inscribed on the list in 2006, is divided between the Henan Provincial Administration of Cultural Heritage and the Anyang City Bureau of Culture.[11] And for the Paotala Palace in Tibet, its administration is shared between the Lhasa Bureau of Culture, which similarly reports directly to the State Administration for Cultural Heritage, and in particular to a special team in the Department of Conservation and Archaeology dedicated to all issues in relation to World Heritage.

The State Administration for Cultural Heritage is responsible for all administrative matters, including the handling of new applications for national nomination. The same authority is also taking charge of the general promotion of heritage sites. A new color monthly magazine entitled *China Cultural Heritage* was launched in 2005. It provides updates about important sites and finds, and periodically covers special topics in relation to conservation plans and urban development around the country.[12] In fact the State Administration for Cultural Heritage has primary responsibility when it comes to education, so a number of similar provincial magazines dedicated to cultural heritage have been launched, following the national example, whereas individual sites included in the World Heritage List are being held responsible for the daily operation and promotion through, for example, the use of official websites. That explains why the websites of Yinxu and the Longmen grottoes, which are also located in Henan Province, are not easily recognizable as sharing the same status as World Heritage sites. At the same time, the promotion of heritage sites is also a special responsibility for the provincial tourist bureaus. So although both sites are located in the same province, they do not share resources and to a large extent their developments are independent.[13]

Altogether there are not many direct connections between UNESCO and the individual agencies, museums and universities in China with regard to world heritage. The only connecting point is the World Heritage Institute of Training and Research established in 2007, which has its main office in Shanghai, attached to the Institute for Urban Planning and Design at Tongji University. Two other offices are in Beijing and Suzhou.[14]

Probably because of the specialities of Tongji University faculty members, the work of the World Heritage Institute of Training and Research focuses mainly on the urban redevelopment of historic towns, cultural landscapes and historical architecture in accordance with the World Heritage

Convention of 1972. The institute has also until recently been active in launching academic seminars which focus on some of the listed sites, such as the Dujiangyan in Sichuan or the classical gardens in Suzhou. It has also provided a couple of two-week-long training camps in conservation and heritage for museum workers and other interested individuals, and the center provides an advisory service and accompanies representatives from UNESCO when they are in China to examine the sites. At the same time the institute acts as an independent, academic agent that supports historic towns and possible future heritage sites to develop in a planned and harmonious manner due to the social and economic welfare gained from the preservation of cultural and natural heritage. Nonetheless, it is a small team compared with the enormous number of existing historic sites in China. Moreover, its role is only advisory and it has no authority or commitment to any official agents directly responsible for world heritage.

In 2007 the Chinese Government made another move when a new Chinese Academy of Cultural Heritage (CACH) was set up.[15] Its main tasks, similar to those of the institute, are offer give guidance regarding conservation plans, provide conservation training and support international exchange projects. Practically speaking, however, neither the State Administration for Cultural Heritage nor its subordinate academy gets involved in the operation of individual World Heritage or national heritage sites. The state administration is purely an administrative unit, whereas the academy is purely an academic body. The management of the heritage sites therefore relies entirely on the experience and exposure of the corresponding staff members. At the moment the academy is the only unit that organizes seminars to gather staff members working at different World Heritage sites in the country. Having said that, "World Heritage" is still understood rather differently in the various provinces of China due to the fact that the connections between UNESCO and the different cultural units in China are both indirect and rather sporadic.

Heritage fever and Chinese society

China has "heritage fever", it is often claimed. Its symptoms appeared in around 2004 when scholars and critics noted that the officials were active, almost restless, in having heritage sites in their corresponding province inscribed on the World Heritage list.[16]

When an inspection was made of the so-called Stone Forest in the Yunnan Province in the south-western part of China, the representatives of ICOMOS made a few remarks that give some idea of the extent of this "fever". First, the Chinese officials had arranged that more than a thousand people, each holding a flower in their hands, would receive the investigators on the site, and the subsequent reception was later described by the visitors as "presidential". Second, although an enormous number of people were involved in the investigation, the ICOMOS experts soon recognized that the locals had only a little knowledge about what the site represented. Finally, the conservation measures at the Stone Forest proved to be far from satisfactory – hardly any money had been spent on it.[17] The report of the Stone Forest investigators exposes a serious problem in China: that there is often a discrepancy between the overall will to conserve important heritage sites and how the heritage sites may benefit from the people living around them. One of the reasons for this is that world heritage often provides a credit-bearing opportunity for Chinese officials in their career. The country is privileged to own many important sites of great aesthetic and historical values, and to have a site inscribed is a seemingly straightforward project to complete in order to win national pride in an international setting.[18] There is, of course, a strong economic reason too. A much-quoted example is the ancient city of Lijiang, also in Yunnan Province, which was inscribed in 1997. According to the Yunnan Provincial Bureau of Tourism, one of the two main counties of Lijiang received 109,000 visitors in the first quarter of 2010 and made CNY10.48 billion.[19] "World Heritage" is in other words unquestionably an international brand that assures constant income from tourism. Indeed, "heritage fever" has even spread so that it now includes a competition toward the candidacy on the tentative list pending national nomination.

In fact, the World Heritage Programme is more or less an official campaign in China. As noted by the investigators of the Stone Forest program, local participation in heritage conservation was observably weak, and scholars often criticize the overexploitation of cultural sites, seeking to put a halt to the applications. The site of Yinxu cost CNY230 million to get prepared for the application.[20] Lijiang witnessed such an incredible influx of immigrants in search of fortune that it forced the inhabitants, who were mainly the non-Han population, to move out of the town. The demographic change consequentially gave rise to the issue of the authenticity of the so-called ancient city, which is now packed with modern pubs, guesthouses and shops.[21]

Worse still, such enlisted sites as the Mountain Tai and Zhangjiajei Geopark of Hunan received warnings from UNESCO about destroying the scenery of the landscape.[22] The many negative comments suggest that "heritage fever" has not yet found a cure to rectify the existing problems on the one hand, and the Chinese officials continue to compete at the expense of the locals on the other. It is perhaps not surprising to find that while scholars and critics urge for the refining of the guidelines regarding conserving World Heritage sites, the official media continue to deliver optimistic messages about World Heritage sites in China.

As suggested by an anonymous author of an entry for "heritage fever" in the widely used Baidu web-encyclopedia, China will in the long run benefit from the sites registered due to the steps taken to develop sustainable measures to protect those sites, and the practice of submitting applications for enlistment is, after all, also evidence of a revived Chinese tradition to respect cultural heritage. At least this more or less official view is obviously the dominant one, despite the many criticisms and questions mentioned above, and it is to this Chinese learning of cultural heritage that we shall now turn.

"China Dream" and heritage

Xi Jinping, president of the People's Republic of China, seems more interested in the collaboration with UNESCO than any of his predecessors. On his trip to France in spring 2014, he delivered a speech at UNESCO's headquarters, announcing that China was a country with a civilization made up of "one hundred flowers in full blossom" such that it "brings spring to the garden". In the same month, China sent USD8 million to support a UNESCO project in Africa. In other words, China has not only a civilization but also resources.[23]

It was by no means a coincidence that the new Chinese leader chose to demonstrate his national strength at a cultural occasion. Two years earlier, as soon as he took up the presidency, Xi visited an exhibition entitled the Road to Revival in the newly renovated National Museum of China. Afterwards he told the media that China would enable everyone to make his dream real, and that the key could be found in the "great revival of the Chinese nation".[24] "China Dream" has thereafter been the key concept in his policy.

Many foreign columnists have been critical about what "China Dream" – or "the Chinese Dream" – means or consists of. The term implies wealth and a middle-class lifestyle, but to the Chinese leaders it seems also to mean a revival of the Chinese nation, culture and civilization. Experts on contemporary Chinese politics are not unaware that the Chinese leaders are rather self-conscious about their country's culture. As Ai Jiawei points out, while the previous term of office headed by Hu Jintao employed Confucianism to revive the Chinese nation, the present leaders expand the definition to include the entire "cultural tradition", and heritage too plays an important role in this political scheme.[25]

Indeed, this political intention has alarmed scholars, especially those who are interested in the cultural heritage sites in the provinces near the borders, because those areas are inhabited by different groups of people of non-Han ethnic origins. When, for example, Robert Shepherd gathered fieldwork data for his study on cultural heritage in China, a number of the interviewed Han Chinese tourists in Tibet admitted that they had little knowledge of the other ethnic groups or minorities in the country. In fact, the sites had been turned into tourist resorts, and there was hardly any program set up for tourists from the country or overseas to understand the values and meanings of the staged performances within the World Heritage sites. Apart from tourism, the case of Tibet also faces an additional problem of the political agenda of the Chinese Government that Tibetan features of the site are distorted, suppressed or stereotyped in a way to portray the image of the minorities as opposed to the Han Chinese culture.[26]

The concept of heritage ownership is something that most Han Chinese use as a reason to explain the lack of motivation to improve the site, because a sense of ownership must come first from people at the site, then from the management and only then from the Han Chinese. The idea that World Heritage consists of values that can be shared equally by all people on the planet exists, but the idea that it is something that belongs to the entire world also causes hesitation. More often, the existing cultural treasures are seen in China as something that tell them about the glorious past, and the efforts made by people in order to contribute to present-day China. When, for example, Anyang, or Yinxu, was inscribed on the World Heritage List in 2006, fulfilling four of the ten prescribed criteria, the Chinese representatives wrote in the document submitted to UNESCO:

> The bronze culture of Yin Xu is a reflection of the artistic level and social customs of the Late Shang Period and represents the highest level of development in China's ancient bronze culture ... This interaction of several Bronze Age cultures within the area of modern day China paved way to the formation of the Chinese nation with its own unique characteristics.[27]

This and other heritage sites are in other words heritage that represents the present achievement of the country, and first and foremost something for the Chinese to be proud of.[28]

Unless this concept is enriched by the ideas that cultural treasures are something to be appreciated and respected across all nations, the World Heritage Programme will probably never be in China as originally intended by UNESCO. And if China, or the Han Chinese at large, refuse to decrease the amount of cultural pride attached to the sites and continue to embrace the idea of Han superiority over other ethnic groups in the country, it will never be possible to offer a justified or fair understanding of the heritage sites in the many different parts of the country. The opportunity is even smaller to promote cultural appreciation and respect for other ethnic groups. Altogether, UNESCO has had an enormous impact when it comes to appreciating heritage, but the barrier to accepting the universal view of heritage is not only a contemporary but also a historically rooted fact very much present in today's China.

Looking forward

In this chapter I have reviewed the administrative structure for World Heritage sites in China, the conflicting views about heritage between the officials and the citizens, as well as the different understanding of heritage in China compared with that of UNESCO, even though the bottom line is that there is not a coherent view of World Heritage, or heritage in general, in Chinese writing or scholarship.

It will indeed be difficult to urge China to revise the existing administrative system and the way of thinking about heritage. Nonetheless, it is not an entirely impossible task moving ahead toward the ultimate mission of the original purpose of the 1972 convention, but it should be expected that the process of revision or improvement would not be a quick and easy one. The next campaigns would have to consider including different members of Chinese society to become familiar with the practice of setting up an international scheme to conserve cultural heritage sites, and UNESCO should further encourage such countries as China to make use of heritage to nurture a sense of cultural appreciation and mutual respect, and the importance of local participation. This heritage campaign has only just begun.

To a large extent the exercise of the World Heritage Programme has in fact already successfully encouraged China to redefine its heritage. For the sites that have been registered, China does indeed experience problems, but many of them are problems similar to those in Italy and other countries, and they consist of the handling of tourism, cultural authenticity and ethnic crisis over cultural properties. However, in the case of China, the exercise of preserving heritage sites and turning them into articles to serve social and economic welfare is inevitably intertwined with the century-old need of restoring national pride.

Inscribed or not, the cultural heritage sites in China are likewise imbued with specific historical significance to emphasize the linearity of Chinese civilization and the integrity of the present territorial boundary. In the foreseeable future, the possibilities to alter such discourse promoted by the Chinese state are few. Nevertheless, as the practice of preserving heritage, and World Heritage, was developed only recently, there is room to empower the practitioners with skills and techniques, and to enrich the concept of developing heritage for social welfare through new means of administration and promotion. As demonstrated by the fairly new English name of the State Administration for Cultural Heritage, the country is after all ready to learn something new. And in October 2014, when the first draft of this chapter was completed, the Chinese Government celebrated the country's success in collaborating with Kazakhstan and Kyrgyzstan to register the 8,700 km-long Silk Route. The scale of the World Heritage "site" for the Chinese is in other words substantially expanded and will thus involve even greater complexities in administration and the division of work between the two countries. Before China has come up with a scheme to promote the idea of heritage and conservation effectively, a new set of challenges toward the conservation and management of the extensive sites along the

Silk Route – and the collaboration across countries and cultural boundaries – will eventually emerge. It could prove to be very positive. No matter what, as Xu Ming, a columnist for *Globe News*, remarks in the title of an article, "winning world heritage status is not the end of the race".

Notes

1. Helaine Silverman and Tami Blumenfield, "Cultural Heritage Politics in China: An Introduction" in *Cultural Heritage Politics in China*, ed. Helaine Silverman and Tami Blumenfield (New York: Springer, 2012), 5–7.
2. Rana Mitta, *Modern China: A Very Short Introduction* (Oxford: Oxford University Press, 2008), 55–73; Robert J. Shepherd and Larry Yu, *Heritage Management, Tourism, and Governance in China: Managing the Past to Serve the Present* (New York: Springer, 2013), 13–22.
3. Peter Stearns, *Western Civilization in World History* (New York: Routledge, 2003).
4. Julian Huxley, *UNESCO: Its Purpose and Its Philosophy* (Paris: UNESCO, 1946); William Logan, "States, governance and politics of culture" in *Routledge Handbook of Heritage of Asia*, ed. Patrick Daly and Tim Winter (Oxford and New York: Routledge, 2012), 113–115.
5. Kwang-chich Chang, "China on the Eve of the Historical Record" in *The Cambridge History of Ancient China*, ed. Michael Loewe and Edward Shaughnessy (Cambridge: Cambridge University Press, 1999), 54–73; Li Liu and Chen Xingchan, *The Archaeology of China: From the Late Paleolithic to the Early Bronze Age* (Cambridge: Cambridge University Press, 2012), 350–360.
6. Ge Jiaoguang, *Zezi zhongguo: zhongjian youguan "Zhongguo" de lishi lunshu* [Making Residence in the Centre of States: Reconstructing the Historical Descriptions about "China"] (Taipei: Linking Publishing, 2011).
7. *Oxford Learners' Dictionaries*, http://www.oxfordlearnersdictionaries.com/definition/english/heritage?q=heritage (accessed 25 March 2015).
8. It should be noted that until around 2000, China was not an active member in the area of heritage conservation and promotion under the lead of UNESCO. Xie Zheping and other scholars in international politics and affairs consider that the country participated more as a supporter than a partner throughout the 1980s and 1990s. So the history of partnership between China and UNESCO is rather short in the area of heritage protection. For heritage and other cultural developments, such as Education, see Xie Zheping, *China and UNESCO: An Empirical Study of International Organizations' Impact on Member States* (Beijing: Jiaoyu kexue, 2010), 111–172; Xie Jiping and Zhang Xiaojing, "Chuanshou yu xuexi: Zhongguo canyu lianheguo jiaokewen zuji de jingyan yanjiu (Learning and Following: A Study on China's Experience in Participating in UNESCO)", *Foreign Affairs Review* 1 (2011): 48–59.
9. *Rules on the Implementation of the Law of the People's Republic of China on the Protection of Cultural Relics*, promulgated on 5 May 1992, appendix in the application documents of the site of Yinxu submitted to UNESCO, *World Heritage Scanned Nomination: 1114 Yinxu*, 16 July 2006, http://whc.unesco.org/uploads/nominations/1114.pdf (accessed 25 November 2014).
10. China ICOMOS, *Principles for the Conservation of Heritage Sites in China* (Los Angeles: Getty Conservation Institute, 2004), 97, http://www.getty.edu/

conservation/publications_resources/pdf_publications/pdf/china_prin_heritage_sites.pdf (accessed 25 November 2014).

11. *World Heritage Scanned Nomination: 1114 Yinxu*, section 4.
12. The official website of the magazine can be found at http://www.weibo.com/p/1002061181023032/home?from=page_100206&mod=TAB#_loginLayer_1418616720631 (accessed 14 December 2014).
13. The Yinxu site is usually addressed as Anyang Yinxu in the Chinese language, which is the name of the official website, http://www.ayyx.com (accessed 14.12.2014). The official website of the Longmen Grottoes is http://www.lmsk.cn (accessed 14 December 2014).
14. The background and tasks of the WHITR can be found at its official website at http://www.whitr-ap.org/index.php?classid=1459 (accessed 25 November 2014).
15. CACH was restructured from a former Chinese Academy of Cultural Relics, which traced its history back to the 1930s. In my opinion, CACH is a new establishment, which deals with heritage and conservation that have never been attended to before. The background and tasks of CACH can be found on its official website at http://www.cach.org.cn (accessed 25 November 2014).
16. The columnist Zhou Weihung was one of the first who raised the issue of "heritage fever" in "Shenyi ru yu tong" (Heritage fever and hurt), *Chinese Times* 12 (2004), available at *Wisenews* (accessed 20 November 2014). See also Wei Houkai, "UNESCO tag: a double-edged sword", *China Daily* 6.9.2010, available at *Wisenews* (accessed 20 November 2014); and Annie Lee, "China obsessed with World Heritage", *The Asia Magazine* 3.9.2010, available at *Wisenews* (accessed 20 November 2014).
17. Wen Yu, "Shijie yizhan: tuidong yu congji" ["World Heritage: Launching and Impacts"], *Jinwan Bao*, 10 November 2013, available at *Wisenews* (accessed 20 November 2014).
18. Raymond Li, "Warning over Commercialising Heritage Sites", *South China Morning Post*, 1.1.2013, available at *Wisenews* (accessed 20 November 2014).
19. The data were retrieved from the Yuannan Bureau of Tourism at http://www.ynta.gov.cn/Item/2177.aspx (accessed 14 December 2014).
20. Zheng Zhi, "World Heritage Fever: Good or Bad?", *CRIENGLISH.com*, online newspaper, 24.8.2010, http://english.cri.cn/6909/2010/08/24/2321s590609.htm (accessed 20.11.2014).
21. Xiaobo Su, "Tourism, Migration and the Politics of Built Heritage in Lijiang, China" in *Cultural Heritage Politics in China*, ed. Helaine Silverman and Tami Blumenfield (New York: Springer, 2012), 101–114.
22. Xu Ming, "Winning World Heritage Status is not the End of the Race", *Globe News* 22 January 2013, available at *Wisenews* (accessed 20 November 2014).
23. Speech by Xi Jingping, President of the People's Republic of China at UNESCO Headquarters, 28 March 2014, http://www.fmprc.gov.cn/mfa_eng/wjdt_665385/zyjh_665391/t1147894.shtml (accessed 20 November 2014).
24. "Chasing the Chinese Dream: China's New Leader Has Been Quick to Consolidate His Power. What Does He Now Want for His Country?", *The Economist*, 4 May 2013, http://www.economist.com/news/briefing/21577063-chinas-new-leader-has-been-quick-consolidate-his-power-what-does-he-now-want-his (accessed 25 November 2014).
25. Jiawen Ai, "Selecting the Refined and Discarding the Dross: The Post-1990 Chinese Leadership's Attitude towards Cultural Tradition" in *Routledge Handbook*

of Heritage of Asia, ed. Patrick Daly and Tim Winter (Oxford and New York: Routledge, 2012), 129–138.

26. Robert Shepherd, "UNESCO and the Politics of Cultural Heritage in Tibet", *Journal of Contemporary Asia* 36:2 (2006): 243–257; Robert Shepherd, "Cultural Heritage, UNESCO, and the Chinese State", *Heritage Management* 2:1 (2009): 55–80; and Shepherd and Yu, *Heritage Management*, 20–30.
27. *World Heritage Scanned Nomination: 1114 Yinxu*, 14.
28. Magnus Fiskesjo, "Politics of Cultural Heritage" in *Reclaiming Chinese Society*, ed. You-tien Hsing and Ching Kwan Lee (New York and Oxford: Routledge, 2010), 225–245.

Index

Lightning Source UK Ltd.
Milton Keynes UK
UKHW020145121021
392024UK00012B/2809